AF773009

ADVANCES IN SYSTEM RELIABILITY ENGINEERING

ADVANCES IN SYSTEM RELIABILITY ENGINEERING

Edited by

MANGEY RAM

J. PAULO DAVIM

ACADEMIC PRESS

An imprint of Elsevier

Academic Press is an imprint of Elsevier
125 London Wall, London EC2Y 5AS, United Kingdom
525 B Street, Suite 1650, San Diego, CA 92101, United States
50 Hampshire Street, 5th Floor, Cambridge, MA 02139, United States
The Boulevard, Langford Lane, Kidlington, Oxford OX5 1GB, United Kingdom

Notices

Knowledge and best practice in this field are constantly changing. As new research and experience broaden our understanding, changes in research methods, professional practices, or medical treatment may become necessary.

Practitioners and researchers must always rely on their own experience and knowledge in evaluating and using any information, methods, compounds, or experiments described herein. In using such information or methods they should be mindful of their own safety and the safety of others, including parties for whom they have a professional responsibility.

To the fullest extent of the law, neither the Publisher nor the authors, contributors, or editors, assume any liability for any injury and/or damage to persons or property as a matter of products liability, negligence or otherwise, or from any use or operation of any methods, products, instructions, or ideas contained in the material herein.

Library of Congress Cataloging-in-Publication Data
A catalog record for this book is available from the Library of Congress

British Library Cataloguing-in-Publication Data
A catalogue record for this book is available from the British Library

ISBN: 978-0-12-815906-4

Publisher: Matthew Deans
Acquisition Editor: Brian Guerin
Editorial Project Manager: Thomas Van Der Ploeg
Production Project Manager: Sruthi Satheesh
Cover Designer: Harris Greg

Typeset by SPi Global, India

CONTENTS

CONTRIBUTORS

Mohini Agrawal
School of Business, Galgotias University, Greater Noida, India

Adarsh Anand
Department of Operational Research, University of Delhi, New Delhi, India

Alessandro Barbiero
Department of Economics, Management and Quantitative Methods, Università degli Studi di Milano, Milan, Italy

Navneet Bhatt
Department of Operational Research, University of Delhi, New Delhi, India

F. De Caro
Department of Engineering, University of Sannio, Benevento, Italy

Lirong Cui
School of Management & Economics, Beijing Institute of Technology, Beijing, China

Antônio Carlos Lopes da Costa
Centro de Desenvolvimento da Tecnologia Nuclear—CDTN, Belo Horizonte, Brazil

Vanderley de Vasconcelos
Centro de Desenvolvimento da Tecnologia Nuclear—CDTN, Belo Horizonte, Brazil

Chen Fang
School of Management & Economics, Beijing Institute of Technology, Beijing, China

Soufiane Gasmi
Optimization, Modeling and Decision Support (OMAD), University of Tunis—Tunis National Higher School of Engineering, Tunis, Tunisia

Miroslav Kvassay
Department of Informatics, University of Zilina, Zilina, Slovakia

Dongjin Lee
School of Computing, Informatics, and Decision Systems Engineering, Arizona State University, Tempe, AZ, United States of America

Preeti Malik
Department of Computer Science and Engineering, Graphic Era Deemed to be University, Dehradun, India

Monika Manglik
Department of Mathematics, University of Petroleum & Energy Studies, Dehradun, India

Lata Nautiyal
Department of Computer Science and Engineering, Graphic Era Deemed to be University, Dehradun, India

Rong Pan
School of Computing, Informatics, and Decision Systems Engineering, Arizona State University, Tempe, AZ, United States of America

Brian A. Polin
Jerusalem College of Technology, Jerusalem, Israel

Ludmila Pustylnik
Department of Mechanical Engineering, Afeka Tel-Aviv Academic College of Engineering, Tel Aviv, Israel

Jan Rabcan
Department of Informatics, University of Zilina, Zilina, Slovakia

Mangey Ram
Department of Mathematics, Computer Science & Engineering, Graphic Era University, Dehradun, India

Amanda Laureano Raso
Centro de Desenvolvimento da Tecnologia Nuclear—CDTN, Belo Horizonte, Brazil

Alexander Rotshtein
Department of Industrial Engineering, Jerusalem College of Technology, Jerusalem, Israel

Patrik Rusnak
Department of Informatics, University of Zilina, Zilina, Slovakia

Luis Mejia Sanchez
Cummins Inc., Columbus, IN, United States of America

Wellington Antonio Soares
Centro de Desenvolvimento da Tecnologia Nuclear—CDTN, Belo Horizonte, Brazil

A. Vaccaro
Department of Engineering, University of Sannio, Benevento, Italy

D. Villacci
Department of Engineering, University of Sannio, Benevento, Italy

Sylwia Werbińska-Wojciechowska
Department of Operation and Maintenance of Logistic Systems, Transportation Systems and Hydraulic Systems, Wroclaw University of Science and Technology, Wroclaw, Poland

Petek Yontay
Intel, Inc., Chandler, AZ, United States of America

EDITORS BIOGRAPHY

Dr. Mangey Ram received a Ph.D. degree, majoring in Mathematics and minoring in Computer Science, from G. B. Pant University of Agriculture and Technology, Pantnagar, India. He has been a faculty member for around 10 years and has taught several core courses in pure and applied mathematics at undergraduate, postgraduate, and doctorate levels. He is currently a professor at Graphic Era Deemed to be University, Dehradun, India. Before joining Graphic Era, he was a deputy manager (probationary officer) with Syndicate Bank for a short period. He is editor-in-chief of *International Journal of Mathematical, Engineering and Management Sciences* and the guest editor and member of the editorial board of various journals. He is a regular reviewer for international journals, including IEEE, Elsevier, Springer, Emerald, John Wiley, Taylor & Francis, and many other publishers. He has published 131 research publications in IEEE, Taylor & Francis, Springer, Elsevier, Emerald, World Scientific, and many other national and international journals of repute, and also presented his works at national and international conferences. His fields of research are reliability theory and applied mathematics. He is a senior member of the IEEE, life member of Operational Research Society of India, Society for Reliability Engineering, Quality and Operations Management in India, Indian Society of Industrial and Applied Mathematics, member of International Association of Engineers in Hong Kong, and Emerald Literati Network in the UK. He has been a member of the organizing committee of a number of international and national conferences, seminars, and workshops. He has been conferred with "Young Scientist Award" by the Uttarakhand State Council for Science and Technology, Dehradun, in 2009. He has been awarded the "Best Faculty Award" in 2011 and recently Research Excellence Award in 2015 for his significant contribution in academics and research at Graphic Era.

Dr. J. Paulo Davim received a Ph.D. degree in Mechanical Engineering in 1997, an M.Sc. degree in Mechanical Engineering (materials and manufacturing processes) in 1991, a Mechanical Engineering degree (5 years) in 1986 from the University of Porto (FEUP), the Aggregate title (Full Habilitation) from the University of Coimbra in 2005, and a D.Sc. from London Metropolitan University in 2013. He is Eur Ing by FEANI-Brussels and senior chartered engineer by the Portuguese Institution of Engineers with an MBA and Specialist title in Engineering and Industrial Management. Currently, he is Professor at the Department of Mechanical Engineering of the University of Aveiro, Portugal. He has more than 30 years of teaching and research experience in Manufacturing, Materials and Mechanical & Industrial Engineering with special emphasis in Machining & Tribology. He also has an interest in Management, Engineering Education, and Higher Education for Sustainability. He has guided large numbers of postdoctorate, Ph.D., and Masters students as well as coordinated and participated in several research projects. He has received several scientific awards. He has worked as evaluator of projects for international research agencies as well as examiner of Ph.D. theses for many universities. He is the editor-in-chief of several international journals, guest editor of journals, book editor, book series editor, and scientific advisory for many international journals and conferences. Presently, he is an editorial board member of 25 international journals and acts as reviewer for more than 80 prestigious Web of Science journals. In addition, he has also published as editor (and coeditor) more than 100 books and as author (and coauthor) more than 10 books, 80 book chapters, and 400 articles in journals and conferences (more than 200 articles in journals indexed in Web of Science core collection/h-index 45+/6000+ citations and SCOPUS/h-index 52+/8000+ citations).

PREFACE

Advances in modeling and simulation is arguably one of the most multidimensional topics that one can face in system reliability engineering today. This rapid development also creates many opportunities and challenges for both industrialists and academics, and has completely changed the global design and system engineering environment. More modeling tasks can now be undertaken within a computer environment using simulation and virtual reality technologies.

Through this book, *Recent Advances in System Reliability Engineering*, engineers and academicians will gain the knowledge to help them in system reliability engineering. This book is meant for those who are pursuing reliability engineering as a subject of study. The material is intended for an audience at the level of postgraduate or senior undergraduate students. That's why system reliability engineering is now a well-recognized and rapidly developing branch of engineering.

Topics of focus include:

- Delay-time-based maintenance modeling for technical systems
- Probabilistic and deterministic safety analysis
- Reliability-game theory
- Integrating reliability models and adaptive algorithms for wind power forecasting
- Time-dependent analysis of series-parallel multistate systems using structure function and Markov processes
- Modeling-correlated counts in reliability engineering
- Statistical inference of an imperfect repair model with uniform distributed repair degrees
- Method of fuzzy perfectness in human reliability analysis; selection of performance conditions
- System reliability assessment through Bayesian network modeling
- Multistate multifailures system analysis with reworking strategy and imperfect fault coverage
- Software patch scheduling policy incorporating functional safety standards
- Tools and techniques in software reliability modeling

Through this book, the undergraduate and postgraduate students of engineering, engineers, research scientists, and academicians will gain the knowledge to help them in their reliability courses. This book is meant for those who take reliability and safety as a subject of study.

The editors acknowledge Elsevier for this opportunity and for their professional support. Finally, we would like to thank all the chapter authors for their availability for this work.

Mangey Ram
Dehradun, India

J. Paulo Davim
Aveiro, Portugal

ACKNOWLEDGMENTS

The editors acknowledge Elsevier and the editorial team for their adequate and professional support during the preparation of this book. Also, we would like to acknowledge all of the chapter authors and reviewers for their availability for work on this book project.

Mangey Ram
Graphic Era Deemed to University, Dehradun, India

J. Paulo Davim
University of Aveiro, Aveiro, Portugal

CHAPTER 1

Delay-Time-Based Maintenance Modeling for Technical Systems—Theory and Practice

Sylwia Werbińska-Wojciechowska
Department of Operation and Maintenance of Logistic Systems, Transportation Systems and Hydraulic Systems, Wroclaw University of Science and Technology, Wroclaw, Poland

Abstract

This chapter will be a valuable resource for understanding the latest developments in maintenance modeling issues in the field of delay-time approach implementation. This chapter presents the literature review on delay-time modeling for single- and multiunit systems. The maintenance models for single-unit systems assume two-stage or three-stage failure processes. In the case of complex systems, the discussed problems are with regard to models' parameters, estimation issues, case studies analysis, or hybrid modeling approach implementation. The maintenance models for multiunit systems examine the known models for nonseries systems. A case study for maintenance modeling for multiunit systems based on a delay-time approach is provided. The directions for further research work are defined.

Keywords: Delay-time concept, Multiunit systems, Literature review, Maintenance modeling, Optimization

Abbreviations

AIC	Akaike information criterion
BI	Block Inspection
CM	corrective maintenance
DT	delay time
DTA	delay-time approach
DTM	delay-time models
ELECTRE	elimination and choice expressing the reality
HPP	homogeneous Poisson process
MAUT	multiattribute utility theory
MC	Monte Carlo
MLE	maximum likelihood estimation
MRT	mean repair time
MTBF	mean time between failures
NHPP	nonhomogeneous Poisson process
PAR	proportional age reduction
PM	preventive maintenance
PROMETHEE	preference ranking organization method for enrichment evaluations

Advances in System Reliability Engineering
https://doi.org/10.1016/B978-0-12-815906-4.00001-4

Important Notations

$c(T_{in})$	expected cost over each inspection cycle
$c(T_{ini})$	expected cost over *i*th inspection cycle
$C(T_{in})$	total costs resulting from chosen maintenance policy
c_{cs}	clean-up cost connected with cleaning up any spillage that may result in a failure
c_{in}	cost of inspection action performance
c_p	cost of preventive replacement of a unit
c_r	cost of failed unit replacement
d_{in}	time of single inspection action performance
d_p	the time of preventive replacement of a system
d_r	time of corrective replacement of a system (after a failure)
$E[x]$	expected value of random variable x
$E_d(T_{in})$	expected downtime in an inspection cycle of length T_{in}
$F(t)$	probability distribution function of system/unit lifetime; $\overline{F}(t)=1-F(t)$
$F_h(h)$	probability distribution function of system delay time
$f_h(h)$	probability density function of system delay time
$F_{hi}(h)$	probability distribution function of *i*th element delay time
$f_{hi}(h)$	probability density function of *i*th element delay time
$G_h(t)$	probability distribution function of the initial time u, which elapses from the beginning of operation by "as good as new" elements of a system until the moment of first symptoms of failure occurrence
$g_h(t)$	probability density function of the initial time u, which elapses from the beginning of operation by "as good as new" elements of a system until the moment of first symptoms of failure occurrence
$G_{hi}(t)$	probability distribution function of initial time u of *i*th element in a system
$g_{hi}(t)$	probability density function of initial time u of *i*th element in a system
h	delay time of a defect, denoting the period between the moment of appearance of the first symptoms of potential failure and the moment of an object's failure occurrence
k_u	constant rate for fault's arrival in a technical system for any inspection period
M_{cs}	measure of possible impact of a failure of a piece of equipment on an environment
$M_{env}(T_{in})$	environmental model of a technical object
m	number of failures until element replacement
n	number of elements in a system
n_k	minimum number of operating components to make a system function (performance in n_k-*out-of-n* reliability structure)
$N_{rin}(t^{i-1}_{in}, t^{i}_{in})$	expected number of failures over inspection interval $(t^{i-1}_{in}, t^{i}_{in})$
p_{cs}	probability of a failure resulting in a spillage requiring clean up
p_{in}	probability of imperfect inspection occurrence
$P_{uF}(T_{in})$	probability of a fault arising as a breakdown in a system
p_w	probability that, during system inspection, performance symptoms of forthcoming failures (if they occur in a system) are identified
$R(t)$	reliability function of a system

$R_{Tin}(t)$	component reliability at time t
R_{Tini}^{c}	a cubic approximation of average reliability over an ith inspection
R_{Tini}^{L}	a linear approximation of average reliability over an ith inspection
r_{mr}	repair cost rate
$r_{Tin}^{(i)}(t)$	reliability function for a component at time t
$\dot{r}_{Tin}^{(k)}(kT_{in})$	left-hand derivative of reliability function at $t=kT_{in}$
t_f	random moment of failure occurrence in a system
T_{in}	time between inspection actions' performance
T_{ini}	time of ith inspection cycle performance (between two consecutive inspection actions' performance)
$T_M(T_{in})$	total expected length of a renewal cycle
$T_M(T_{ini})$	expected length of ith inspection cycle
T_{OP}	random variable denoting system's lifetime
$T_{OP}(T_{in})$	expected length of element/system's lifetime
t_u	moment of the first symptoms of potential failure occurrence
u	initial time of a defect
δ_h	standard deviation for normal distribution of a delay time h
δ_r	standard deviation for normal distribution of a time to failure
$\lambda_h(t)$	intensity function of a system delay time
$\lambda_u(u)$	random rate for fault's arrival in a technical system for any inspection period

1 INTRODUCTION

Many real-life systems may display some symptoms of forthcoming failure. One example may be a production process, which may start producing defective items after some random amount of time. If the situation is not corrected, product quality gradually deteriorates to a level where it is self-evident to an operator that the system has failed. By inspecting the product quality at some intervals, the operator may be able to reduce the cost incurred with the system lifetime [1]. These types of components may benefit from an inspection policy whereby a component is inspected for a defect and consequently replaced at inspection to prevent failure [2]. Recent reviews on inspection maintenance modeling issues are presented in the literature [3–9].

The basic inspection models are extended by using a technique called delay-time analysis (DTA), which was developed for modeling the consequences of an inspection policy for any system [10]. This approach was first described and investigated by Christer et al. [10–16].

This maintenance concept is based on the assumption that before a system breaks down, there will be some signs of its reduced performance. The time between the first identification of abnormalities (called initial point)

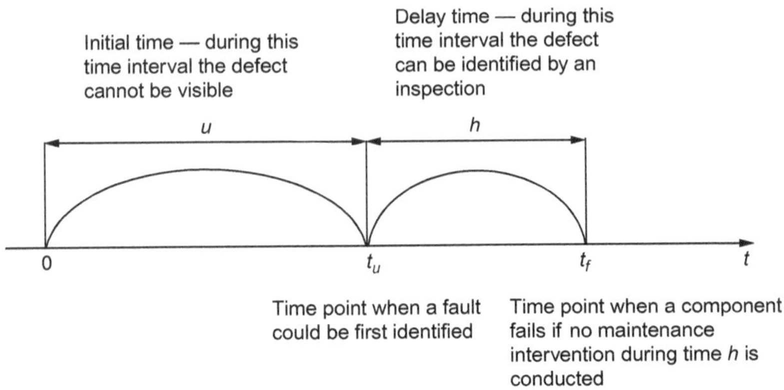

Fig. 1 Time-delay-modeling concept.

and the actual failure time (called failure point) is called a delay time and determines the best opportunity to carry out maintenance or an inspection (Fig. 1). As a result, the delay-time concept defines a two-stage process for three-stated systems (being either in upstate, partially upstate, or in down-state). For more information, the author recommends reading Wang et al. [17].

The inspection schemes in delay-time models may be periodic or based on condition-based maintenance implementation [18]. In this chapter, the author focuses on periodic inspection maintenance modeling issues. More information about condition-based delay-time models may be found in the literature [19–25].

The interest in DT modeling issues has increased significantly in recent years. Literature reviews, in which delay-time models are investigated along with other preventive maintenance models, are given in several studies [26–32]. The states of art works, dedicated to DT modeling, were mostly developed in the 1990s. One of the first literature reviews was given by Christer et al. [19], where the authors discussed a basic DT model for a single-unit case taking into account model parameters estimation problems. Later, the state of art was updated in Christer et al. [33]. In this work, mathematical methods for updating delay-time models of industrial inspection practice are proposed. They presented a linear delay-time update method and model parameter variation. Moreover, the authors also provide a combination of these methods in the maintenance area and defines the criteria for choosing an updating method.

Christer later presented a basic delay-time model [34] and discussed the main development directions, including perfect/nonperfect inspection cases. There have been simple DT models investigated for a repairable component and for a complex plant. Moreover, the authors focused on the problems of a parameter estimation process and DT models' implementation possibilities. The work ends with a discussion of further developments in delay-time modeling. The presented state of art was further extended by Christer [35]. In this work, the author focuses on investigation of industrial maintenance problems and presents basic maintenance models for a repairable component and a complex plant. The main extensions of the basic models include, for example, nonperfect inspection case and nonsteady-state conditions.

One of works that provides recent advances in delay-time-based maintenance modeling is given by Wang [36]. In this work, the author focuses on the main development directions in delay-time-based models. As a result, he investigates the delay-time models for optimized inspection intervals (distinguishing, for example, imperfect inspection and maintenance issues or multiple inspection intervals problem) and DT modeling application areas (including e.g., the problem of spare part inventory and maintenance modeling or maintenance outsourcing issues). The author also defines the main directions for further research in the analyzed maintenance area.

For more information, the author recommends further reading of the literature [2, 6, 37, 38]. Some recent developments in this area are also available for further reading [17, 21, 39].

Following these considerations, in Fig. 2, the main classification of existing delay-time models is presented. The proposed classification divides the known models into two main groups of inspection strategies: single- and multiunit systems.

To sum up, the author summarizes and reviews the existing literature on periodic delay-time maintenance modeling and clusters the existing works into several fields, mostly based on the classification on single- and multiunit models.

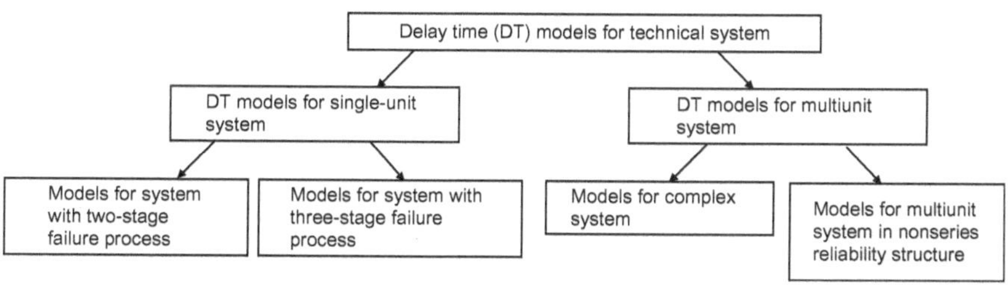

Fig. 2 Classification scheme of delay-time models (periodic inspection).

The articles referred to in this study were found using Google Scholar as a search engine and ScienceDirect, JStor, SpringerLink, SAGEJournals, and Taylor & Francis as online databases. The author primarily searched the relevant literature based on keywords, abstracts, and titles. The following main terms and/or a combination of them were used for searching the literature: *delay time maintenance* or *inspection maintenance*. As a result, 130 papers from 1982 to 2018 are reviewed in this study.

The applicability of the given maintenance approach is presented in a case study. The case study is developed based on historical data of maintenance and operation processes of four production plants (injection molding machines) operated by an international company located in Poland.

These production plants have defined inspection policies based on service regulations defined by a producer. However, a company finds that some developments in this area may be useful to make further decisions.

Following this, this chapter is organized as follows. Sections 2 and 3 examine various types of delay-time modeling approaches for single- and multiunit systems, which are the most widely known in the literature. In Section 4, a case study is proposed to investigate the optimal inspection interval for a four-unit system performing in a series reliability structure. Section 5 concludes the chapter and highlights future research in DT modeling.

2 DELAY-TIME MODELS FOR SINGLE-UNIT SYSTEMS

First, there are delay-time maintenance models for one-unit stochastically failing or deteriorating system that were investigated in which the system's defective state can be detected by actual inspection. In this area, the known models have been explored in two main dimensions: optimization of inspection policy for systems with a two-stage failure process and for systems with a three-stage failure process. Moreover, some research works can be defined that are aimed at investigation and development of special problems (e.g., case studies, comparative studies).

2.1 Delay-Time Models for Single-Unit Systems With a Two-Stage Failure Process

The basic delay-time concept considers that the degenerative process of a system can be divided into two stages: a normal working stage and a delayed defect stage. For this modeling case, the main developments are given in Fig. 3.

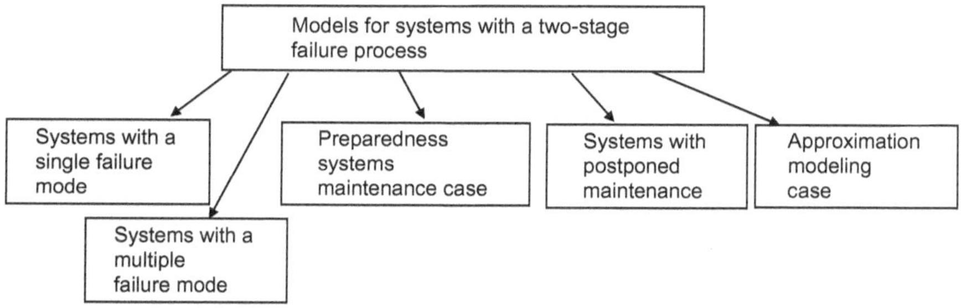

Fig. 3 Classification scheme of delay-time models for systems with a two-stage failure process.

2.1.1 DTM for a Single-Defect Case

One of the first developed DT models for single-unit systems is presented by Christer [11]. In his work, the author develops the reliability model for a single component subject to one type of inspectable defect, which will subsequently lead to a failure. In the presented model, under the policy of inspecting every T_{in} time unit, the author develops a function of component reliability at time t $R_{Tin}(t)$.

Taking into account the following main assumptions:

- a fault may arise randomly at time u since new with *pdf* $g_h(u)$,
- a perfect maintenance policy,
- the time required for inspection is negligible,
- variables u and h are assumed to be independent,
- a infinite horizon case,

the required reliability function $R_{Tin}(t)$ is given by [11]:

$$R_{Tin}(t) = r_{Tin}^{(i)}(t), \quad \text{where } (i-1)T_{in} \le t < iT_{in} \tag{1}$$

where i is a positive integer and $r_{Tin}^{(i)}(t)$ is a reliability function for the component at time t, given by the formula [11]:

$$r_{Tin}^{(i)}(t) = \left\{ \sum_{j=1}^{(i-1)} \left[\left(\int_{u=(j-1)T_{in}}^{jT_{in}} g_h(u) M_h(jT_{in} - u) du \right) r_{Tin}^{(i-j)}(t - jT_{in}) \right] \right.$$
$$\left. + \int_{u=t}^{\infty} g_h(u) du + \int_{u=(i-1)T_{in}}^{t} g_h(u) M(t-u) du \right\},$$
$$\text{where } (i-1)T_{in} \le t < iT_{in} \tag{2}$$

where $M_h(x)$ is defined as [11]:

$$M_h(x) = \int_x^\infty f_h(u)du \tag{3}$$

In the absence of an inspection procedure ($T_{in} = \infty$), the reliability function is given by the following formula [11]:

$$R(T_{in}) = R_\infty(t) = r_{Tin}^{(1)}(t) = 1 - \int_{u=0}^{t} F_h(t-u)g_h(u)du \tag{4}$$

For such a model, a numerical example is provided.

DT models for single-unit systems are later investigated in works by Redmond and Wang [2, 17]. According to these works, there is a possibility to define the *C.d.f.* of time to failure, $F(x)$, as the convolution of u and h such that $u + h \leq x$ [2, 17]:

$$F(x) = \int_{u=0}^{x} g_h(u)F_h(x-u)du \tag{5}$$

and the reliability function, $R(x) = 1 - F(x)$.

Taking into account the following additional assumptions and notation [2, 17]:

- the system is renewed at either a failure repair or at a repair done at an inspection if a defect is identified,
- after either a failure renewal or inspection renewal, the inspection process restarts,
- maintenance actions restore maintained components to good as new condition,
- failures of the system are identified immediately, and repairs or replacements are made as soon as possible,
- perfect inspections case, that is, every defect is assumed to be identified during inspection action performance,

the expected cost over each cycle, $c(T_{in})$, may be defined by the following equation [2, 17]:

$$c(T_{in}) = c_r F(T_{in}) + (c_p + c_{in})\int_{u=0}^{T_{in}} g_h(u)(1 - F_h(T_{in} - u))du + c_{in}(1 - G_h(T_{in})) \tag{6}$$

The expected downtime $E_d(T_{in})$ of an inspection cycle of length T_{in} is given as [34]:

$$E_d(T_{in}) = (d_r - d_p - d_{in})F(T_{in}) + d_p G_h(T_{in}) + d_{in} \tag{7}$$

Assuming instantaneous inspection and replacement times, the expected cycle length, $T_M(T_{in})$, is given by [2, 34]:

$$T_M(T_{in}) = \int_{x=0}^{T_{in}} xF'(x)dx + T_{in}R(T_{in}) \tag{8}$$

Examples of such model implementations are given in the literature [2, 17], as well as an analysis of the problem of model parameter estimation [17].

This model is later extended by Jodejko-Pietruczuk et al. [40, 41]. In one study, the authors focus on the development of a mathematical model for a technical object maintenance, which extends the approach used thus far and allows analyzing the long-term operation time period of a single-unit system [40]. The model gives the possibility to find a constant time period between the inspection actions' performance that is optimal due to minimal maintenance costs or maximal availability criterion satisfaction. The solution is based on Block Inspection policy implementation. The authors present analytical and Monte Carlo simulation models with comprehensive sensitivity analysis performance. This model is an extension of a preliminary analytical maintenance model [42].

Another extension is presented in work by Jodejko-Pietruczuk et al. [41], where the assumption of imperfect inspections is introduced. In this work, it is assumed that the system inspection actions are imperfect, which means that the defect occurrence in the system is identified with the given probability p_w. In such a situation in the analyzed system, the following maintenance actions may occur:

- preventive replacement during inspection action performance, provided that the defect has occurred and been identified with the probability p_w,
- inspection action performance without preventive replacement, provided that the defect has not occurred nor been identified with probability $(1 - p_w)$.

Based on the main assumptions given earlier [40], the long-term expected maintenance costs per unit time function is defined.

2.1.2 DTM for a Multidefect Case

One of the extensions in delay-time modeling regards consideration of more than one type of inspectable and independent defect occurrence in a system. The first model for single-unit systems with n different types of inspectable defects is developed by Christer [11] and later extended by Wang [43]. Wang [43] investigated the production process subjected to two types of deterioration. The delay-time-based inspection model was developed for a component with a minor defect and a major defect. It is assumed that the minor defect is fixed by minor perfect inspections and repairs, whereas the major defect is dealt with by major imperfect inspections and repairs.

The major defects, if they occur, may not be identified by minor inspections and repairs. They have to be detected and rectified by major inspections and repairs, otherwise they can lead to a breakdown of the process. The model also takes into account the possible relationship between the two defined defects due to the possible influence of major defects on the minor defects occurring in the system. The authors aimed at finding the optimal values for both types of inspection intervals by minimizing the long-run expected cost per unit time.

The case of imperfect maintenance model for single components with multiple failure modes was later analyzed by Li et al. [44]. The authors in their work describe an accumulative age concept and imperfect inspection maintenance due to insufficient allocation of maintenance resources. The model is based on the optimization of the average cost per unit time over an infinite period and average reliability function.

2.1.3 Approximation Modeling Case

Another extension of the basic delay-time-based maintenance model for single-unit systems regards approximation modeling. The inspection-maintenance model given earlier [11] was simplified by Cerone [45]. The author in his work developed an average reliability over an ith inspection period ${R_{Tini}}^c$ as a cubic approximation:

$$R^c_{Tini} = \frac{1}{2}\left[r^{(i)}_{Tin}(iT_{in}) + r^{(i)}_{Tin}((i-1)T_{in})\right] - \frac{T_{in}}{12}\left[\dot{r}^{(i)}_{Tin}(iT_{in}) + \dot{r}^{(i)}_{Tin}((i-1)T_{in})\right] \quad (9)$$

and a linear approximation:

$$R^L_{Tini} = \frac{1}{2}\left[r^{(i)}_{Tin}(iT_{in}) + r^{(i)}_{Tin}((i-1)T_{in})\right] \quad (10)$$

The author shows that the linear approximation gives a relative error of the order of 10% and by a cubic resulting in a relative error of less than 1%.

In another study [46], the author investigated a problem named "the Converse problem". The model given by Christer [11] is here extended by the development of a cost model, which involves determining both the number of inspections and the inspection interval that will produce the maximum reliability at some future point in time at minimum cost. There is also a numerical example provided.

In another work [47], a new simplified delay-time model of the reliability function with inspections was obtained. Taking into account the same assumptions as Cerone [46], the author formulated the reliability functions when the distributions of time of defective u and delay time h are respectively negative exponential distributions with parameters λ_u and λ_h. The reliability function for the component is given by Attia [47]:

$$r_{Tin}^{(i)}(t) = \sum_{j=1}^{(i-1)} k_j(T_{in}) r_{Tin}^{(i-j)}(t - jT_{in}) + k_x(t) \tag{11}$$

where:

$$k_j(T_{in}) = \frac{\lambda_u}{\lambda_h - \lambda_u}\left(e^{-\lambda_u t T_{in}} - e^{-\lambda_h T_{in}}\right)e^{-(j-1)\lambda_u T_{in}} \tag{12}$$

and:

$$k_x(t) = \frac{1}{\lambda_h - \lambda_u}\left(\lambda_2 e^{-\lambda_u t} - \lambda_1 e^{-\lambda_h t} e^{(\lambda_h - \lambda_u)(i-1)T_{in}}\right) \tag{13}$$

The author also provided a numerical example, where for different values of λ_u and λ_h, the reliability function is given.

Another approximation model was presented [48], where the author presents the model and method to find the timeliness-based optimal inspection interval. The proposed solution is based on Monte Carlo simulation use, and the optimization process uses the cost rate function.

2.1.4 Preparedness Systems Maintenance Case

The issues of a preparedness system maintenance with DT concept use were investigated by Jia and Christer [49]. The authors developed the long-run availability for a preparedness system based on the renewal theory use. Moreover, the authors in their work investigated various variants of the basic model, including the situations, when (1) a delay-time period exists, but the technology to detect a defect is not available; (2) the delay time is zero, so that only failures are detected; or (3) the system is replaced on a regular basis without any state testing.

The problem of preparedness system maintenance optimization was further investigated [50]. The authors in their work assume that a component may be either weak or strong, so that the time in the good state has a distribution that is a mixture. Following this, the main extensions of another study [49] regards the implementation of a two-phase inspection policy with an anticipated high inspection frequency in early life and low inspection frequency in later life of the component. The cost and reliability models are developed for finite and infinite horizon cases.

Another extension of the previously discussed model [49] is presented by Wang et al. [51]. The authors in their work introduced the availability DT model with imperfect maintenance at inspection. The proportional age reduction (PAR) model was used to present that the accumulative age (wear) caused by previous imperfect maintenance will affect the instantaneous rates of defect and failure at the next maintenance stage. The solution is based on renewal theory use. Moreover, the authors presented the model parameters' estimation method based on maximum likelihood estimation (MLE) use.

An interesting block-based inspection model for a single component was presented by Zhang et al. [52]. The authors in their work focused on the effect of the possible overlapping of inspection span with failure renewal on determining the inspection interval. Based on the renewal theory, the s-expected interval availability function is developed for the infinite horizon case. The solution is also provided with MC simulation implementation. Moreover, the presented model was also compared with the age-based inspection model to show its practical sense.

2.1.5 Postponed Maintenance Case

The last problem regards postponed replacement introduction. Van Oosterom et al. [53] developed a DT model to determine the optimal maintenance policy relaxing assumption of instantaneous replacement enforced as soon as a defect is detected during inspection. The authors' key motivation is to achieve better utilization of the system's useful life and to reduce replacement costs by providing a sufficient time window to prepare necessary maintenance resources. As a result, the PM replacement cost is modeled as a nonincreasing function of the postponement interval. Moreover, there are analyzed two cases: when delay time is either a deterministic or a random variable. This model was later extended by Yang et al. [54], where the authors investigated a system that successively executes missions with random durations. Following this, the inspections are carried out periodically and immediately after the completion of each mission (random inspection). As a result, the replacement is immediate if the system fails or is defective at a periodic inspection. However, in the situation when the system is defective at a random inspection, then replacement will be postponed if the time to the subsequent periodic inspection is shorter than a predetermined threshold, and immediate otherwise. Based on the main assumptions given by Wang [55], the cost model is derived to obtain an optimal periodic inspection interval and postponement threshold. Moreover, the continuation of the investigations on postponed replacement is presented in further study [56]. The authors in their work investigated imperfect inspection performance and opportunity replacements that arise after a positive inspection.

2.2 Delay-Time Models for Single-Unit Systems With a Three-Stage Failure Process

In the known literature, there are some research that introduces a two-level inspection policy model for single-component systems based on a three-stage failure process. Such a failure process divides the system's life into three stages: good, minor defective, and severe defective stages.

One of the first work that investigates DTA implementation for systems with a three-stage failure process is given by Wang [57]. In the developed maintenance model, the inspection actions may not be perfect, in the sense that the minor defective stage may be missed with a given probability, but the severe defective stage of the item is always identified perfectly. As a measure to optimize the inspection interval, the author uses the long-run expected cost function for two main models with respect to the two options connected with maintenance actions taken. The solution is also presented on the example of maintenance optimization for the cold water pumps used in a soft drink company.

Yang et al. [58] extended the model discussed earlier [57] by assuming the proportion of shortening the inspection interval, when the minor defective stage is identified as a decision variable to be optimized. The imperfect maintenance was introduced by Yang et al. [59]. The developed model uses an age reduction concept and is based on MC simulation method implementation.

Zhao et al. [60] introduced the three-stage failure process model with inspection and age-based replacement. According to their assumptions, the system needs to be repaired immediately in case of a severe defective stage identified during an inspection, at a failure, or when it reaches a certain age. The authors proposed two cost models and a numerical example to show the procedure of the model to solve the optimal inspection and age-based replacement intervals.

This model was later extended by Wang et al. [61], where the authors introduced a two-phase inspection schedule and an age-based replacement policy for a single-plant item with a three-stage degradation process. The solution is based in the implementation of the hybrid bee colony algorithm.

2.3 Other Maintenance Models for Single-Unit Systems Where DTA Has Been Used

Other delay-time maintenance modeling problems are mostly aimed at case studies introduction, comparative studies investigation, or hybrid maintenance implementation. Following this, most of the works are aimed at, among others, optimization of inspection and maintenance decisions for infrastructure facilities (see Yamashina and Otani [62] for maintenance optimization for single elevator performance) or transportation means (see Christer et al. [63] for modeling ship reliability optimization). Moreover, Christer et al. [64] compared semi-Markov and delay-time single-component inspection models. The authors' main interest was to investigate to what extent the results of a semi-Markov decision model remain valid, when the Markov property is assumed but is not valid. In another study [65], the authors presented two alternative policies for preventive replacement of a component—age replacement during delay-time policy and opportunistic age replacement policy—and compared them to define the possibilities of their implementation for real-life systems. The last interesting problem regards the integration of multicriteria decision-making techniques with a delay-time model for optimal inspection maintenance policy definition [66, 67]. In one study, Emovon et al. [67] focused on marine machinery system maintenance, and the solution based on ELECTRE and MAUT methods implementation. In a separate study [66], the PROMETHEE decision-making technique was used. The problem of supplier choice in preventive maintenance, including inspection and replacement, was investigated in another study [68].

Delay-time modeling, in the context of risk analysis of maintenance problems, was investigated by Wang [55]. The presented model regards single- and multiunit cases. It is worth mentioning that the developed prototype software packages give a possibility of automating the delay-time modeling process. The software is based on the multiple component system DTM with perfect inspection, and the model parameters are estimated using the moment matching method.

The safety constraints are considered by Aven and Castro in their work [69]. In their paper, the authors considered the basic delay-time model for the system, whose failures are safety critical, thus the risk should be controlled. The model is aimed at determining optimal inspection intervals, T_{in}, minimizing the expected discounted costs under the safety constraints.

Lastly, an interesting problem is presented by Mahfoud et al. [70], where the authors present the proportional delay-time model, which includes parameters related to maintenance effectiveness and working conditions (utilization rate) of the maintained medical equipment. The continuation of maintenance modeling for medical equipment is later continued in a follow-up study [71].

3 DELAY-TIME MODELS FOR MULTIUNIT SYSTEMS

In this chapter, the author presents the multiunit system case. Usually, in the literature, there are developed DT models for complex systems, where generally it is assumed that a system is comprised of many independent component parts, and a breakdown can be caused by any one component (a series structure). The arrival pattern of defects within the system is modeled by an instantaneous arrival rate parameter $\lambda_h(u)$ at time u. If $\lambda_h(u)$ is constant, the model is a homogeneous Poisson process type (HPP), otherwise it is of a nonhomogeneous Poisson process type (NHPP) [6].

In the known literature, there are also some developments for delay-time-based maintenance of systems with nonseries reliability structures. The known analytical solutions mostly regard systems with parallel reliability structures and perfect inspection case. In the next sections, the main achievements in this maintenance modeling research area are presented.

3.1 Delay-Time Models for Complex Systems

First, the complex system maintenance case is analyzed. The general classification of the main delay-time-based maintenance models for complex systems is presented in Fig. 4. The author defines four main groups of DT models according to the main issues considered in the investigated research

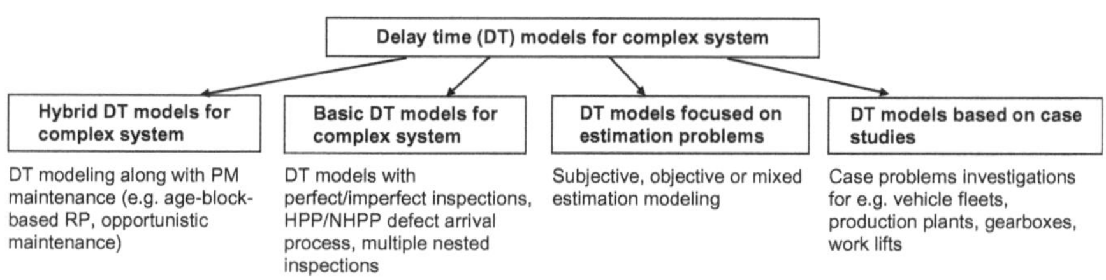

Fig. 4 The general classification of delay-time-based maintenance models for complex system.

works. Usually in the given research works, the authors focus on DT models development together with estimation method investigation or for the known DT models providing comprehensive case studies. The main known research findings are summarized in the following sections.

3.1.1 Delay-Time-Based Models With Parameters Estimation

The first models developed in the '80s and '90s of the 20th century are based mostly on estimation modeling issues and present simple DT models with very strong assumptions.

One of the first models of an inspection maintenance policy was developed for building complexes [15]. The authors in their work developed the cost-based system model for perfect inspection cases. The main assumptions given in this model include perfect inspection case or HPP process of faults arising within the technical system. For such assumptions, it is possible to estimate the probability of a fault arising as a breakdown $P_{uF}(T_{in})$ [15]:

$$P_{uF}(T_{in}) = \int_0^{T_{in}} \left(\frac{T_{in} - h}{T_{in}} \right) f_h(h) dh \tag{14}$$

The expected downtime per unit time to be incurred operating an inspection policy of period T_{in} is given by $E_d(T_{in})$, where [15]:

$$E_d(T_{in}) = \frac{k_u T_{in} d_r P_{uF}(T_{in}) + d_{in}}{T_{in} + d_{in}} \tag{15}$$

With average breakdown and inspection repair costs, c_r and c_{in}, respectively, the expected cost per unit time of maintaining the plant on an inspection system of period T_{in} is $C(T_{in})$ [15]:

$$C(T_{in}) = \frac{1}{T_{in} + d_{in}} \{ k_u T_{in} [c_r P_{uF}(T_{in}) + c_{in}(1 - P_{uF}(T_{in}))] + 1 \} \tag{16}$$

For the developed basic delay-time model, the authors focused on the issues connected with model parameters' estimation process based on subjective estimation method use. They reported on a study of snapshot modeling applied to model the downtime consequences of a high-speed production line maintained under an inspection system. They also proposed the structure of a questionnaire that may be used during the subjective information gathering process (survey research). This problem was later continued and extended in the literature [14, 72], where the authors investigated the problem of vehicle fleet maintenance.

The subjective estimation issues were also investigated by Christer [10]. In the presented research work, the author proposes a technique for assessing the subjective information connected with the expected number of defects identified at an inspection and the consequential cost-saving. The presented approach is later discussed in the context of two industrial case studies [16].

A DT model where inspections may be imperfect was also analyzed [73]. In the presented work, a probability p_w was introduced that a specific defect will be identified at *n*th inspection, and a corresponding probability $(1-p_w)$ that it will not. Moreover, the authors in their work relaxed another assumption connected with the uniform spread of the initiation of defects after an inspection (NHP process).

The fully Bayesian approach in relation to an inspection maintenance decision problem was discussed by Apeland and Scarf [74]. The authors in their work presented the classical approach to delay-time modeling and compared it with the new fully subjective approach, investigating its key features, such as integration of engineering judgments, uncertainty treatment, and the type of performance measures to be used.

The issues of objective estimation of models parameters were also analyzed [75]. The authors in their work considered a repairable machine that may fail or suffer breakdown many times during the course of its service lifetime. The authors presented the DT models for single- and multiunit systems, and for perfect/imperfect inspection cases. The model parameters are fitted by the method of maximum likelihood, and selection of the "best" model is made with the use of the Akaike information criterion (AIC). Later, this research work was continued by Baker and Wang [76], where several extensions to the basic DT model were derived. The main extensions include: (1) allowing the age of a machine to influence both the period u and period h, (2) allowing an inspection to have a hazardous or beneficial effect on the lifetime of a component, and (3) allowing several mechanisms that induce a correlation between the two independent periods of u and h.

Since 1995, the works of Baker et al. [77, 78] have been a continuation of prior research developments [15, 73, 76]. In one work, the authors discussed the problem of sample size influence on the possibility of determining cost-based maintenance policies [77]. In another work, the authors' focus was also placed on the estimation of the model parameters, their errors from records of failure times, and the number of defects found during inspections of a machine [78].

In the previous studies, the parameter estimation of the DTM was solved by using subjective data or objective data. However, there is a majority of research works that use a combination of subjective data and objective data, mainly because there is rarely sufficient maintenance data to allow the use of fully objective data to solve it.

In another study, Wang and Jia [79] developed an approach using a standard hierarchical Bayesian method and the likelihood function. The new estimation method assumes that the initial estimates are made using the empirical Bayesian method matching with few subjective summary statistics provided by the experts. The considered basic delay time inspection model of a complex plant is based on the assumptions given in another work [15].

In the next work [80], the authors presented the DT model and estimation procedure based on the assumption that the historic data exist for failure time points and PM times, but no records exist for the number of defects identified and removed at PM. Thus the authors obtained a subjective estimate of the mean number of defects identified and removed at PM from the factory engineers, who maintained the plant. The analyzed model is mostly based on the assumptions and modeling solutions given earlier [15]. This problem was later continued by the authors in another study [81]. In this work, the authors based the main assumptions of the presented model (given in their work [80]), providing more comprehensive case study of a production plant maintenance interval optimization. The simulation model use confirms the validity of the presented estimation method.

Another interesting research work in this area is given by Jones et al. [82], where the authors developed a methodology of applying delay-time analysis to a maintenance and inspection department. The considered cost and downtime models were based on the assumptions given in an earlier study [15]. The developed environmental model is shown by:

$$M_{env}(T_{in}) = \frac{1}{T_{in} + d_{in}} \{k_u T_{in} [M_{cs} P_{uF}(T_{in}) + c_{cs} p_{cs} (1 - P_{uF}(T_{in}))]\} \quad (17)$$

To demonstrate the applicability of the developed methodology, the authors provided the case study of a factory producing carbon black in the United Kingdom.

3.1.2 Basic DT Models for Complex Systems

Here, the authors mostly developed the new DT models providing some numerical examples and case studies that confirm their applicability.

The estimation problem is usually not considered, or authors base it upon the estimation methods later introduced in this chapter.

One of the first well-known DT models for complex systems is given by Christer and Wang [83]. The authors in their work extended the DT models given in previous studies [14, 15, 75, 76] into two main ways. The first extension regards the implementation of the opportunistic inspection at component failure. Thus the system is inspected on a planned basis and when a component fails. The second extension is the use of a nonhomogeneous Poisson process to describe defect arrivals in the system. For the perfect inspection case, the total expected cost per unit time is minimized with respect to the inspection intervals and the system replacement time:

$$C(T_{in}) = \frac{1}{T_{OP}} \sum_{i=1}^{m+1} \left[c_r N_{rin}\left(\mathrm{t}_{\mathrm{in}}^{\mathrm{i}-1}, \mathrm{t}_{\mathrm{in}}^{\mathrm{i}}\right) + c_p \int_{\mathrm{t}_{\mathrm{in}}^{\mathrm{i}-1}}^{\mathrm{t}_{\mathrm{in}}^{\mathrm{i}}} \lambda_u(u)du + c_{in} \right] + \frac{c_{in}}{T_{OP}} \qquad (18)$$

The developed model is later analyzed by Wang and Christer [84], where the authors provided three solution algorithms that use the recursive procedure to determine the replacement time and reduce the number of decision variables to one, namely the first inspection interval. The first algorithm is developed for obtaining the system replacement time for NHPP defect arrival process. The second algorithm extends the first one by providing the possibility of obtaining the nonconstant optimal inspection intervals. The last algorithm is a numerical algorithm for solving an integral equation arising within the model in the case of opportunistic inspection at failures.

In another work [85], the authors analyzed the DT model for aircraft structure maintenance optimization to decrease maintenance cost rates. The model assumes implementation of imperfect inspection, minimal repair of detected defects, and a finite horizon case. In the proposed model, the NHP process is studied and adopted to obtain the renewal probabilities due to defects or failures between the two different successive inspections. The proposed algorithm is based on the Nelder-Mead downhill simplex method use.

Another extension of the basic DT models given by Christer and Waller [14, 15] is presented by Christer and Lee [86]. The main extension of the developed model is made in the context of downtime modeling. The paper presents the revised models with nonnegligible system downtime for perfect and nonperfect homogeneous processes.

Moreover, an interesting maintenance model for a multicomponent complex engineering system with many components is later given by Wang

[87]. The author in his work presents a delay-time-based maintenance model, where the probability of successful defect identification is an increasing function of the delay time. The expected downtime per unit time and the expected cost per unit time functions are developed based on the assumption of NHPP process of defect arrival.

A prototype software package, developed to automate the delay-time modeling process, is discussed by Wang [55]. The software is developed based on the complex system DTM with perfect inspection and subjective data use [73]. The model parameters are estimated using the moment matching method, where the experts are supposed to provide three pieces of information: mean number of failures over a given time horizon, the mean number of defects identified at T_{in}, and the mean delay time.

A model of two-type inspection services and maintenance was developed by Wang et al. [88]. The main extensions regard a finite horizon case implementation and a dynamic inspection policy introduction due to technological insertions occurrence. Thus the authors developed the model, where upgrading activities and technological insertions are regarded as part of the maintenance planning process, and capability enhancement trade-off decisions are readily incorporated. The solution procedure is based on two stages: first, the model for the infinite horizon case is discussed, and later, the finite horizon modeling scenario is introduced.

A multicomponent monotone safety system inspection issues was by Aven [89]. The developed model is based on the DT concept use for systems comprised of components having an arbitrary finite number of states. The DT models for simple two- and three-unit systems are provided as special cases.

3.1.3 Hybrid Delay-Time Based Models

There have also been a few hybrid delay-time-based maintenance models identified that combine various replacement and inspection policies. One of the developments in this area was presented by Scarf et al. [90]. The authors considered a hybrid maintenance policy for components that arise from a mixture of two subpopulations. The firs subpopulation represents weak, low-quality components, whereas the second represents stronger, more long-lived components. Thus a hybrid maintenance policy based on the assumption that inspections are carried out over the early life of the system (to mitigate early failures of weak components) and age-based replacement is performed during later life of the system (to reduce wear-out failures in later life). With the use of renewal-reward theorem, the long-run cost per unit time is obtained.

Scarf and Cavalcante [91] investigated the problem of a hybrid maintenance policy. The authors proposed hybrids of inspection maintenance and block-based and age-based replacement for multicomponent series system. The developed models based on the assumptions and model of distribution of time to defect arrival were later presented by Scarf et al. [90].

In another study [92], the authors considered a block-based inspection model and focused on the problem of a forward time distribution definition. Based on the general Markov chain implementation, the authors proposed a recursive algorithm for determining a limiting distribution of forward time. The simulation verification based on the assumptions and main model parameters taken from a study by Baker and Wang [75].

The last hybrid maintenance model was presented by Flage [93] and considers preventive age replacement issues. In this work, the author presents the inspection-based maintenance optimization model, where inspections are imperfect and potentially failure-inducing. The main maintenance problem is determining the optimal inspection interval T_{in} and preventive age replacement limit that jointly minimizes the long-run expected cost per unit time.

3.1.4 Delay-Time-Based Model Applications

In this area, the main research works are aimed at implementation of the known DT models and their parameters' estimation methods into a case study. In the first work [94], the authors applied the objective estimation approach to delay-time modeling into a case study. The authors focused on the performance of a copper products manufacturing company, providing the delay-time maintenance model of a plant. This work is a continuation of research problems analyzed earlier [75], based on the assumptions given by Christer and Waller [15]. Later, Christer et al. [95] continued their investigations on DT modeling issues by introducing the subjective estimation method for PM and DT modeling based on the same data study from earlier [94], adopting their earlier assumptions [15]. They presented the implementation of DT model to extrusion press maintenance. In another work [96], the authors also based their assumptions on the literature [14, 73, 94]. The paper presents the reliability data analysis and modeling of an offshore oil platform plant.

Later, Wang focused on the problem of maintenance and inspection for tea bag production lines with a local company [97]. The general model assumptions were adopted from a study by Christer [34].

The next two works reviewed [98, 99] incorporated the DTA methodology presented by Jones et al. [98] with the use of both subjective and objective estimation methods. In the first work [82], the authors provided a simple cost analysis for a factory producing carbon black in the United Kingdom. In the second paper [99], the authors modeled the maintenance and inspection for the same case company [82] but incorporated Bayesian network modeling.

The extension of the maintenance methodology given by Jones et al. [82] was also presented by Werbińska-Wojciechowska and Zając [100]. The authors in their work presented the implementation algorithm for DTA use in the area of logistic systems maintenance performance. The case study was provided for 10 forklifts' performance, based on the objective estimation of CM and PM data and subjective estimation of the delay time parameter.

The subjective estimation method is used in works by Leung et al. [101, 102]. In the first work [101], the authors focused on improving the effectiveness of maintenance policy for the gearboxes operated by a franchised bus company from Hong Kong. The main modeling methodology was based on the assumptions given by Christer and Waller [15]. The next work [102] focuses on applying the DT analysis to model the inspection frequency of fresh water pumping systems. The model based on the developments of modeling approaches are given in the literature [15, 19, 33].

The subjective methods implementation for the case of bus fleet maintenance is presented by Desa and Christer [103]. Due to the problem with inadequate maintenance data for gaining an overview and understating of the nature of the actual maintenance problems, the snapshot modeling approach was applied based on the developments discussed in the literature [15, 16, 86, 94].

Later, Akbarov et al. [104] focused on investigating maintenance-related problems in a chocolate cake manufacturing plant, taking into account production downtime and equipment failure analyses. The data collection phase bases on subjective estimation of the delay time distribution, presented by Wang [110]. The maintenance modeling uses snapshot modeling (according to Christer and Whitelaw [16]) and assumptions presented in the literature [86, 94, 103, 105].

The last work, given by Christer et al. [106], reports on a case study of delay-time modeling of maintenance applied to a key subsystem of a complex machine used in manufacturing brake linings. In the presented paper, they implemented the case of a mixture of the objective method for the failure data (according to developments presented by Baker and Wang [75]) and the subjective method for the PM data (according to developments presented by Christer et al. [94]). The adopted DT model was based on the imperfect inspection case and HPP/NHPP arrival rate of failures.

Other research works were aimed at case studies' implementation with regard to, among others, production system maintenance planning [107], maintenance and inspection policies development for marine systems [108], or optimal inspection policy determination for protection devices of electronic equipment [109].

3.1.5 Other Maintenance Models Where DTA Has Been Used

In this area, one paper of note was presented by Wang [110], where the author presents a model for maintenance service contract design, negotiation, and optimization. In the presented paper, the three contract options were considered with regard to the extent of outsourced maintenance activities. The aim of this paper was to develop a model to obtain the optimal strategies for the customer (to optimize the contract parameter under each option) with a required availability and reliability level and fixed budget.

A different problem was considered by Scarf and Majid [111], where the authors focused on the problem of modeling the cost of warranty extensions for a particular automotive supplier in Malaysia. To model the effect of services during the warranty period, an inspection maintenance model based on the delay time concept was used.

Moreover, joint analysis of spare part provisioning and maintenance planning issues should be investigated. In this area, valuable contributions are given in the literature [112–114]. Wang [113] proposed a component-based DTM to model the connection between spare parts and maintenance of n identical items installed in a production line or a fleet of transportation vehicles. The objective function is the long run-expected cost per unit time in terms of the three decision variables to be optimized: the ordering quantity, ordering interval, and inspection interval. These problems are later continued in another study by Wang [112], where the author introduces an enumeration algorithm with stochastic dynamic programming for finding the joint optimal solutions over a finite-time horizon. The problem of demand for spare parts forecasting is later investigated by Wang in another work [114]. The author compares demand forecasting obtained from a statistical time-series forecasting method and maintenance-based methods using simulated data from a well-known maintenance models given in the literature [17, 75].

3.2 Delay-Time Models for Systems With Nonseries Reliability Structures

In the given research area, there are few papers that investigate DT modeling for systems with nonseries reliability structures. One of the first works in this area was by Jodejko-Pietruczuk and Werbińska-Wojciechowska [115]. The authors in their work focused on the comparison of two common maintenance models taking into account the same assumptions for both of them. The first model regards the group maintenance policy, and the second model is based on the DT approach (BIP implementation). The investigated system is a three-component system with an n_k-out-of-n reliability structure. For perfect inspection case and an infinite planning horizon, the simulation model was given. Later, a more detailed analysis of the BI maintenance model was given by Jodejko-Pietruczuk and Werbińska-Wojciechowska [116], and problems with the model's parameter's estimation were also investigated by the authors [117].

The developed simulation models [116] were later extended in future studies [118, 119]. In both papers, the authors introduced imperfect inspection cases. Thus any component's defect, which occurred in the system until the moment of inspection, will be unnoticed with probability p_{in} or correctly identified with probability $1 - p_{in}$. Following this, when p_{in} is equal to 0, it means that all components are correctly diagnosed (perfect inspection case). All elements with identified defects will be replaced within the inspection period.

In the same year, Wang et al. [120] considered a two-component system in parallel. They presented an analytical model for perfect inspection and maintenance with exponentially distributed time to defect. They analyzed eight different inspection renewal scenarios and one failure renewal scenario that gave them the possibility to estimate the expected cost per unit time.

This model was later extended by Liu et al. [121], where the authors proposed a new delay-time-based inspection model for an n-component parallel system and study optimal inspection interval that minimizes the long-term expected cost per unit time. The inspection and maintenance actions are perfect. For the defined assumptions, the authors considered two maintenance policies. The first maintenance policy assumes that the defective or failed components are replaced only when all components are either defective or failed at an inspection before a system failure. This maintenance policy is estimated analytically. The second, more general one is solved with the use of simulation modeling.

Jodejko-Pietruczuk and Werbińska-Wojciechowska [122] presented the cost delay-time-based maintenance model for a multiunit system performing in an n_k-out-of-n reliability structure and a simple method of determining the near-optimum inspection period length T_{in}. The solution is based on the renewal theory use (the analytical model of the first inspection period) and simulation procedure implementation for an infinite time horizon case. The problem of the model's parameter's estimation was later analyzed by the authors [123].

In their last work [124], the authors presented the maintenance policy based on inspections and replacement, taking into account opportunities for a system composed of two components, whose failure characteristics are different. They used a simulation to obtain the minimal expected maintenance costs. Moreover, this model is the first attempt to model economic and structural dependency with the use of DT modeling.

The main target of the developed literature review was to provide the synthetic summary of the results on delay-time-based maintenance. The conducted research gives the possibility to identify the potential future research directions on one side. The applicability of the delay-time approach is presented via a case study, given in Section 4.

4 CASE STUDY

The applicability of DT modeling is presented on the basis of chosen enterprise from the plastics industry, located in the center of Poland. Currently, the company hires more than 300 people and produces more than 200 articles distributed on the worldwide markets. The company produces, among other things, covers and plant pots, watering systems, serving bowls, dish drainers, vases, and boxes. All the products are categorized into three main groups: products for kitchen, garden, and cemetery line.

The chosen production company offers services of plastic processing with an injection molding method almost in all respects. They offer full service when it comes to injection molding: starting from the project, through its construction to the mold, technological trials, and finally production.

The company also offers production of injection molds based on client's samples, drawings, projects, or even just rough conceptions.

The production processes are performed with the use of energy- and work-efficient injection molding machines with a clamping force ranging from 86 to 1200 tons. All of the machines have coworking robots, and the company is mostly aimed at large volume serial production. Recently, the production company operated 33 injection molding machines situated in two main production halls.

Maintenance of plants' park is one of the basic conditions for the production company to be able to meet customers' expectations and to provide the required quantity of products, consistent with quality requirements. In the analyzed production company, the core production processes are performed with the use of injection molding machines. Thus any hazard event occurrence during the operation of these machines may result in higher maintenance costs, lower product quality, or delays in master production scheduling.

Following this, the author focused on the maintenance of these kind of machines. These production plants have defined inspection policy based on service regulations defined by a producer. However, the company finds that some developments in this area may be useful to make further decisions.

Injection molding machines are used to form a thermoplastic material by injection molding under pressure. The injection molding machine consists of the following elements:

- granulate dispenser,
- an injection cylinder, in which the material melts,
- a screw used to press the material,
- mold form that cools the plastic, resulting in a molder obtaining.
- The plant's main components are:
- injection system (hopper, cylinder, piston, or screw),
- drive system (electric motor with hydraulic pump or electric drive),
- closing and opening system (hydraulic cylinders),
- regulation and control system,
- cooling systems (hydraulic oil cooling system and mold cooling system).

In the given company, maintenance processes are based on operators' control carts and maintenance cards implementation. This system assumes that operators may perform some of simple maintenance actions. These are simple activities that do not require specialized tools. These activities are primarily connected with cleaning, lubrication, and observation of equipment during operation.

The task of each operator after finishing his work shift is to fill in the control card according to the developed instructions for autonomous maintenance. This employee's duty helps to detect faults, which can be a symptom of failure. These activities increase the safety, cleanliness, and reliability of workstations in a plant.

The control card should be checked every day, at least once during the shift. The obligation to complete it belongs to an operator of an injection molding machine (confirmed by operator's signature). If any irregularities are noticed, a repair order should be issued. The main symptoms of forthcoming failure that are checked by an operator includes:

- plastic leak occurrence,
- oil leak occurrence,
- any custom noises occurring during machine operation,
- visual inspection of electrical plugs,
- control of a mold and oil cooling system temperature.

Moreover, maintenance inspection operations are performed periodically. In the company can be distinguished two types of inspection cards that are related to periodic testing. The first of them concern the periodical maintenance of injection molds. The injection mold subjected to periodic testing is transferred to the tool shop. During an inspection, mechanics check a cooling system, a hydraulic system, a pneumatic system, an intake system, a dismantling system, and forming sockets. At the same time, an electrician checks all electric wires looking for abrasions or loose connections. The second inspection card concerns injection molding machines. This card is divided into six parts. Each part concerns a different system (connected with plant functional decomposition), which consists of several subitems that should be periodically inspected.

A plant inspection schedule includes weekly (or after 145 working hours) and quarterly (or after 2500 working hours) maintenance. The activities performed every week include, for example, checking relays or air conditioning units. A maintenance mechanic performs these checks.

Activities that are carried out once every 3 months include, for example, checking of limit switches. During this inspection, a plant has to be stopped because not all operations can be performed on a running injection molding machine.

Analysis of failure data obtained from the company gives a possibility to define which injection molding machines generate the highest maintenance costs and time loss in production. The gathered data includes such information as:

- failure type (failure of machine, injection mold, or other plants in a production line),
- date of failure occurrence,
- number of production plant,
- comment (characteristic of occurred failure),
- repair order number,
- number of product being manufactured in a time of failure occurrence,
- information if a failure is removed,
- repair time.

Following exemplary results (Fig. 5), the most maintenance time consumed in August 2017 were four injection molding machines marked with numbers 1, 2, 3, and 4. The reason for this is the dependence of these machines on each other, because together they form a manufacturing cell and perform in a series reliability structure. The second reason is connected with a significant increase of production processes performance with the use of these four machines. Moreover, unavailability of this manufacturing cell increased a failure rate of cooperating machines by as much as 18h only in August 2017. This generates considerable losses.

Following this, the author investigated an optimal inspection interval for the given manufacturing cell composed of a four-unit system performing in a series reliability structure.

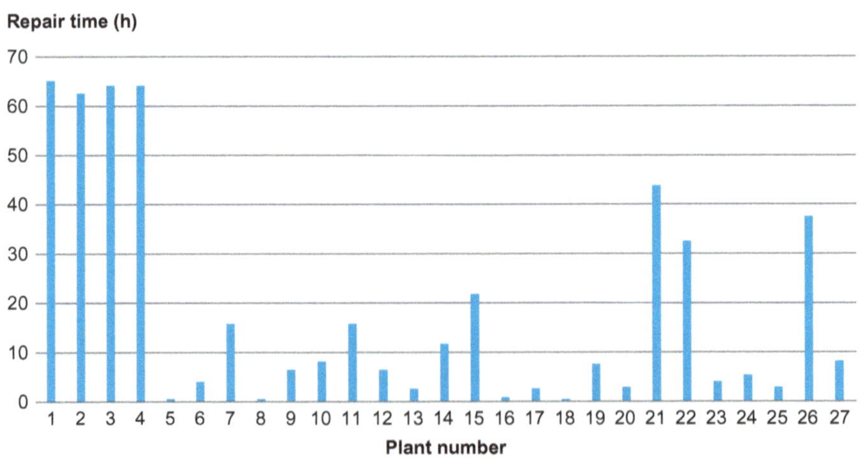

Fig. 5 Total repair time for injection molding machines in August 2017.

To find an optimal (or best) inspection interval for the analyzed system, delay time modeling may be implemented. The chosen analytical DT model is based on Block Inspection policy (BI) which assumes that inspections take place at regular time intervals of T_{in}, and each requires a constant time. This simple DT model was investigated in more depth [125]. Moreover, the detailed sensitivity analysis of the developed analytical DT model is presented by Jodejko-Pietruczuk and Werbińska-Wojciechowska [123], where the authors also focused on the problem of the model's parameters estimation process.

The main assumptions include:

- the system is a three-state system where, over its service life, it can be either operating, operating acceptably, or down for necessary repair or planned maintenance,
- failures of the system are identified immediately, and repairs or replacements are made as soon as possible,
- inspections are assumed to be perfect; any components' defect, which occurred in the system until the moment of inspection, will be identified,
- inspection action performance begins the new maintenance cycle for the analyzed system,
- all elements with identified defects will be replaced within the inspection period,
- maintenance actions restores system to good as new condition.

As a result, one of the two maintenance operations may be performed in the analyzed system: failure repair or inspection together with replacement of elements with defects. Following this, it is assumed that, when a system failure occurs, there is only performed replacement of failed components without additional inspection action performance. However, in the case of planned inspection action performance, the replacement will be performed only for those elements with visible symptoms of their future damage.

According to the renewal theory [126], the expected maintenance costs per unit of time may be estimated based on the equation:

$$C(T_{ini}) = \frac{c(T_{ini})}{T_M(T_{ini})}, \quad \text{where } T_{ini} = T_{in1} \tag{19}$$

The expected costs of four-element system maintenance in the first inspection cycle are defined as [125]:

$$c(T_{in1}) = c_r F(T_{in1}) + c_p(1 - F(T_{in1}))\left(\sum_{i=1}^{4} G_{hi}(T_{in1})\right) + c_{in}(1 - F(T_{in1})) \tag{20}$$

The maintenance cost expressed in Eq. (20) presents the sum of possible cost of system failure, replacement cost of working elements with observable defects, and inspection costs per a single inspection period. It should be underlined here that the developed mathematical model gives the possibility for an estimation of expected maintenance costs for the system, whose elements are as good as new at the beginning of an inspection cycle of length T_{in} (e.g., first inspection cycle performance being a part of a single renewal cycle).

Moreover, the expected system lifetime T_M in the same period [125]:

$$T_M(T_{in1}) = \int_0^{T_{in1}} R(u)du \tag{21}$$

and the system reliability function $R(x)$ is estimated according to the equation [125]:

$$R(x) = 1 - F(x) = 1 - \left\{1 - \prod_{i=1}^{4}\left[1 - \int_{u=0}^{x} g_{hi}(u)F_{hi}(x-u)du\right]\right\} \tag{22}$$

To implement this model, there is an assumption that all plants may be defined by the same economic and reliability parameters. It is possible because these injection molding machines operate together in one manufacturing cell and usually have similar degradation processes that results from the performance of operational processes and operational conditions.

Maintenance data obtained from a company and experts' opinion gave the possibility to estimate the maintenance models main parameters (Table 1).

Following the obtained information, preventive maintenance action is connected with the expected cost of 500 zl (cost includes labor and new element purchasing) and remaining useful life wasting. However, if a failure occurs, the consequences may be very harmful. The main consequences are connected with production delays or manufacturing of products with

Table 1 Parameters of the analyzed production system

Element	MTBF	*E* (h)	δ_h	MRT	Inspection time	c_r	c_p	c_{in}
Injection molding machine	15 h	2 h	0.5 h	2.5 h	0.5 h	1500 zl	500 zl	50 zl

poor quality. Thus basic inspection actions are performed in every 145 working hours (performed by a service man) and per every shift (performed by a plant operator).

Based on historical data and many years of experts' experience, it is estimated that time to failure and delay time may follow a normal distribution with parameters $N(15,2)$ and $N(2,0.5)$, respectively.

For the given parameters, the author discusses the optimal inspection interval to minimize cost based on the simple DT model.

Following this, the author first bases assumption on Eqs. (19)–(22), where the initial time and delay time follow a normal distribution. Numerical results for applying the analytical DT model are presented in Figs. 6 and 7.

As a result, Fig. 6 shows the expected costs as a function of the optimal inspection interval for the first inspection period. For the given assumptions and parameters values, the optimal T_{in} should be equal to 13 h (the minimum cost per unit of time is equal then to 0.03846). The obtained result is close to the value of MTBF, what is obvious for the first inspection cycle. Inspection action performance at time moment of 13 working hours of production plant could maximize operational time of analyzed units and provide the shortest remaining useful life wasting. This result is comparable to the

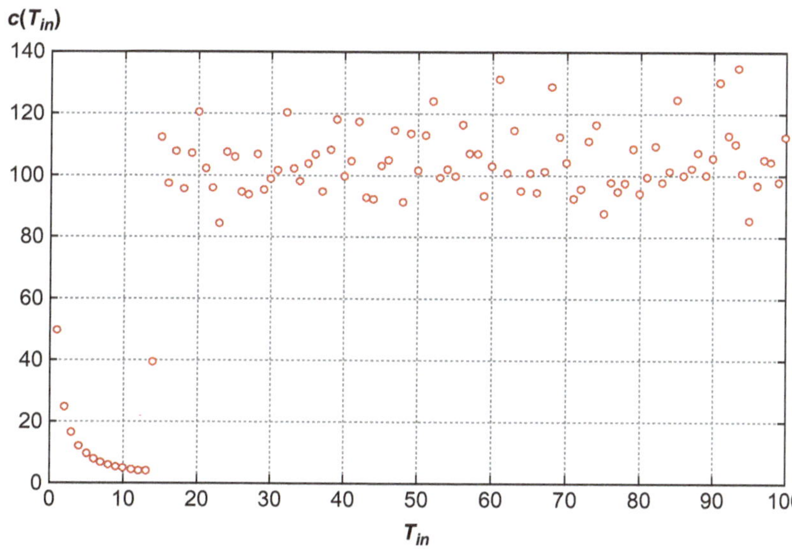

Fig. 6 Expected maintenance costs per unit of time ($C(T_{in})$) for the first inspection period.

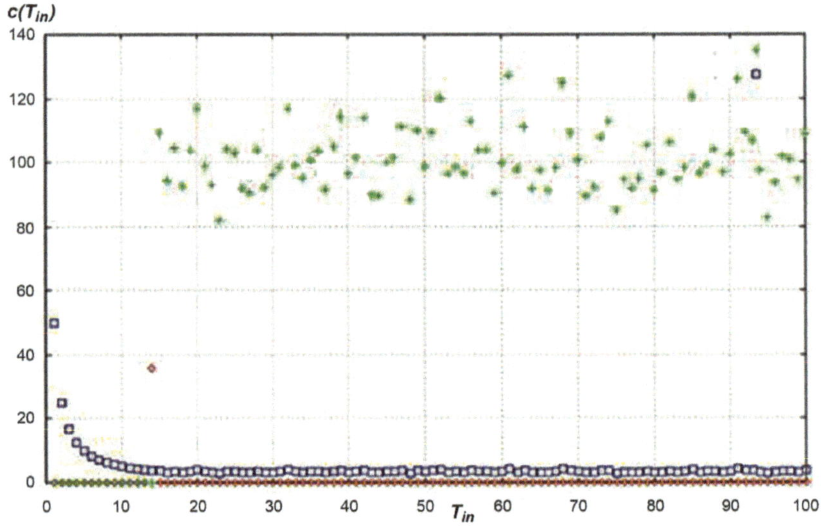

Fig. 7 Expected costs per unit of time for the first inspection period for preventive maintenance performance *(red color)*, consequences costs *(green color)*, and inspection action performance *(blue color)*.

maintenance time period-connected production company procedures related to simple maintenance performance by an operator (per every working shift). However, taking into account a failures' history, it seems to be reasonable to verify if the defined checklist is sufficient. What is also visible, for relation $T_{in} \approx \text{MTBF} + \delta_r$, the expected maintenance costs function is determined by consequences cost occurrence (all units fail until that time).

The process of searching of the best inspection period in an infinite time horizon is based on the simulation process use. The used simulation model is presented, e.g. in [115,125]. The obtained results are presented in Figs. 8 and 9.

When analyzing an infinite time horizon, the optimal time between inspections is not so obviously visible. For the given assumptions, the optimal time between inspections T_{in} is equal to 10h (at minimum cost level=4.425) and is consistent with the results obtained for the short time horizon (the same T_{in} value).

Such time period is more than 10 times shorter than the inspection interval defined for weekly inspections. As for the previous calculations, this result is significantly determined by consequences costs and should be revised carefully, taking into account working conditions, cost requirements, and failure history.

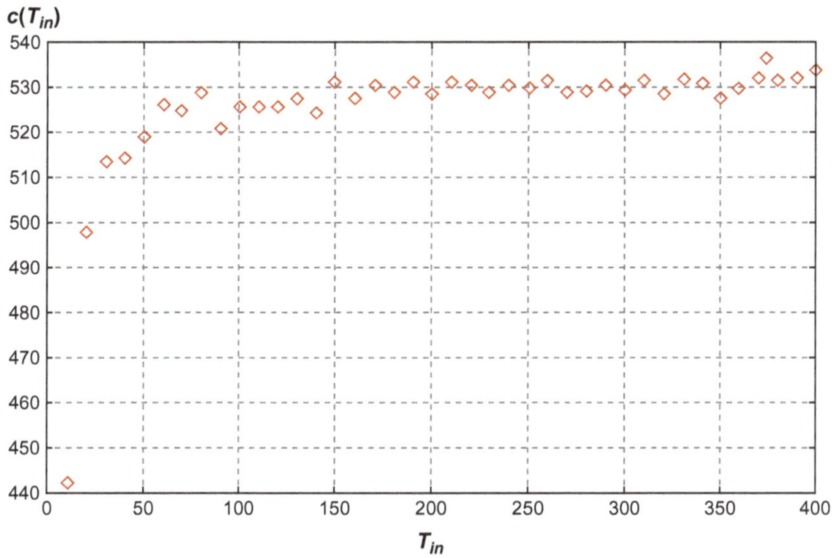

Fig. 8 Expected maintenance costs per unit of time ($C(T_{in})$) (simulation results for an infinite time horizon, series structure).

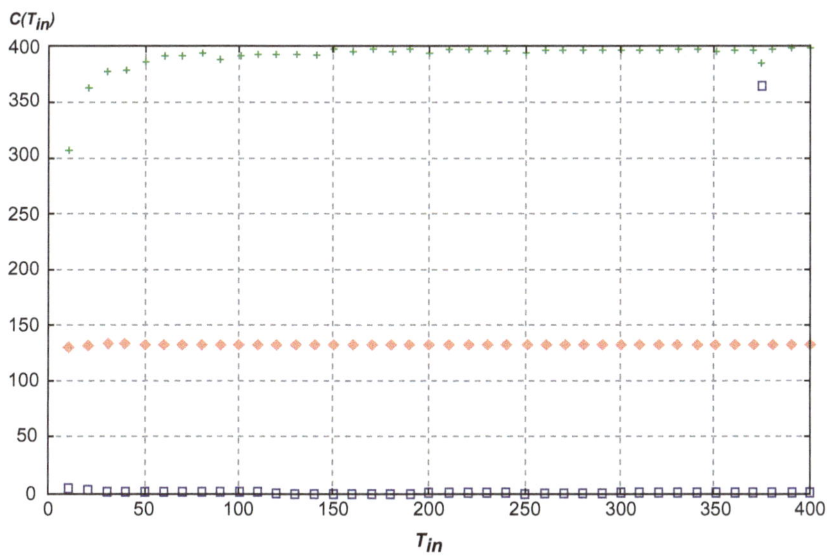

Fig. 9 Expected costs per unit of time for preventive maintenance performance *(red color)*, consequences costs *(green color)*, and inspection action performance *(blue color)* (simulation results for an infinite time horizon).

5 SUMMARY AND DIRECTIONS FOR FUTURE RESEARCH

The delay time concept is a convenient approach to maintenance modeling and optimization of various types of systems, such as production plant, transportation systems, or civil engineering. This modeling technique is specifically for inspection-based optimization in terms of inspection intervals, but it has a close connection with condition monitoring and spare part inventory control [36, 127].

The presented literature review mainly focuses on two research areas: maintenance modeling and applications. The presented theory is still developing, and the discussion of further developments in delay-time modeling is given in the literature [17, 34, 36].

The case study confirms the ease of applying the developed DTMs, and the results for analytical models (first inspection cycle) seem logical in relation to the given assumptions, and reliability and cost parameters' levels. In general, the current inspection policy for the production system is acceptable. The production operator may reschedule inspections to shorter time intervals depending on the working conditions and cost requirements.

Based on the presented material in this chapter literature review, some thoughts and conclusions regarding the future research directions and topics are presented here. They partially confirm research findings given in the literature [17, 34, 36]:

- *Single defect case*—Much work assumes a single and specific type of inspection. Few works regard multiple nested inspections, and they are usually developed for a single-unit case [43, 44]. In practice, inspections can be hierarchical or synchronized, where several systems are inspected and maintained at the same time. This research direction can be viewed as an extension of the model given by Wang et al. [88], where dynamic inspection policy is investigated.
- *Components dependency*—The known delay-time-based maintenance models omit the systems' elements dependency (stochastic, economic, or structural), which may not reflect real life. The main DT modeling issue, which may be treated as an extension in this area, regards a system's multiple failure mode investigation [128, 129].
- *Reliable maintenance data availability*—In the literature, there exists three main techniques to estimate the delay-time parameters given objective data, subjective data, or a mixture. A set of known works with applications have given consistent results. However, the author puts an emphasis on problems by gathering well-tested methods to collect adequate data to enable the calculation of optimum maintenance policies for dedicated

systems. The problem of effective estimation procedures for models' parameters is also indicated as to be worth investigating [17, 36].

- *Simplified assumptions of maintenance policy parameters' probability density functions estimation*—The well-known models assume exponential approximation of model parameters' distributions and Poisson defect arrival time. It is a convenient situation from a modeling point of view but usually not sufficient to real-life system's performance.
- *Imperfect inspection case*—Most works assume that the defective and failed state can be always revealed if they were there [36, 130]. However, some defects may be missed during inspections due to, for example, working conditions, human factors, or unreliable facility use. Based on the developed literature review, there is a gap in this research area on joint analysis of imperfect inspections performance and delay-time modeling for multicomponent systems performing in various reliability structures.
- *Dynamic inspection policy*—In many systems, the inspection interval may not be constant, and therefore the consideration of a dynamic inspection policy is required. Some developments in this area may be found where the authors propose an algorithm for the optimal dynamic determination of the sequential inspection intervals using dynamic programming [88]. They consider a complex system case.
- *Linking of maintenance with spare part inventory modeling*—The analyzed models are based on the assumption that, once the system fails, all the resources needed for its renewal process are immediately available for use (logistics availability). This simplified assumption usually cannot be satisfied due to some randomness of logistic processes and failure processes performed in real-life systems. Some recent developments in this area are presented in the literature [112–114].

Other research for future directions are presented in the literature [17, 36].

The discussions in this chapter literature review and research directions show that this research area is still growing, and the opportunities and challenges of today and future development (e.g., computer science, system complexity) will need further investigation.

REFERENCES

[1] B. Senegupta, Inspection procedures when failure symptoms are delayed, Oper. Res. 28 (3) (1980) 768–776.

[2] D.F. Redmond, Delay Time Analysis in Maintenance (Ph.D. Thesis), University of Salford, Salford, 1997.

[3] A. Chelbi, D. Ait-Kadi, Inspection strategies for randomly failing systems, in: M. Ben-Daya, S.O. Duffuaa, A. Raouf, J. Knezevic, D. Ait-Kadi (Eds.), Handbook of Maintenance Management and Engineering, Springer, London, 2009.

[4] A. Chelbi, D. Ait-Kadi, H. Aloui, Optimal inspection and preventive maintenance policy for systems with self-announcing and non-self-announcing failures, J. Qual. Maint. Eng. 14 (l) (2008) 34–45.
[5] H. Guo, F. Szidarovszky, A. Gerokostopoulos, P. Niu, On determining optimal inspection interval for minimizing maintenance cost, Proc. of 2015 Annual Reliability and Maintainability Symposium (RAMS), 2015, pp. 1–7.
[6] C. Lee, Applications of Delay Time Theory to Maintenance Practice of Complex Plant (Ph.D. Thesis), University of Salford, UK, 1999.
[7] T. Nakagawa, Random Maintenance Policies, Springer-Verlag, London, 2014.
[8] Y.S. Sheriff, Reliability analysis: optimal inspection & maintenance schedules of failing equipment, Microelectron. Reliab. 22 (1) (1982) 59–115.
[9] T. Tang, Failure Finding Interval Optimization for Periodically Inspected Repairable Systems (Ph.D. Thesis), University of Toronto, 2012.
[10] A.H. Christer, Modelling inspection policies for building maintenance, J. Oper. Res. Soc. 33 (1982) 723–732.
[11] A.H. Christer, Delay-time model of reliability of equipment subject to inspection monitoring, J. Oper. Res. Soc. 38 (4) (1987) 329–334.
[12] A.H. Christer, Comments on finite-period applications of age-based replacement models, IMA J. Math. Manag. 1 (1986) 111–124.
[13] A.H. Christer, E. Keddie, Experience with a stochastic replacement model, J. Oper. Res. Soc. 36 (1) (1985) 25–34.
[14] A.H. Christer, W.M. Waller, An operational research approach to planned maintenance: modelling P.M. for a vehicle fleet, J. Oper. Res. Soc. 35 (11) (1984) 967–984.
[15] A.H. Christer, W.M. Waller, Reducing production downtime using delay-time analysis, J. Oper. Res. Soc. 35 (6) (1984) 499–512.
[16] A.H. Christer, J. Whitelaw, An operational research approach to breakdown maintenance: problem recognition, J. Oper. Res. Soc. 34 (11) (1983) 1041–1052.
[17] W. Wang, Delay time modelling, in: A.H. Kobbacy, D.N. Prabhakar Murthy (Eds.), Complex System Maintenance Handbook, Springer, London, 2008.
[18] T. Nowakowski, S. Werbińska-Wojciechowska, in: T. Berenguer, A. Grall, A. Guedes-Soares (Eds.), Developments of time dependencies modelling concept, Advances in Safety, Reliability and Risk Management—Proc. of European Safety and Reliability Conference ESREL, Taylor and Francis Group, London, 2011.
[19] A.H. Christer, D.F. Redmond, A recent mathematical development in maintenance theory, IMA J. Math. Appl. Bus. Ind. 2 (1990) 97–108.
[20] A.H. Christer, W. Wang, A model of condition monitoring of a production plant, Int. J. Prod. Res. 30 (9) (1992) 2199–2211.
[21] T.F. Lipi, J.-H. Lim, M.J. Zuo, W. Wang, A condition- and age-based replacement model using delay time modelling, Proc. Inst. Mech. Eng. O: J. Risk Reliab. 226 (2) (2012) 221–233.
[22] S. Okumura, An inspection policy for deteriorating processes using delay-time concept, Int. Trans. Oper. Res. 4 (5–6) (1997) 365–375.
[23] S. Okumura, A.K.S. Jardine, H. Yamashina, An inspection policy for a deteriorating single-unit system characterized by a delay-time model, Int. J. Prod. Res. 34 (9) (1996) 2441–2460.
[24] W. Wang, Modelling Condition Monitoring Inspection Using the Delay-Time Concept (Ph.D. Thesis), University of Salford, Salford, 1992.
[25] W. Wang, P.A. Scarf, M.A.J. Smith, On the application of a model of condition-based maintenance, J. Oper. Res. Soc. 51 (11) (2000) 1218–1227.
[26] R. Dekker, P.A. Scarf, On the impact of optimisation models in maintenance decision making: the state of the art, Reliab. Eng. Syst. Saf. 60 (1998) 111–119.

[27] R. Guo, H. Ascher, E. Love, Towards practical and synthetical modelling of repairable systems, Economic 16 (2) (2001) 147–182.
[28] R. Guo, H. Ascher, E. Love, Generalized models of repairable systems a survey via stochastic processes formalism, ORiON 16 (2) (2000) 87–128.
[29] T.A. Mazzuchi, J.M. Van Noortwijk, M.J. Kallen, Maintenance Optimization (Technical Report, TR-2007-9), (2007).
[30] P.A. Scarf, On the application of mathematical models in maintenance, Eur. J. Oper. Res. 99 (1997) 493–506.
[31] L.C. Thomas, D.P. Gaver, P.A. Jacobs, Inspection models and their application, IMA J. Math. Appl. Bus. Ind. 3 (1991) 283–303.
[32] C. Valdez-Flores, R. Feldman, A survey of preventive maintenance models for stochastically deteriorating single-unit systems, Nav. Res. Logist. 36 (1989) 419–446.
[33] A.H. Christer, D.F. Redmond, Revising models of maintenance and inspection, Int. J. Prod. Econ. 24 (1992) 227–234.
[34] A.H. Christer, Developments in delay time analysis for modelling plant maintenance, J. Oper. Res. Soc. 50 (1999) 1120–1137.
[35] A.H. Christer, Review of delay time analysis for modelling plant maintenance, in: S. Osaki (Ed.), Stochastic Models in Reliability and Maintenance, Springer, Berlin, Heidelberg, 2002.
[36] W. Wang, An overview of the recent advances in delay-time-based maintenance modelling, Reliab. Eng. Syst. Saf. 106 (2012) 165–178.
[37] R.D. Baker, A.H. Christer, Review of delay-time OR modelling of engineering aspects of maintenance, Eur. J. Oper. Res. 73 (1994) 407–422.
[38] K.M. Choi, Semi-Markov and Delay Time Models of Maintenance (Ph.D. Thesis), University of Salford, UK, 1997.
[39] J.R. Jiang, Relationship between delay time and gamma process models, Chem. Eng. Trans. 33 (2013) 19–24.
[40] A. Jodejko-Pietruczuk, S. Werbińska-Wojciechowska, Development and sensitivity analysis of a technical object inspection model based on the delay-time concept use. Maint. Reliab. 19 (3) (2017) 403–412, https://doi.org/10.17531/ein.2017.3.11.
[41] A. Jodejko-Pietruczuk, S. Werbińska-Wojciechowska, Block inspection policy model with imperfect maintenance for single-unit systems, Procedia Eng. 187 (2017) 570–581.
[42] A. Jodejko-Pietruczuk, T. Nowakowski, S. Werbińska-Wojciechowska, Time between inspections optimization for technical object with time delay, J. Pol. Saf. Reliab. Assoc. 4 (1) (2013) 35–41 (Summer Safety and Reliability Seminars).
[43] W. Wang, An inspection model for a process with two types of inspections and repairs, Reliab. Eng. Syst. Saf. 94 (2) (2009) 526–533.
[44] X. Li, R. He, Z. Yan, H. Hu, G. Cheng, A new imperfect maintenance model based on delay-time concepts for single components with multiple failure modes, Int. J. Syst. Assur. Eng. Manag. 6 (4) (2015) 479–486.
[45] P. Cerone, On a simplified delay time model of reliability of equipment subject to inspection monitoring, J. Oper. Res. Soc. 42 (6) (1991) 505–511.
[46] P. Cerone, Inspection interval for maximum future reliability using the delay time model, Eur. J. Oper. Res. 68 (1993) 236–250.
[47] A.F. Attia, Estimation of the reliability function using the delay-time models, Microelectron. Reliab. 37 (2) (1997) 323–327.
[48] J.R. Jiang, in: A timeliness-based optimal inspection interval associated with the delay time model, Prognostics and System Health Management Conference (PHM-2012 Beijing), 2012, pp. 1–5.
[49] Z. Jia, A.H. Christer, A periodic testing model for a preparedness system with a defective state, IMA J. Manag. Math. 13 (2002) 39–49.

[50] C.A.V. Cavalcante, P.A. Scarf, A.T. De Almeida, A study of a two-phase inspection policy for a preparedness system with a defective state and heterogeneous lifetime, Reliab. Eng. Syst. Saf. 96 (2011) 627–635.
[51] L. Wang, H. Hu, Y. Wang, W. Wu, P. He, The availability model and parameters estimation method for the delay time model with imperfect maintenance at inspection, Appl. Math. Model. 35 (2011) 2855–2863.
[52] Z. Zhang, M. Chen, D. Zhou, Inspection models considering the overlapping of inspection span and failure downtime, Chin. J. Chem. Eng. 20 (6) (2012) 1167–1173.
[53] C.D. Van Oosterom, A.H. Elwany, D. Celebi, G.J. Van Houtum, Optimal policies for a delay time model with postponed replacement, Eur. J. Oper. Res. 232 (2014) 186–197.
[54] L. Yang, X. Ma, Q. Zhai, Y. Zhao, A delay time model for a mission-based system subject to periodic and random inspection and postponed replacement, Reliab. Eng. Syst. Saf. 150 (2016) 96–104.
[55] W. Wang, A delay time based approach for risk analysis of maintenance activities, Saf. Reliab. 23 (1) (2002) 103–113.
[56] M.D. Berrade, P.A. Scarf, C.A.V. Cavalcante, A study of postponed replacement in a delay time model. Reliab. Eng. Syst. Saf. (2017), https://doi.org/10.1016/j.ress.2017.04.006.
[57] W. Wang, An inspection model based on a three-stage failure process, Reliab. Eng. Syst. Saf. 96 (2011) 838–848.
[58] R. Yang, F. Zhao, J. Kang, X. Zhang, An inspection optimization model based on a three-stage failure process, Int. J. Perform. Eng. 10 (7) (2014) 775–779.
[59] R. Yang, Z. Yan, J. Kang, An inspection maintenance model based on a three-stage failure process with imperfect maintenance via Monte Carlo simulation, Int. J. Syst. Assur. Eng. Manag. 6 (3) (2015) 231–237.
[60] F. Zhao, R. Peng, W. Wang, in: An inspection and replacement model based on a three-stage failure process, IEEE Conference on Prognostics and System Health Management (PHM), 23–25 May 2012, Beijing, China, 2012, https://doi.org/10.1109/PHM.2012.6228928.
[61] H. Wang, W. Wang, R. Peng, A two-phase inspection model for a single component system with three-stage degradation, Reliab. Eng. Syst. Saf. 158 (2017) 31–40.
[62] H. Yamashina, S. Otani, Cost-optimized maintenance of the elevator—single unit case, J. Qual. Maint. Eng. 7 (1) (2001) 49–70.
[63] A.H. Christer, C. Lee, Modelling ship operational reliability over a mission under regular inspections, J. Oper. Res. Soc. 48 (1997) 688–699.
[64] A.H. Christer, W. Wang, K. Choi, The robustness of the semi-Markov and delay time single-component inspection models to the Markov assumption, IMA J. Manag. Math. 12 (2001) 75–88.
[65] A. Das Nath, D. Acharya, Age replacement of components during IFR delay time, IEEE Trans. Reliab. 53 (3) (2004) 306–312.
[66] I. Emovon, Inspection interval determination for mechanical/service systems using an integrated PROMETHEE method and delay time model, Mech. Eng. Technol. 8 (1) (2016) 13–29.
[67] I. Emovon, R.A. Norman, A.J. Murphy, An integration of multi-criteria decision making techniques with a delay time model for determination of inspection intervals for marine machinery systems, Appl. Ocean Res. 59 (2016) 65–82.
[68] P.A. Scarf, C.A.V. Cavalcante, Modelling quality in replacement and inspection maintenance, Int. J. Prod. Econ. 135 (2012) 372–381.
[69] T. Aven, I.T. Castro, A delay-time model with safety constraint, Reliab. Eng. Syst. Saf. 94 (2009) 261–267.
[70] H. Mahfoud, A. El Barkany, A. El Biyaali, Cost effectiveness of healthcare maintenance policy: a new proportional delay time model, Int. J. Perform. Eng. 12 (5) (2016) 433–450.

[71] H. Mahfoud, A. El Barkany, A. El Biyaali, Reliability assessment of degradable systems under imperfect maintenance and utilisation rate: a case study, Int. J. Eng. Res. Afr. 26 (2016) 184–194.
[72] W. Wang, Subjective estimation of the delay time distribution in maintenance modelling, Eur. J. Oper. Res. 99 (1997) 516–529.
[73] A.H. Christer, W.M. Waller, Delay time models of industrial inspection maintenance problems, J. Oper. Res. Soc. 35 (5) (1984) 401–406.
[74] S. Apeland, P.A. Scarf, A fully subjective approach to modelling inspection maintenance, Eur. J. Oper. Res. 148 (2003) 410–425.
[75] R.D. Baker, W. Wang, Estimating the delay-time distribution of faults in repairable machinery from failure data, IMA J. Math. Appl. Bus. Ind. 3 (1992) 259–281.
[76] R.D. Baker, W. Wang, Developing and testing the delay-time model, J. Oper. Res. Soc. 44 (1993) 361–374.
[77] R.D. Baker, P.A. Scarf, Can models fitted to small data samples lead to maintenance policies with near-optimum cost? IMA J. Math. Appl. Bus. Ind. 6 (1995) 3–12.
[78] R.D. Baker, P.A. Scarf, W. Wang, A delay-time model for repairable machinery: maximum likelihood estimation of optimum inspection intervals, IMA J. Math. Appl. Bus. Ind. 8 (1997) 83–92.
[79] W. Wang, X. Jia, An empirical Bayesian based approach to delay time inspection model parameters estimation using both subjective and objective data, Qual. Reliab. Eng. Int. 23 (2007) 95–105.
[80] L. Wen-Yuan, W. Wang, in: Modelling preventive maintenance of production plant given estimated PM data and actual failure times, Proc. of International Conference on Management Science and Engineering, 2006 ICMSE '06, IEEE, 2006, pp. 387–390, https://doi.org/10.1109/ICMSE.2006.313857.
[81] W. Lv, W. Wang, Modelling preventive maintenance based on the delay time concept in the context of a case study, Maint. Reliab. 3 (2011) 5–11.
[82] B. Jones, I. Jenkinson, J. Wang, Methodology of using delay-time analysis for a manufacturing industry, Reliab. Eng. Syst. Saf. 94 (2009) 111–124.
[83] A.H. Christer, W. Wang, A delay-time-based maintenance model of a multi-component system, IMA J. Math. Appl. Bus. Ind. 6 (1995) 205–222.
[84] W. Wang, A.H. Christer, Solution algorithms for a nonhomogeneous multi-component inspection model, Comput. Oper. Res. 30 (2003) 19–34.
[85] J. Cai, L. Zhu, in: A delay-time model with imperfect inspection for aircraft structure subject to a finite time horizon, Proc. of IEEE International Conference on Grey Systems and Intelligent Services Nanjing, China, 15–18 September, 2011, pp. 716–719, https://doi.org/10.1109/GSIS.2011.6044102.
[86] A.H. Christer, C. Lee, Refining the delay-time-based PM inspection model with non-negligible system downtime estimates of the expected number of failures, Int. J. Prod. Econ. 67 (2000) 77–85.
[87] W. Wang, Modeling planned maintenance with non-homogeneous defect arrivals and variable probability of defect identification, Maint. Reliab. 2 (2010) 73–78.
[88] W. Wang, J. Carr, T.W.S. Chow, A two-level inspection model with technological insertions, IEEE Trans. Reliab. 61 (2) (2012) 479–490.
[89] T. Aven, Optimal test interval for a monotone safety system, J. Appl. Probab. 46 (2009) 330–341.
[90] P.A. Scarf, C.A.V. Cavalcante, R.A. Dwight, P. Gordon, An age-based inspection and replacement policy for heterogeneous components, IEEE Trans. Reliab. 58 (4) (2009) 641–648.
[91] P.A. Scarf, C.A.V. Cavalcante, Hybrid block replacement and inspection policies for a multi-component system with heterogeneous component lives, Eur. J. Oper. Res. 206 (2010) 384–394.

[92] W. Wang, D. Banjevic, Ergodicity of forward times of the renewal process in a block-based inspection model using the delay time concept, Reliab. Eng. Syst. Saf. 100 (2012) 1–7.
[93] R. Flage, A delay time model with imperfect and failure-inducing inspections, Reliab. Eng. Syst. Saf. 124 (2014) 1–12.
[94] A.H. Christer, W. Wang, R.D. Baker, Modelling maintenance practice of production plant using the delay-time concept, IMA J. Math. Appl. Bus. Ind. 6 (1995) 67–83.
[95] A.H. Christer, W. Wang, J. Sharp, R.D. Baker, A case study of modelling preventive maintenance of production plant using subjective data, J. Oper. Res. Soc. 49 (1998) 210–219.
[96] W. Wang, H.B.A. Majid, Reliability data analysis and modelling of offshore oil platform plant, J. Qual. Maint. Eng. 6 (4) (2000) 287–295.
[97] W. Wang, Delay time modelling for optimized inspection intervals of production plant, in: M. Ben-Daya, S.O. Duffuaa, A. Raouf, J. Knezevic, D. Ait-Kadi (Eds.), Handbook of Maintenance Management and Engineering, Springer-Verlag, London, 2009.
[98] B. Jones, I. Jenkinson, J. Wang, in: Application of the delay-time concept in a manufacturing industry, Proc. of the 6th International Conference on Manufacturing Research, 9–11 September, Brunel University, London, UK, 2008, pp. 23–28.
[99] B. Jones, I. Jenkinson, Z. Yang, J. Wang, The use of Bayesian network modelling for maintenance planning in a manufacturing industry, Reliab. Eng. Syst. Saf. 95 (2010) 267–277.
[100] S. Werbińska-Wojciechowska, P. Zając, Use of delay-time concept in modelling process of technical and logistics systems maintenance performance, Maint. Reliab. 17 (2) (2015) 174–185.
[101] F.K.N. Leung, M. Kit-Leung, Using delay-time analysis to study the maintenance problem of gearboxes, Int. J. Oper. Prod. Manag. 16 (12) (1996) 98–105.
[102] F.K.N. Leung, T.W. Ma, A study on the inspection frequency of fresh water pumps, Int. J. Ind. Eng. 4 (1) (1997) 42–51.
[103] M.I. Desa, A.H. Christer, Modelling in the absence of data: a case study of fleet maintenance in a developing country, J. Oper. Res. Soc. 52 (2001) 247–260.
[104] A. Akbarov, A.H. Christer, W. Wang, Problem identification in maintenance modelling: a case study, Int. J. Prod. Res. 46 (2) (2008) 1031–1046.
[105] A.H. Christer, C. Lee, W. Wang, A data deficiency based parameter estimating problem and case study in delay time PM modelling, Int. J. Prod. Econ. 67 (2000) 63–76.
[106] A.H. Christer, W. Wang, K. Choi, The delay-time modelling of preventive maintenance of plant given limited PM data and selective repair at PM, IMA J. Math. Appl. Med. Biol. 15 (1998) 355–379.
[107] X. Liu, W. Wang, R. Peng, An integrated production and delay-time based preventive maintenance planning model for a multi-product production system, Maint. Reliab. 17 (2) (2015) 215–221.
[108] D. Mcnamara, A. Cunningham, R. Riahi, I. Jenkinson, J. Wang, Application of Monte Carlo techniques with delay-time analysis to assess maintenance and inspection policies for marine systems, Proc. Inst. Mech. Eng. E: J. Process Mech. Eng. 231 (2) (2015) 89–106.
[109] X.-I. Jiang, J.-M. Zhao, Z.-W. Li, An optimal inspection policy for protection devices of electronic equipment using delay time model, IJUNESST 8 (8) (2015) 169–178.
[110] W. Wang, A model for maintenance service contract design, negotiation and optimization, Eur. J. Oper. Res. 201 (2010) 239–246.
[111] P.A. Scarf, H.A. Majid, Modelling warranty extensions: a case study in the automotive industry, Salford Business School Working Paper Series, Paper No. 339/10, 2010.

[112] W. Wang, A stochastic model for joint spare parts inventory and planned maintenance optimisation, Eur. J. Oper. Res. 216 (2012) 127–139.
[113] W. Wang, A joint spare part and maintenance inspection optimisation model using the delay-time concept, Reliab. Eng. Syst. Saf. 96 (2011) 1535–1541.
[114] W. Wang, A.A. Syntetos, Spare parts demand: linking forecasting to equipment maintenance, Transp. Res. E 47 (2011) 1194–1209.
[115] A. Jodejko-Pietruczuk, S. Werbińska-Wojciechowska, in: Analysis of block-inspection policy parameters from economical and availability point of view, 11th International Probabilistic Safety Assessment and Management Conference & The Annual European Safety and Reliability Conference, PSAM 11 & ESREL 2012, Helsinki, Finland, 25–29, June, 2012, , pp. 1–10.
[116] A. Jodejko-Pietruczuk, S. Werbińska-Wojciechowska, in: Block inspection policy for non-series technical objects, in: Safety, reliability and risk analysis: beyond the horizon, Proc. of the European Safety and Reliability Conference, ESREL 2013, Amsterdam, The Netherlands, 29 September–2 October, CRC Press/Balkema, Leiden, 2014, pp. 889–898.
[117] A. Jodejko-Pietruczuk, S. Werbińska-Wojciechowska, Analysis of maintenance models' parameters estimation for technical systems with delay time, Maint. Reliab. 16 (2) (2014) 288–294.
[118] A. Jodejko-Pietruczuk, S. Werbińska-Wojciechowska, A delay-time model with imperfect inspections for multi-unit systems, J. Konbin 3 (23) (2012) 157–172.
[119] A. Jodejko-Pietruczuk, T. Nowakowski, S. Werbińska-Wojciechowska, Block inspection policy model with imperfect inspections for multi-unit systems, Theory Appl. 8 (3) (2013) 75–86. http://gnedenko-forum.org/Journal/2013/032013/RTA_3_2013-08.pdf.
[120] W. Wang, X. Liu, R. Peng, L. Guo, in: A delay-time-based inspection model for a two-component parallel system, International Conference on Quality, Reliability, Risk, Maintenance, and Safety Engineering (QR2MSE), 15–18 July 2013, Chengdu, China, 2013.
[121] X. Liu, W. Wang, R. Peng, F. Zhao, A delay-time-based inspection model for parallel systems, Proc. Inst. Mech. Eng. O: J. Risk Reliab. 229 (6) (2015) 556–567.
[122] A. Jodejko-Pietruczuk, S. Werbińska-Wojciechowska, in: Expected maintenance costs model for time-delayed technical systems in various reliability structures, Proc. of Probabilistic Safety Assessment and Management, PSAM 12: Honolulu, Hawaii, USA, 22–27 June, 2014, pp. 1–8. http://psam12.org/proceedings/paper/paper_572_1.pdf.
[123] A. Jodejko-Pietruczuk, S. Werbińska-Wojciechowska, in: Influence of data uncertainty on the optimum inspection period in a multi-unit system maintained according to the block inspection policy, Dependability Engineering and Complex Systems: Proceedings of the Eleventh International Conference on Dependability and Complex Systems Dep CoS-RELCOMEX, June 27–July 1, Springer International Publishing, Brunów, Poland, 2016, pp. 239–256.
[124] M.F. Marsaro, C.A.V. Cavalcante, Random preventive maintenance policy based on inspection for a multicomponent system using simulation, Maint. Reliab. 19 (4) (2017) 552–559.
[125] S. Werbińska-Wojciechowska, Technical system maintenance. Delay-time-based modelling (in press).
[126] W.R. Blischke, D.N. Prabhakar Murthy, Reliability: Modelling, Prediction and Optimization, John Willey & Sons, Inc., New York, 2000.
[127] T. Nowakowski, S. Werbińska-Wojciechowska, Uncertainty problem in decision support system for means of transport maintenance processes performance development, J. KONBiN 3 (2012) 173–192.

[128] W. Wang, D. Banjevic, M. Pecht, A multi-component and multi-failure mode inspection model based on the delay-time concept, Reliab. Eng. Syst. Saf. 95 (2010) 912–920.
[129] L. Zhu, H.-F. Zuo, The delay-time maintenance optimization model with two failure modes, Adv. Mater. Res. 452–453 (2012) 190–194.
[130] H. Wang, H. Pham, Optimal imperfect maintenance models, in: H. Pham (Ed.), Handbook of Reliability Engineering, Springer-Verlag London Limited, London, 2003, pp. 397–414.

CHAPTER 2

Deterministic and Probabilistic Safety Analyses

Vanderley de Vasconcelos, Wellington Antonio Soares, Antônio Carlos Lopes da Costa, Amanda Laureano Raso
Centro de Desenvolvimento da Tecnologia Nuclear—CDTN, Belo Horizonte, Brazil

Abstract

This chapter introduces deterministic safety analysis (DSA) and probabilistic safety analysis (PSA) as applied in risk assessment during the lifetime of complex industrial facilities. It starts by presenting some safety analysis concepts and their applications to hazardous facilities with regard to nuclear power plants. A framework to implement DSA and PSA, their approaches, steps, and complementarities are presented. Available techniques needed for these analyses are described, and their advantages and disadvantages for qualitative and quantitative assessments are discussed. A case study using consequence assessment and fault tree analysis is presented to illustrate some important issues of the two approaches. It is concluded that the conservative features of DSA, based on worst-case scenarios, and the more realistic analyses of PSA, selecting alternative scenarios, are complementary and suitable, depending on scope of analysis, to ascertain the compliance of many complex and hazardous industrial facilities with safety requirements and acceptance criteria.

Keywords: Deterministic safety analysis, Probabilistic safety analysis, Risk assessment, Hazardous facility, Nuclear power plant, Fault tree analysis

1 INTRODUCTION

Safety analysis can generally be considered as the "evaluation of potential hazards associated with operation of a facility or the conduct of an activity" [1]. It is carried out during the lifetime of complex industrial facilities, for example, nuclear power plants (NPPs). For these facilities, safety analysis is relevant in design, licensing, operation, and life extension. It includes analytical evaluations of physical phenomena with the purpose of demonstrating that safety requirements are met for the postulated accidents that could occur, including the actuating of safety systems. The main goal of safety analysis is to verify if the risks associated with the facility have acceptable levels. At this viewpoint, this concept overlaps with the concept of risk analysis,

Advances in System Reliability Engineering
https://doi.org/10.1016/B978-0-12-815906-4.00002-6

seen as a process of understanding the nature of risk, determining the level of risk, and providing the basis for risk treatment [2]. Two types of safety analysis are available: deterministic and probabilistic.

Deterministic safety analysis (DSA) for an industrial facility has the main objective of predicting its response to postulated initiating events, using some acceptance criteria, based on physical and structural features. Initiating events are events that upset the normal operations of the facility that may require response from systems and operators to avoid undesirable outcomes. The results of DSA are spatial and time-dependent calculations of parameters that govern the physical processes in the facility or estimate the consequence to workers or public. DSA is characterized by conservative assumptions in terms of minimum safety margins related to the acceptance criteria for the postulated accidents [3].

Deterministic (or nonprobabilistic) methods consider the consequences of well-defined events or combinations of events but do not necessarily consider the probability of these events or assure that all possible events are included in the deterministic analysis. Often, this is the starting point for the safety analysis. On the other hand, probabilistic (or stochastic) analysis tends to include all possible outcomes looking for their probabilities. Requiring much more component and facility data, as well as advanced risk assessment methods, probabilistic safety analysis (PSA) is currently recognized as a mature safety analysis methodology and is widely used, in a complementary way with DSA, for safety and risk assessments of NPPs [4].

The examples mentioned in this chapter are mainly from the nuclear industry. However, the lessons learned from this industry can be applied to other areas, as the process industry, from which the case study presented at the end of this chapter was selected, involving major flammable liquid leakage from a storage vessel. In principle, the DSAs and PSAs for process industries (for example, chemical or petrochemical) are similar to these studies for NPPs, but they differ in some important features. Process industries have a diversity of technologies and processes, where the dangerous materials are often transferred from one vessel to another and are used in different processes. On the other hand, the nuclear and radioactive materials in NPPs are mostly concentrated in the solid fuel in the reactor core. In addition, in process industries, there are more frequent changes in operations and systems, which require continuous improvements. Thus the sources of hazards in these facilities are diverse, distributed through the process, and dependent on the type of plant. For this reason, safety analysis for process industries usually have to be first screened by DSA and PSA qualitative analysis techniques before carrying out more detailed quantitative analyses.

2 CONCEPTS ON SAFETY ANALYSIS

Some concepts on safety analysis are presented here in alphabetical order.

2.1 As Low as Reasonably Practicable (ALARP)

"As Low as Reasonably Practicable" is a principle used in some areas, for example, radiation protection and chemical prevention. It recognizes that not all risks can be eliminated, and "there are no other practicable options that could reasonably be adopted to reduce risk further" [5]. Fig. 1 illustrates the ALARP principle. The triangle represents an increasing level of cumulative risk (all risks to which workers or public are exposed) from the lower vertex of triangle, to a high risk at the top of the triangle. The equivalent term used in the nuclear industry is ALARA ("As Low As Reasonably Achievable"). ALARA means "making every reasonable effort to maintain exposure to ionizing radiation as far below the dose limits as practical, consistent with the purpose for which the licensed activity is undertaken, taking into account the state of technology, economic factors, and public interest" [6].

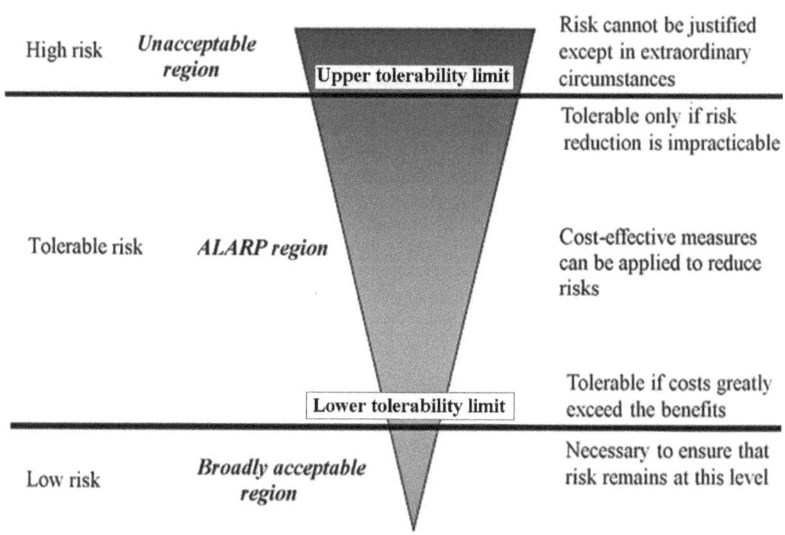

Fig. 1 Illustration of the ALARP principle. *(Based on NOPSEMA, National Offshore Petroleum Safety and Environmental Management Authority (ALARP Guidance Note. N-04300-GN0166 Revision 6), NOPSEMA, Melbourne, 2015.)*

2.2 Defense-in-Depth

It is a safety philosophy for designing and operating plants focusing on features that prevent and mitigate events with potential undesirable consequences. The key point is to create multiple independent and redundant layers of protection, so as not to rely exclusively on a single layer. This safety philosophy applied to a process plant is illustrated in Fig. 2. This example presents the measures for preventing and mitigating accidents are presented as layers of protection or lines of defense [7].

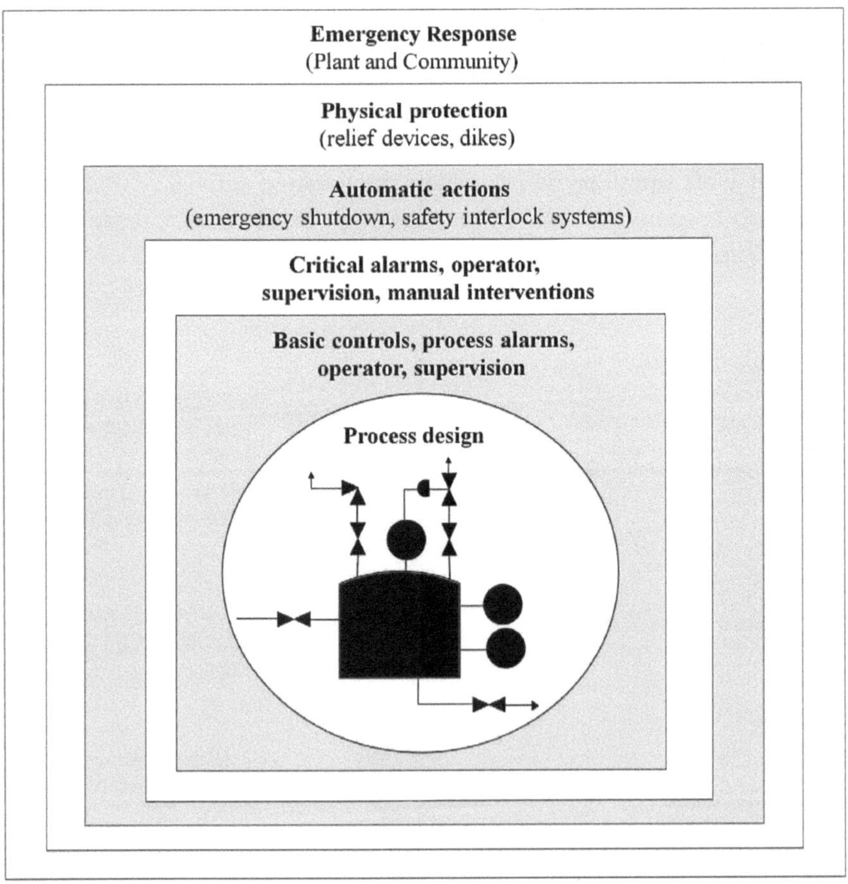

Fig. 2 Illustration of defense-in-depth safety philosophy. *(Based on A. Franks, Lines of Defence/Layers of Protection Analysis in the COMAH Context, Amey VECTRA Limited, Westbrook, Warrington, 2017.)*

2.3 Design Basis Accidents (DBAs)

They are "postulated accidents that a facility must be designed and built to withstand without loss to systems, structures, and components" [6]. The selection of DBAs when designing an NPP takes into account that the plant should withstand and recover effectively from these types of accidents.

2.4 Deterministic Safety Analysis (DSA)

It is the "engineering analysis of a plant response using validated models, calculations, and data, which predict transient response of the plant to event sequences" [8]. Typically, DSA uses conservative estimates, safety margins, and DBAs.

2.5 Endpoint Distance

For flammable substances, it is defined as the "distance where the consequence analysis results in an overpressure of 1.0 psi for vapor cloud explosions, or in the lower flammability limit for the released substance" [9]. This is the maximum distance at which people can suffer serious injury. The threshold of 1.0 psi was chosen because of the potential injuries caused by explosions, for example, flying glass from shattered windows.

2.6 Exclusion Area

According to U.S. Nuclear Regulatory Commission, exclusion area is the "area surrounding a nuclear facility, where the plant owner has the authority to determine all activities, including exclusion or removal of people and property in case of accidents" [6]. This concept is used in the licensing of nuclear facilities, especially in site selection.

2.7 Hazard

It is an event that poses some risk to a facility or activity. It includes internal hazards, such as equipment and process failures, and external hazards, such as flooding, earthquakes, and aircraft crashes. Hazard analysis identifies materials, systems, and processes that can produce undesirable consequences. According to American Nuclear Society [8], the hazard analysis should examine the "complete spectrum of potential events that could expose members of the public, workers, and the environment to hazardous materials."

2.8 Probabilistic Safety Analysis (PSA)

PSA comprises "qualitative and quantitative assessments of the risk associated with plant operation and maintenance" [8]. In case of NPPs, PSA outcomes are given in terms of frequency of occurrence and severity of consequences of undesired events, such as core damage or radioactive material release.

2.9 Risk

There are many concepts of risk used in different technological or scientific areas. A general approach considers risk as a quantitative assessment of frequency of occurrence and severity of an undesired event sequence, including the evaluation of the uncertainties [8]. Eq. (1) expresses mathematically the concept of risk [10].

$$Risk\left[\frac{consequence}{time}\right] = frequency\left[\frac{event}{time}\right] \times severity\left[\frac{consequence}{event}\right] \tag{1}$$

Risk assessment is the general term that encompasses risk identification, risk analysis, and risk evaluation. Risk identification is the first step of risk assessment process. It identifies hazards, areas of impacts, and undesired events. Risk analysis involves developing an understanding of risk, its causes, occurrence probabilities, and potential consequences. The analysis can be qualitative, semiquantitative, or quantitative. Risk evaluation involves "comparing the level of risk found during the analysis process with risk criteria established when the context of analysis was considered" [2]. Based on this comparison, risk management can be needed. Fig. 3 illustrates three basic questions concerning the concept of risk assessment for an undesired event [8].

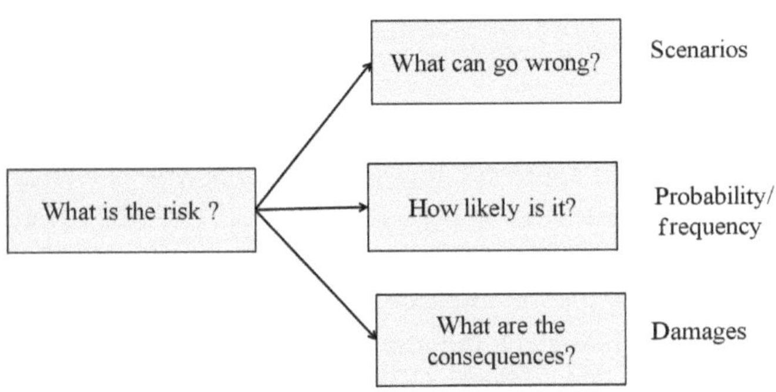

Fig. 3 Illustration of the concept of risk assessment.

2.10 Safety

It can be seen as a "practical certainty that adverse effects will not result from exposure to an agent under defined circumstances" [11]. The concepts of safety analysis and safety assessment can be considered equivalent to risk analysis and risk assessment, respectively [2]. Even the concepts of safety analysis and safety assessment are frequently used interchangeably. More specifically, safety analysis is considered the documented process for the study of safety, and safety assessment is the quantitative evaluation of safety, for example, the performance assessment of safety and mitigation measures, as well as the impact of abnormal or accidental conditions of a facility [1].

2.11 Single Failure

It is a "failure that results in the loss of capability of a single system or component to perform its intended safety function, and any consequential failure which results from it" [1]. Compliance with the single failure criterion requires, for instance, that a safety system must be capable of continuing to perform its function in the presence of a single failure. This usually requires that two or more redundant systems be provided.

3 DETERMINISTIC SAFETY ANALYSIS (DSA)

DSA approach generally considers the worst-case scenarios. Traditionally, safety analyses of industrial facilities, mainly in the scope of licensing processes, are based on deterministic approaches. To verify compliance with safety standards, deterministic safety margins are introduced in the design of systems important to safety, largely based on expert judgment, to take into account the uncertainties and reduce failure risks. DSA often produces an "all or nothing" type of assessment, considering a system "safe" or "unsafe." Each engineering team may use different requirements and modeling parameters in DSA, so that expected effects from an event may be significantly different between analysts [3].

Depending on the type of facility and requirements of regulatory bodies, the main objectives of DSA are: understanding the design basis of facilities or equipment, as well as their safety concepts; verifying the compliance with safety goals, principles, and criteria; and demonstrating the safety of facilities and activities. A simplified approach for a deterministic safety analysis is shown in Fig. 4. Each of the steps shown in the figure is described as follows.

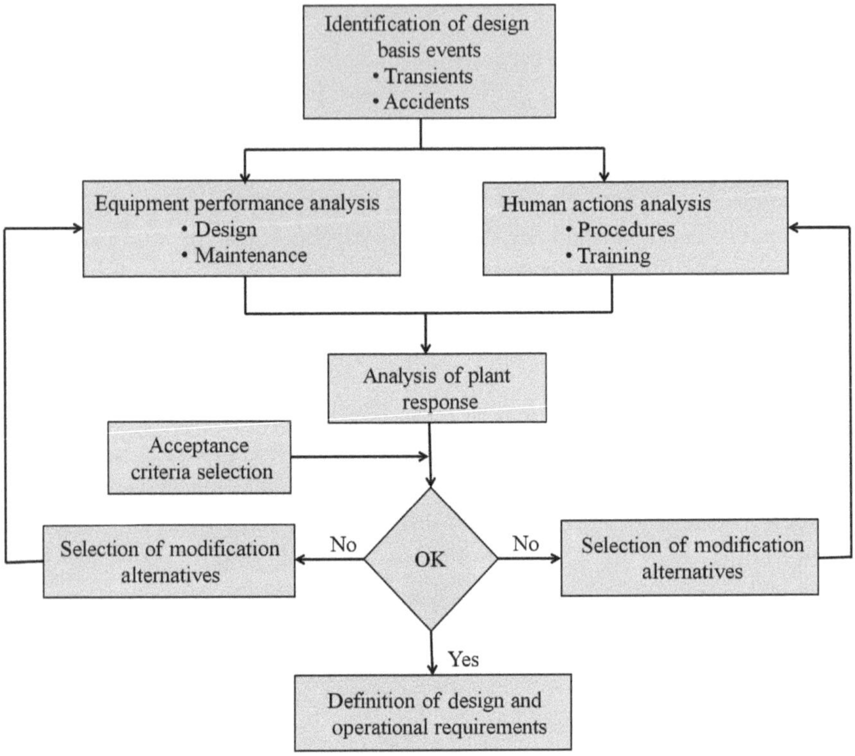

Fig. 4 Simplified approach for a deterministic safety analysis. *(Adapted from International Atomic Energy Agency (IAEA), Needs for Safety Assessment & The Safety Assessment Process, 2009, http://www-ns.iaea.org/downloads/ni/training/specific_expert_knowledge/safety%20assessment/II%202%20and%20II%203_2%20-%20Needs%20for%20Safety%20Assessment%20&%20the%20Safety%20.pdf (Accessed 06.02.18).)*

3.1 Identification of Design Basis Events

Normal operation is a plant state within specified operational limits and conditions. As normal operating deviations can occur during the lifetime of a facility, then appropriate design provisions (safety margins) must be provided to items important to safety. In this case, DBAs are defined through analysis of postulated initiating events. This includes operator errors, equipment failures, and human-induced or natural events that challenge the systems required to maintain safety [3].

3.2 Equipment Performance Analysis

The equipment performance should be analyzed to prove that the design agreed with the requirements of regulatory bodies through inspections, testing, and maintenance. Reliability, availability, and performance of equipment and control systems are analyzed, taking into account conservative initial and boundary conditions of the plant [12].

3.3 Human Actions Analysis

Action and diagnosis errors must be taken into account, when analyzing operator reliability responding to an accident sequence. Under accident conditions, the first step is the diagnosis of the nature of the accident, before selecting the appropriate procedures and recovery actions. Diagnosis errors are usually more frequent than action errors. Human-induced actions required to maintain the safety of the plant should be included in the analysis of the plant response. Operational procedures should be analyzed, and training of operators in using them should be carried out. For severe accidents that involve facility damage and potential consequences to the environment, the operator emergency procedures should be also analyzed [13, 14].

3.4 Analysis of Plant Response

DSA should predict the plant response to postulated initiating events, like transients, postulated accidents, DBAs, and severe accidents. In the case of NPPs, calculations of neutronic, thermohydraulic, and structural plant responses are analyzed often using analytical or computational models. Typically, the parameters of interest to assess radiological consequences are the radiation doses to workers, public, and environment. Computer codes are available for analyzing equipment performance or specific phenomena such as reactivity excursions and dynamic loads on components. Computational fluid dynamics codes and computer codes for estimating the consequences of fire, explosion, and release of dangerous materials are also usually required [3, 9].

3.5 Acceptance Criteria Selection

Acceptance criteria are limits and conditions set by regulators with the purpose of ensuring an adequate level of safety. Licensees carry out their duty to protect workers, public, and environment by establishing safety standards to ensure that consequences of normal operations and accidents comply with the ALARP principle. Regulatory bodies verify if licensees develop, achieve, and maintain standards, ensuring that any necessary safety measures are taken, in addition to enforcing safety and standards. In general, regulators do not prescribe in detail how the licensees should comply with their safety requirements. The licensees should select the acceptance criteria, apply the

ALARP principle, carry out safety analyses, and demonstrate to regulators that the risks are acceptable. The requirements of regulators cover, among other things: application of the defense-in-depth philosophy; the single failure criterion; requirements for redundancy and diversity; the preference for passive over active systems; criteria related to human factors and cognitive ergonomics; and risk criteria related to dangerous materials [5, 9, 13].

3.6 Selection of Modification Alternatives

DSA should take into account cost-effectiveness requirements, to improve safety and support modifications. A risk analysis is useful when several alternatives are available, to support the decision makers, verifying the impact of modifications on safety and demonstrating the compliance with acceptance criteria. The following examples require revision of DSAs in a facility: implementation of modification programs on the plant or on operating procedures; availability of new technical knowledge of physical phenomena; implementation of life management programs (aging of the plant); and changes of applicable standards and regulations [3].

3.7 Definition of Design and Operational Requirements

DSA helps in analyzing events that can occur during plant operation with the purpose of: checking the adequacy of postulated initiating events; providing additional information on the time-dependency of parameters that are not directly observable during operation; checking the performance of operators and plant systems; reviewing emergency procedures; supporting the troubleshooting of potential safety issues identified in the analysis; and verifying and validating models and computer codes used in analyses and in training simulators [3]. DSA is an important approach for identifying design deficiencies, and optimizing design and operational procedures, thorough evaluation of operation practice and improving the understanding of the plant behavior. In routine operation, the regulatory bodies can require the operator to report periodically on the compliance with safety goals, specified regulatory requirements, and on efforts to enhance safety [12].

4 PROBABILISTIC SAFETY ANALYSIS (PSA)

PSA is a stochastic risk modeling, also commonly named probabilistic risk assessment (PRA) [15] or quantitative risk assessment (QRA) [16]. Both industry and regulators currently use PSAs to analyze individual and societal risks, to assess the hazard consequences to workers, public, and environment. Many industrial requirements worldwide include some PSA level in their license processes, to support risk assessment and management processes. PSA seeks to understand and model potential undesired events with their associated probabilities and consequences, providing a sophisticated analysis of different risk management strategies, supported by quantitative outcomes [4, 13].

The main advantage of these probabilistic methods is that they consider frequency and severity together in a more comprehensive and complex way than the deterministic ones. The main problem is the difficulty in obtaining the data required by quantitative assessments. In addition to data uncertainties, the uncertainties inherent to the suitable models and the completeness of the analyses may not be apparent or properly appreciated. Taking into account uncertainty assessments and applying probabilistic data and methods in a nonbiased way, PSA approach can be a cost-effective tool to support decision-makers and communication of risks to the public [17]. Fig. 5 shows a simplified approach for a PSA. Each of the steps shown in the figure is described as follows.

4.1 Scope of PSA

Depending on the scope of the safety analysis, the PSA can encompass the identification of initiating events, the estimation of the frequency of potential accident sequences, and the calculation of their consequences to workers, public, and environment. In NPPs, the PSAs are usually categorized as: Level 1 PSA, including estimation of the core melt frequency per reactor-year and a risk assessment for workers; Level 2 PSA, including the assessment of radioactivity released to the environment; and Level 3 PSA, including the risk assessment for the population [4]. Sometimes a simplified PSA intends only to demonstrate the compliance with safety criteria. If more comprehensive analyses are required, individual and societal risks may be developed, quantitatively describing and assessing the spectrum of possible scenario outcomes. The internal and external boundaries of the analysis, including human operators and management systems, are important issues when defining PSA scope [18].

4.2 Initiating Event Analysis

The approaches to identify initiating events involve engineering evaluation, a reference to previous sets of initiating events, deductive analyses, and

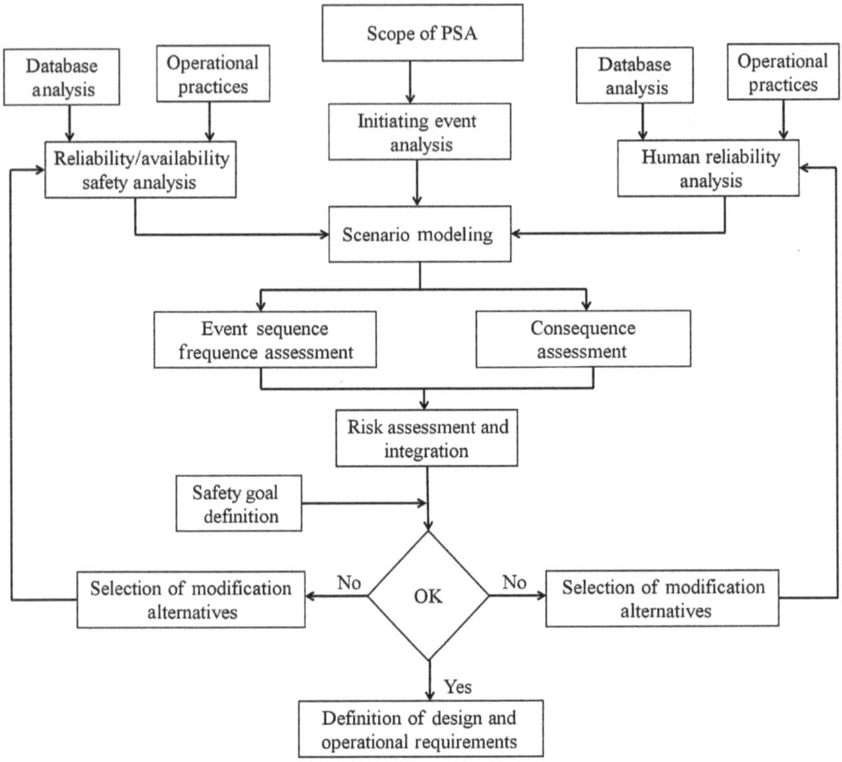

Fig. 5 Simplified approach for a probabilistic safety analysis. *(Adapted from International Atomic Energy Agency (IAEA), Needs for Safety Assessment & The Safety Assessment Process, 2019, http://www-ns.iaea.org/downloads/ni/training/specific_expert_knowledge/safety%20assessment/II%202%20and%20II%203_2%20-%20Needs%20for%20Safety%20Assessment%20&%20the%20Safety%20.pdf (Accessed 06.02.18).)*

consideration of operational experience. Such analyses include the identification of postulated initiating events and their frequencies, as well as evaluating the development and consequences of these events [4].

4.3 Database/Operational Practices Analysis

This step involves familiarization with the facility and gathering the information required in subsequent analyses. The engineering safety features related to initiating events are then identified. This information contributes to a qualitative risk assessment and input data for the frequency and consequence models. Generic and plant-specific data needed to make these estimates and their associated uncertainties should be collected and treated properly. The understanding of the facility behavior under normal and abnormal conditions allows developing the accident scenarios. In addition, toxic, hazardous, or radioactive materials in the facility should also be identified and quantified [4].

4.4 Reliability/Availability Analysis

This step involves acquiring data needed for the quantification of the frequencies and probabilities of accident sequences. Data for reliability and availability assessments of components and systems, common-cause failures (CCFs), and maintenance/repair information should be gathered and analyzed [17]. Sometimes, because of the nature and diversity of the processes and phenomena in complex facilities, the uncertainties of the data and models should be assessed. Information for time-dependent assessment of reliability and availability, for example, the operator performance in repairing a component, should also be analyzed [19].

4.5 Human Reliability Analysis (HRA)

Usually, it involves the evaluation of operator performance, taking into account factors such as task complexity, working conditions, and the physical and cognitive characteristics of operators. Qualitative HRA identifies the possible operator actions that, if not properly performed, will have adverse impact on the accident sequence. Quantitative HRA assesses human error probabilities, taking into account the particular conditions of the human error under consideration, such as the "performance-shaping factors" (PSFs). PSFs typically include everything that influences human performance, such as sociological and psychological issues, workload, stress, and ergonomics. The frequency assessment of each accident sequence where action or diagnosis tasks are involved should consider the PSFs [14, 18].

4.6 Scenario Modeling

The scenario modeling within a PSA consists of the logic modeling of accident sequences and consequence analyses. This step also includes the human performance models analyzed in the HRA step. The logic modeling of an accident sequence starts with an initiating event that places a demand on safety systems (defense-in-depth levels). The scenario modeling combines success and failure of these defense-in-depth levels, which could result in undesired consequences. The consequence analysis involves the use of analytical and simulation models for evaluating the effects of an accident, for instance, in terms of hazardous materials released to a working area or to the environment [18].

4.7 Event Sequence Frequency Assessment

The models constructed during scenario modeling are quantified using generic or plant-specific data. The frequency of accident sequences and potential consequences are estimated. Uncertainties assessments, sensitivity studies, and the relative importance of the various contributors to the outcomes are carried out. Fault trees and event trees are the most common tools used for modeling the accident sequence frequencies. Dependent failures can be important in the frequency assessment of the undesirable events in PSA and should be taken into account. Therefore functional, physical, human interaction, and component failure dependencies should be considered [4].

4.8 Consequence Assessment

One of the most important elements and the initial step of the consequence assessment is the source-term estimation. The source-term is usually defined as the time-dependent release of hazardous material or conditions (e.g., radiation heat or explosion overpressure) from a defined facility boundary. On-site consequences include assessment of exposures of workers to hazardous material and conditions using different models. Off-site consequences include, for example, the estimation of the effects of hazardous materials released to the environment. Once consequence assessments are complex, the use of computer codes may be required, e.g., for transient simulations in a critical system of the facility, or atmospheric dispersion modeling of hazardous materials [18].

4.9 Risk Assessment and Integration

Depending on regulatory and operational requirements, many types of risk measures can be calculated: qualitative risk (use of relative or descriptive scale to measure likelihood and severity); semiquantitative risk (a mix of qualitative and quantitative information to support risk-informed decisions about the facility); and quantitative risk (use of numerical scale, such as exposure of workers or public from the accident scenarios or frequency of fatality per year). Risk integration for a facility or process is carried usually adding the individual risks for all undesired event sequences. Quantitative uncertainty assessments require the knowledge of probability distributions for input parameters and risk results. Importance measures and sensitivity analysis aid interpreting the PSA results. Importance analysis requires estimating the relative importance measures of contributors to event sequence frequencies, system unavailability, and consequences. The sensitivity analysis addresses the modeling assumptions that have a potential significant impact on the results [20, 21].

4.10 Safety Goal Definition

The general aim of safety analysis is to verify if the safety measures are enough to meet the risk criteria for each identified barrier among the hazardous materials and people or environment. Estimated risks are compared with risk criteria, generally derived from standards, which constitute acceptable levels of safety in force for different industrial sectors in different countries. Safety analysis should ascertain the compliance with safety requirements, and if the facility is suitable to continue in operation even in case of changes in regulation and guides. Probabilistic analyses may reveal more information than deterministic ones, because they can explicitly incorporate quantitative uncertainty assessment. Regulators can set numerical criteria, for instance, for acceptable levels of risk to workers and public health and safety, in order to meet the required safety goals. In case of NPPs, risk criteria are expressed in terms of annual average individual probability of fatality due to radiation released under potential accidents. Acceptance criteria can be also defined as exclusion area (for nuclear facilities, in terms of radiation dose exposures) or endpoint distances (for chemical process industries, in terms of concentration of toxic or flammable substances, overpressure, or radiation heat levels) [6, 9].

4.11 Selection of Modification Alternatives

PSA is important to support enhancements of plant safety, assessing the proposed modifications or the way the plant is operated. For example, it can provide plant engineers a ranking of significance of alternative design and operation options in risk-informed approach for safety analysis of NPPs. This approach allows more systematic assessments of alternatives based on risk, operating experience, and engineering judgment. Risk-informed approach is important to reduce unnecessary conservatism of the deterministic approaches, and, on the other hand, it can be used to identify areas with insufficient conservatism. PSA can also support improvements in maintenance, in-service inspections, testing, and technical specifications [4, 20].

4.12 Design and Operational Requirements

The outcomes from PSAs including risk assessments, reliability and availability estimations, relative importance assessments, sensitive studies, and quantification of uncertainty sources provide means to improve design and operational requirements during the lifetime of the analyzed facilities.

5 TECHNIQUES FOR SAFETY ANALYSIS

Techniques adopted, both for DSA and PSA, range from relatively unstructured and qualitative "Checklist" and "What-if" analysis, through more formalized and quantitative (or semiquantitative) tools as "failure mode and effect analysis" (FMEA), "hazard and operability analysis" (HAZOP), and "fault tree analysis" (FTA). These techniques look to identify hazardous events and to find causes, consequences and the existing preventive measures, as well as evaluate qualitatively or quantitatively the safety and risks involved. A brief description of the main available techniques often used in DSA and PSA is given as follows.

5.1 Available Techniques for DSA and PSA

Safety review (SR) typically involves inspections performed by a team nominated to identify and evaluate plant hazards. The safety review is usually a cooperative effort between plant personnel and the inspection team, and the results usually address major risk situations. This review usually begins with a preparatory phase in which a detailed description of the plant and operating procedures is assembled, together with information about materials stored and processed, as well as available records regarding accidents and injuries that have occurred. Particular attention is given to procedures for periodic testing and maintenance of safety-related equipment, as well as emergency response plans. The next steps identify deficiencies and critical areas, as well as develop recommendations for preventive, corrective, and mitigating actions [22]. In some countries, systematic safety reassessments, named "periodic safety review" (PSR), are usually required in life management and plant modifications of NPPs. PSR uses current safety standards and operating practices, ensuring higher levels of safety throughout the plant lifetime [23].

Checklist (CL) is mainly applied to processes covered by standards and engineering practices, for example, pressure vessels designed according to ASME standards. CLs are easy and practical to use, as well as suitable to identify ordinary known hazards. They are highly dependent on the team experience and, depending on their completeness, hazards may not be identified. CLs are generally developed by creating a list of questions that look for nonconformities related to standards and engineering practices. If a particular CL is not available, an experienced team must develop one based on literature and similar processes, equipment, or facilities. Among the hazard assessment CLs available in literature, the following items can be highlighted: general work environment, hazardous chemical exposures, flammable and combustible materials, fire protection equipment and procedures, emergency action plans, electrical, mechanical, ventilation and piping systems, environmental controls, and material handling [22].

What-If (WI) is a kind of structured brainstorming. Once the review team understands the process or system being assessed and the types of hazards involved, each subsystem or step is examined to identify things that can go wrong and to estimate their risks. To carry this out successfully, the team must be properly trained and have a full set of data available about the system being studied. This includes, for instance, operating instructions, process flow sheets, physical and hazardous properties of the materials involved, description of safety systems, and potentially exposed persons and environment. Issues about human errors, process upsets, and equipment failures are then formulated. These possibilities of failures and errors should be considered during normal operation, construction, and maintenance activities, as well as during abnormal or accident conditions. WI has also the benefit of encouraging a wide participation in the risk identification and assessment processes, increasing commitment to the decisions taken [22].

Process hazard analysis (PHA) is a method to identify and analyze hazards associated with the processing of hazardous materials and to make decisions about safety improvements. PHA can be used to look for factors that lead to accidents, such as failure of instrumentation, equipment, system, or human actions. In addition, PHA analyzes the ways or methods of hazard control, focusing on the most hazardous processes [24].

Failure mode and effect analysis (FMEA) is a method designed to identify and understand potential failure modes and their causes, and the effects of failures on systems or processes. It is possible to assess qualitatively the risks, the effects and causes, and then prioritize corrective actions. It is a structured step-by-step process to find out what could go wrong in a task or process, the potential consequences of those failures, and what can be done to prevent them. The likelihood of the failure mode detection is analyzed in some types of FMEA adopted in the industry. Risk Priority Number (RPN) is the index used to prioritize the recommended actions from a FMEA. It is a numerical ranking of the risk of each potential failure mode, resulting from the product of three numbers: the severity of the effect (S), the likelihood of occurrence of the cause (O), and the likelihood of detection of the cause (D). These numbers are obtained from scales based on qualitative, semiquantitative, or quantitative information of the failure modes and vary according to design, process, product, or type of industry [17].

Hazard and operability analysis (HAZOP) has its origin in hazard identification at chemical process plants. The main purpose of a HAZOP is to check systematically a process, identifying if process deviations could lead to undesirable consequences. It can be applied at any stage in the lifetime of a facility, to continuous or batch processes, and to operating procedures. Guidewords are applied to process parameters at defined points in the process. Most used guidewords include "No," "More," "Less," "Part of," "As well as," "Reverse," and "Other than." As the guidewords are applied at the defined points to each process parameter, such as flow, pressure, temperature, level, and composition, the deviations are recorded, including their causes, effects, safeguards, and mitigation actions [22].

Fault tree analysis (FTA) is an analytical technique used to analyze the causes of an undesired state of a system (called top event). The top event is usually a state that is critical to safety or reliability. FTA can be described as a logic and graphic model of parallel and series combinations of faults that will result in the top event. Failures of components or equipment, as well as human errors, are examples of faults. A fault tree thus depicts the logical interrelationships of basic events that lead to the top event. The main FTA qualitative results are the "minimal cut sets" (MCSs). A MCS is the smallest combination of basic events that result in the top event. MCSs represent qualitatively the most important event sets that contribute to the occurrence of top event. Quantitative analysis of fault trees involves assessment of probability of occurrence of the top event based on probability of basic events. Both qualitative and quantitative analyses of complex fault trees, including uncertainty, sensibility, and importance analyses, require the use of computer codes. FTA can support qualitative and quantitative risk assessment tools, as PHA, FMEA, and "event tree analysis" (ETA). Reliability and availability analysis of complex systems can also be carried out with support of FTA [25].

Reliability block diagram (RBD) is a graphical representation of the components of a system and the way they are related or connected. System reliability is computed by using the reliabilities of the components that make up the system. In a series configuration, failure of any component results in failure of the whole system. In parallel configuration, all components must fail for the system to fail. In complex systems, it is not easy to recognize which components are in series and which are in parallel. In such cases, analytical or simulation methods can be used to obtain the reliability characteristics of the system. The primary advantage of analytical solutions is that they produce mathematical expressions that describe the reliability of the system. Once the reliability function of the system has been determined, other reliability characteristics, such as availability estimations, can be obtained. The more complicated a system is, the more difficult it will be to formulate analytical expressions for reliability. In these situations, simulation methods, based on the Monte Carlo method, may be more advantageous than analytical ones [19].

Event tree analysis (ETA) is a branched graph starting from initiating events, showing the possible sequences of plant states, operator performance, or emergency responses, as well as their respective occurrence probabilities. Each branch represents a defense-in-depth level, that is, a layer of protection designed to stop undesired events or to mitigate their consequences. The failure probability of these defense-in-depth levels can be obtained using historical failure data or logic models, such as FTAs or RBDs. Initiating events can be internal or external events, which can create extreme environments to several plant systems. Internal hazards can include fire, explosion, or floods. External hazards can include earthquakes, accidents in neighborhood, tornadoes, or aircraft crashes. An event tree gives a picture of the various plant end-states, each one with an estimated probability. In a PSA approach, ETA is used for estimating the probability and the consequences of each accident scenario caused by an initiating event followed by malfunction of safety and mitigating systems, as well as operating errors [4].

Cause-consequence analysis (CCA) consists of a diagram for graphically displaying the relationships between accident consequences and their basic causes. It has both elements of fault tree and event tree analyses. It can be developed in either direction: toward the consequences (similar to an ETA) or toward the basic causes (similar to a FTA). The outcome consists of a list of MCSs of an accident sequence, which are similar to the MCSs of FTAs. CCA is most frequently applied where the system states change with time. This technique can also include analysis of human errors [22].

Human reliability analysis (HRA) is the study of interactions between human beings and systems to assess system reliability. HRA is a key element of PSAs, particularly for complex systems, which are highly dependent on human-system interactions in many phases of their projects. HRA can model and quantify these human interactions, which will be incorporated as human basic events in the PSA logic models (e.g., FTA, RBD, and ETA). HRA assesses the probabilities of occurrence of errors, the opportunities to recover from them, and their consequences. The most known technique specifically developed for HRA applications is "technique for human error rate prediction" (THERP). THERP uses "performance-shaping factors" (PSFs) to make judgments about particular situations for predicting human errors. THERP has the advantage of simplicity and can be used for design analysis, worker selection, prediction of system effectiveness, and determination of training requirements [17].

Monte Carlo method (MCM), also called "Monte Carlo simulation," is a statistical technique implemented on computers that allows uncertainty assessment in safety analyses and decision-making processes. MCM provides the probabilities of possible outcomes from a set of random input actions, represented by probability functions, analyzing the scenarios of uncertainties. In PSA, MCM can be used for uncertainty assessment of failure rate data, propagation of uncertainty in FTAs, and consequence estimates of accident sequences in ETAs. For instance, MCM can be used to find incidents that lead to a major accident, selecting the probabilistic parameters that affect the accident scenarios. As an example, MCM can be used to estimate the hole size in a pipe or vessel, during an incident, which generates the greatest and most probable damage [26].

Analytical model (AM) in safety analysis uses mathematical expressions that estimates physical consequences of accidents, representing, for instance, plant conditions, dispersion models, discharge models, explosion overpressure, radiation heat or radiological doses, and their variations along time. AMs used in consequence analysis are based on mass, energy, and momentum conservation principles. The consequences can be expressed as fatalities or injuries in humans, damage in physical structures, and environment consequences. Analytical approaches used in PSA include also empirical, statistical, and logical models [26, 27].

Bayes theorem (BT) is used in quantitative safety analysis for specialization ("updating") of failure probabilities when new evidence becomes available. Considering the failure rate (λ) as the parameter under analysis, a mathematical formulation of this theorem could be expressed according to Eq. (2):

$$f(\lambda | E) = \frac{f(\lambda)L(E|\lambda)}{\int_0^\infty f(\lambda)L(E|\lambda)d\lambda}, \tag{2}$$

where $f(\lambda | E)$ is the probability density function of λ given evidence E ("posterior distribution"), $f(\lambda)$ is the probability density prior to having evidence E ("prior distribution"), and $L(E | \lambda)$ is the likelihood function ("probability of the evidence E given λ"). The evaluation of the integral in Eq. (2) cannot, in general, be done analytically and can be carried out with support of computer codes, for specific likelihood function shapes [28].

Life data analysis (LDA) is the study and modeling of observed component lifetime, which can be defined as the length of operation of a component before it fails (measured in terms of time, distance, or cycles). These lifetimes are referred to, generally, as component life data, and are used to estimate the failure rate of components and systems needed in PSAs. When performing LDA, it is necessary to fit a statistical distribution to life data from a representative sample. The life data distribution can be used to estimate reliability, failure probability, mean life, and failure rate of components [29].

Markov model (MM) consists of decision trees used for modeling events that may occur repeatedly. A MM involves the analysis of the possible states of a system, the possible transition paths between the states, and the rate parameters of those transitions. In reliability and safety analysis, the transitions usually consist of failures and repairs. A graphical representation of a MM depicts each state as a circle, and the transition paths between states as arrows. As an example, the MM can be used to model piping failures in locations susceptible to damage mechanisms. This kind of model can be used in PSAs to predict rupture probabilities from occurrence rates of flaw or leak state in piping systems [30].

5.2 Choosing Adequate Techniques for DSA and PSA

Many of the described techniques can be used in each step of DSA and PSA, according to scope of analysis. For example, What-If and HAZOP are better suited to batch processes than FMEA or FTA. Analyses of multiple failure situations are best handled by FTA than HAZOP or FMEA. In addition, FTA and ETA are suitable for quantitative assessments of rare events, when component failure rate or reliability data are available. Safety Review, Checklists, Process Hazard Analysis, and What-If are considered particularly useful when only a screening and a general hazard list are required. HAZOP and FMEA are more suitable for performing detailed analyses of a wide range of hazards, to identify potential accident sequences. Some methods may be used for both qualitative and quantitative risk assessments. FTA, combined with ETA, is widely used in PSA, particularly in nuclear, aerospace, and petrochemical areas [31]. Table 1 shows a list of safety analysis techniques applied to the different steps of DSA and PSA.

The choice of a particular technique or a combination of techniques will depend on the reason for conducting the analysis, the results needed, the data available, the complexity of the systems being analyzed, and the team experience in conducting the analysis and in using the tools, as well as the perceived risks of the processes.

One of the most important factors in determining the choice of such techniques is the scope of analysis. For identifying single failure events, WI, PHA, FMEA, or HAZOP are recommended. If the perceived risks of the potential accident sequences are high, FMEA and HAZOP are indicated. If quantitative assessment is required, HAZOP, FMEA, FTA, RBD, ETA, and HRA are recommended. Some techniques are more systematic than others. For example, the HAZOP technique provides a detailed framework for studying each process, line by line, in an exhaustive manner. This technique analyzes each process and the deviations from normal values and their consequences. Another example of systematic approach is FMEA, which takes into account the various failure modes and evaluates the effects of those failures on the systems or facilities. On the other hand, CL and WI techniques rely on a brainstorming approach to create a list of questions addressing specific events that could produce undesirable consequences. Although the systematic nature of HAZOP and FMEA approaches may compensate for possible weaknesses of the analysis team, CL and WI rely heavily on experience and knowledge of safety analysis team for their success. Finally, the time required to carry out an analysis is also an important factor. Quantitative analysis techniques, such as FTA, RBD, ETA, MCM, and MM are, in general, more time consuming and costly than qualitative ones [31].

6 PSA AS A COMPLEMENT TO DSA

Even though conservative DSAs are widely used as a licensing basis, PSA techniques can be used as complementary tools to support safety analysis. This approach, also named "Risk-Informed Decision-Making" (RIDM), has been recently applied worldwide in licensing processes in nuclear industry, mainly in NPPs [32]. Table 2 summarizes some methodological features of complementarity of both deterministic and probabilistic approaches.

DSA is based on well-defined guidelines stated by the regulatory bodies. It is usually documented in "safety analysis reports" (SARs) that describe and

Table 1 Safety analysis techniques applied to DSA and PSA steps

Safety analysis techniques		SR	CL	WI	PHA	FMEA	HAZOP	FTA	RBD	ETA	CCA	HRA	MCM	AM	BT	LDA	MM
DSA steps	Identification of DBAs	x	x	x	x	x	x				x						
	Equipment performance analysis	x	x	x	x	x	x				x			x			
	Human actions analysis									x	x	x					
	Analysis of plant response				x								x	x			
	Acceptance criteria selection	x	x											x			
	Selection of modification alternatives				x	x	x										
	Definition of design and operational requirements				x	x	x										
PSA steps	Scope of PSA	x	x	x	x	x	x				x						
	Initiating event analysis	x	x	x	x	x	x			x					x	x	
	Database/operational practices analysis	x			x	x	x							x	x	x	x
	Reliability/availability analysis							x	x		x		x	x	x	x	x
	Human reliability analysis							x		x	x	x		x	x		
	Scenario modeling					x	x			x	x	x	x	x			
	Event sequence frequency assessment					x		x	x	x	x	x	x	x	x	x	x
	Consequence assessment					x	x			x	x		x	x	x	x	x
	Risk assessment and integration					x		x	x	x			x	x	x	x	x
	Safety goal definition							x	x	x							
	Selection of modification alternatives				x	x		x	x					x			
	Definition of design and operational requirements				x	x		x	x					x			

Table 2 Complementarity of deterministic and probabilistic approaches

Features	Deterministic approach	Probabilistic approach
Hazard/initiating Event	Limited to consideration of relatively frequent events (DBAs)	All potential credible accidents can be included
Analysis method	Uses conservative rules, standards, guidelines and specialized computer codes	Analysis follows a well-established methodology (realistic, fault trees, event trees, best estimates assumptions, etc.)
Common-cause failure (CCF)	Single failure criterion is used and CCF is not normally addressed in detail	Multiple and common-cause failures usually assessed
Design	Able to support the design process	Aids in identifying cost-effective safety improvements to existing facilities
Licensing	Traditional requirements of regulatory bodies are based on DBAs	Used as support to traditional deterministic licensing
Results	DSA demonstrates compliance with the rules, standards and guidelines	The risk from the facility may be determined and allows the design and operational improvements, as well as planning of inspection, testing and maintenance
Cost/time	Relatively quick and economical	May be time consuming especially if more comprehensive analyses are required

Adapted from A. Weele, Deterministic or probabilistic analysis? RISKWorld, Risktec, Warrington, UK, 2002, p. 2 (issue 1 spring 2002).

present the analyses of the site, the design, and the operation of a facility. The postulated accidents considered in the design basis of the facility are identified, and different scenarios and their consequences are analyzed, taking into account conservative assumptions. Thus the outcomes of the analyses are compared with the acceptance criteria, verifying the adequacy of design of safety systems. Then, DSA considers several safety margins but does not usually take into account low-frequency events.

On the other hand, PSA takes into account the occurrence of rare events, in addition to single failures, multiple failures, "common-cause failures"

(CCFs), and human errors. The failure probabilities are assessed as realistically as possible (best estimates), using both generic and plant-specific data, as well as including uncertainty assessments [33]. Event trees describe the development of different scenarios that either lead to successful sequences or to undesired consequences. Reliability and availability analyses (using RBDs or FTAs) are usually carried out for the safety systems involved. The probabilities of potential human errors important for assessing the accident scenarios are usually considered in PSA. The probabilistic risk assessment of a facility allows the design and operational improvements, as well as planning of inspection, testing, and maintenance [4].

Inputs from deterministic studies (e.g., thermohydraulic behavior for NPPs) with realistic hypotheses are usually required to define the accident scenarios in PSA. The PSA results contain uncertainties arising from various sources: lack of comprehensive data about the area under consideration; incompleteness of analyses; uncertainties of failure rate and reliability data of components; uncertainties of the frequency of initiating events, CCFs, and human errors; and uncertainties related to mechanical behavior of components under accident conditions and understanding of physical phenomena.

It is important to highlight that the uncertainties are not intrinsic to PSAs but may generally be attributed to lack of knowledge (epistemic uncertainties). PSA has the benefits of quantifying the uncertainties and identifying the areas about which more knowledge is required. Despite these uncertainties, the assessments of strengths and weaknesses of the safety systems and mitigation measures are useful for improving design and operation of complex and hazardous facilities. Thus PSA can be considered as a complement to DSA in checking the safety levels of a facility, especially taking into account rare events and severe accidents, improving it from the identification and analysis of the design and operation weaknesses.

7 CASE STUDY

This simple case study presents some important issues on DSA and PSA. Such study, illustrated by the Piping and Instrumentation Diagram (P&ID) shown in Fig. 6, is a simplified version of an example of a leakage from a storage vessel available in the literature [34, 35].

To carry out the quantitative assessments in this case study, a selected storage vessel (VP) was designed to hold 50 tons of a flammable liquid (pentane, an organic compound with the formula C_5H_{12}) under nitrogen positive pressure. In Fig. 6, a control system [pressure indicating controller with

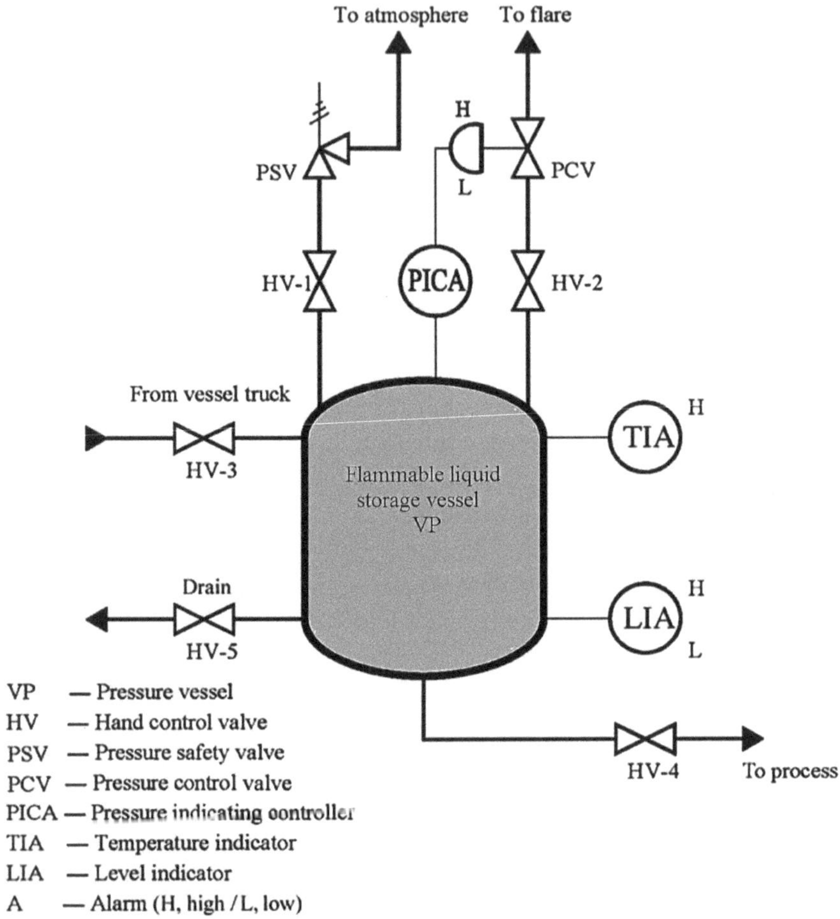

Fig. 6 Simplified flammable liquid storage vessel P&ID. *(Based on American Institute of Chemical Engineers (AIChE), Guidelines for Chemical Process Quantitative Risk Analysis, second ed., Center for Chemical Process Safety (CCPS), AIChE, New York, NY, 2000; H. Ozog, Hazard identification, analysis and control, Chem. Eng. 18 (1985) 161–170.).*

alarm (PICA)] controls pressure through a pressure control valve (PCV) connected to the vessel through a hand control valve (HV-2) and a flare header. A pressure safety valve (PSV) connected to the vessel through a hand control valve (HV-1) is designed to cope with emergencies. Temperature and level of flammable liquid in VP are monitored by temperature indicator with alarm (TIA) and level indicator with alarm (LIA), respectively. Liquid feed from vessel truck and to process flows through the hand control valves HV-3 and HV-4, respectively, while the vessel drain is done through hand control valve HV-5. For illustration purpose, it is considered that the facility boundary is located around 300 m from the vessel site.

7.1 Deterministic Safety Analysis (DSA) Applied to the Case Study

The deterministic (conservative) safety analysis applied to this case study considers the worst-case scenarios of instantaneous release of the largest amount of flammable liquid that could be caused, for instance, by a catastrophic vessel rupture. The consequences of the formation and detonation of the flammable vapor cloud are estimated assuming a "yield factor" of 10% ("fraction of the released substance that participates in blast wave generation"). The selected acceptance criterion to assess if significant consequences of this accident could occur beyond the area of company ownership was the endpoint distance (point where the explosion generates an overpressure of 1.0 psi). The TNT equivalent (analytical/empirical model) was the method selected to estimate the endpoint distance [36].

Eq. (3) estimates the minimum safe distance ($R_{\min}$), according to the TNT equivalent method, based on experimental data from hemispherical charges of TNT [36, 37]:

$$R_{\min} = Z \cdot \sqrt[3]{W_E}, \tag{3}$$

where $R_{\min}$ is the distance from explosion where the overpressure will be 1.0 psi (m), W_E is the mass of TNT equivalent (kg), and Z is the scaled distance (18 $\mathrm{m}/\sqrt[3]{\mathrm{kg}}$). W_E stands for the mass of the flammable liquid that will produce the same blast effect as a unit of mass of TNT and can be calculated by Eq. (4):

$$W_E = \frac{E}{4420}, \tag{4}$$

where 4420 kJ/kg is the value of heat of combustion of TNT, and E is the blast wave energy of flammable liquid (kJ/kg), given by Eq. (5):

$$E = \alpha \, \Delta H_c \, m_F, \tag{5}$$

where, α is the yield factor, ΔH_c is the heat of combustion of flammable liquid (kJ/kg), and m_F is the mass of flammable vapor released (kg).

Assuming $\Delta H_c = 48{,}621$ kJ/kg (heat of combustion of pentane), $m_F =$ 50,000 kg, and $\alpha = 10\%$, Eq. (5) results in a blast wave energy $E =$ 2.43×10^8 kJ that, combined with Eq. (4) results in $W_E = 54{,}977$ kg. Finally, using this value of W_E in Eq. (3) results in $R_{\min} = 684$ m. This means, according to these very conservative calculations, that beyond the facility boundary (300 m from vessel), the overpressure would be greater than 1.0 psi, and the acceptance criterion would not be met.

The possible design and administrative modification alternatives to comply with the acceptance criterion are: increasing the area of ownership of the company, acquiring the land within a radius of 684 m from the vessel site; using a diked area around the vessel with sufficient height to contain all released liquid; replacing the 50-ton vessel with two 25-ton vessels and reassessing the endpoint distance, considering the release from a single vessel. These alternatives, among others, should be analyzed individually or together in a cost-effective basis. For instance, the last alternative of replacing the 50-ton vessel with two 25-ton vessels, besides being very expensive, time-consuming, and with operational implications, would not be enough to meet the acceptance criterion.

7.2 Probabilistic Safety Analysis (PSA) Applied to the Case Study

The case study presented in the previous subitem using a DSA approach estimates the worst-case consequence distances, adopting very conservative assumptions. PSA techniques can support more realistic predictions of possible consequences of vessel releases through the modeling of alternative scenarios. This approach for safety analysis takes into account the active systems, as interlocks, shutdown systems, pressure relieving devices, flares, and emergency isolation systems, as well as passive mitigation systems (e.g., diked areas, blast walls, fire walls, enclosures, and drains). The following analyses consider the specific control, safety, and mitigation devices available in the storage vessel, during the screening of the alternative scenarios.

Fault tree analysis can be used, even in a qualitative way, to illustrate how PSA can help you search for more realistic scenarios, better analyze the risks, and find more cost-effective alternatives of design and operational requirements to meet the safety goals. Fig. 7 shows a fault tree constructed for the top event (T) "major flammable release" for the storage vessel and its control, safety, and mitigation devices shown in Fig. 6. The fault tree was constructed in a top-down process, starting from the top event (T), linked through gates "AND" or "OR," to intermediate events (labeled sequentially with "M"), and to basic or undeveloped events (labeled sequentially with "B").

A qualitative analysis of the fault tree is done searching for the Minimal Cut Sets (MCSs). This can be carried out with support of Boolean algebra,

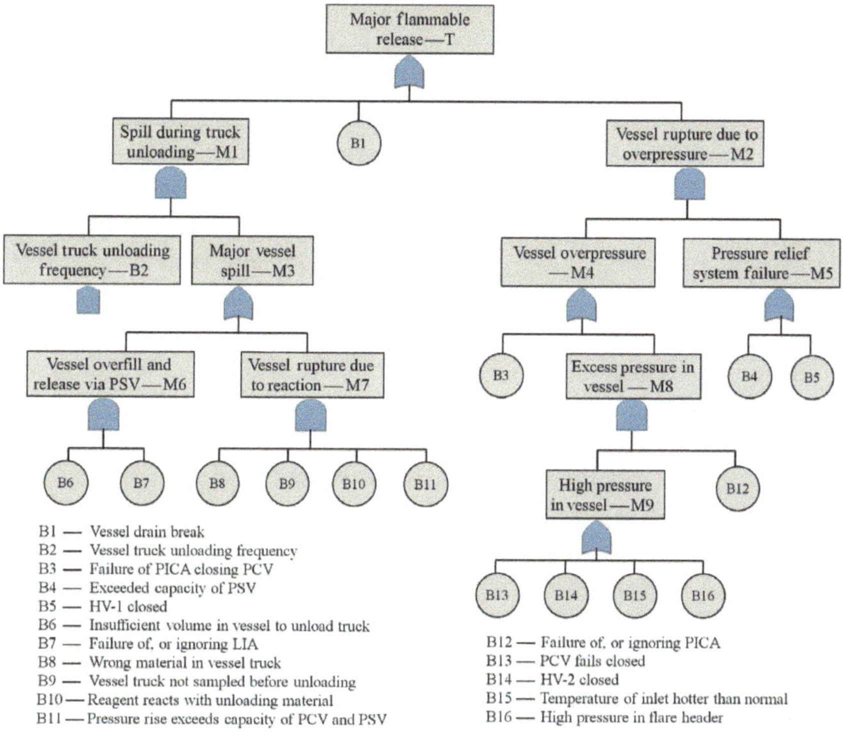

Fig. 7 Fault tree analysis for the case study.

specialized algorithms (e.g., Vesely-Fussell algorithm, described by Ref. [38]), or computer codes [19]. As there are no repeated basic events in the fault tree shown in Fig. 7, the MCS identification can be performed by accurate inspection of the combination of basic events that lead to top event (T). For example, the single event B1 (first-order MCS) and the combinations B3·B4 and B3·B5 (second-order MCSs), lead to T. In addition, MCSs of higher orders also led to occurrence of T (B2·B6·B7, B4·B12·B13, B4·B12·B14, B4·B12·B15, B4·B12·B16, B5·B12·B13, B5·B12·B14, B5·B12·B15, B5·B12·B16, and B2·B8·B9·B10·B11). In a qualitative analysis, the MCSs of lower orders are, in general, the most important contributors for the likelihood of occurrence of top events, irrespective of their probabilities, because a combination of small number of events leads to the occurrence of the top event. Thus vessel drain break (event B1) seems to be the highest contributor for T. Failure of PICA (event B3) seems to be also significant as it is part of two second-orders MCSs, together with B4 (exceeded capacity of PSV) and B5 (V-1 closed). This qualitative analysis is used in this case study to support the selection of more realistic scenarios, consequence assessments, and modification alternatives.

The releases from vessel drain brake and valve failures seem to be more likely than instantaneous releases, taking into account the previous qualitative analyses. Thus a more realistic scenario considered would be the releasing of the entire content of the vessel in a certain time. Because of the complexity of the consequence modeling of this type of alternative scenario, including release rate estimating, evaporation, and atmospheric dispersion models, and influences of surrounding terrain types, as well as passive and active mitigation measures, a computer code is needed to carry out the estimations.

The estimates of consequences of this more realistic scenario is done with support of RMP*Comp, a free browser-based computer code provided by USEPA (U.S. Environmental Protection Agency) to perform off-site consequence analysis required by Risk Management Programs of hazardous chemical facilities [9]. The model for estimating the 1.0 psi overpressure endpoint for vapor cloud explosion assumes a yield factor of 3%, wind speed of 3 m/s, and D atmospheric stability class. Other required parameters are the identification of flammable liquid, liquid temperature, release rate (user-specified or estimated according to storage type, hole or puncture area, and height of liquid column above hole), surrounding terrain type (urban or rural), and release duration (time to shut off flow or time to fully drain the vessel). The computer code also considers the influence of measures of passive mitigation (as diked areas) or active mitigation (as automatic shut off valves) to reduce release rates.

Considering in RMP*Comp as input data a vessel with 50 ton of pentane (a flammable liquid having a boiling point of 36.1°C), liquid temperature of 25°C, release rate of 5 tons/min, release duration of 10 min, vessel site in urban area (many obstacles in the immediate area), and other default parameters assumed by the software in this kind of scenario, the distance to the lower flammability limit is less than 160 m. This means that, even without additional preventive and mitigating measures, the consequences of the postulated accident are restricted to the area of company ownership and can be treated by local emergency planning. In addition, a quantitative assessment of the developed fault tree, using generic or plant-specific failure rates and probabilities of occurrence of the basic events, can be performed to prioritize human and financial resources to improve design and operational requirements for this kind of facility, increasing the reliability of items important to safety, as well as reducing the probability and consequence of the potential accidents.

8 CONCLUSIONS

Concepts, framework, and techniques necessary to carry out DSA and PSA were presented. This chapter leads to the conclusion that, although both approaches for safety analysis of complex and hazardous facilities can provide adequate safety levels, sometimes the PSA results can also be useful to reduce unnecessary conservatism of the DSA approaches. On the other hand, the PSA results can be used to identify insufficient conservatism in DSA results, contributing to reduce possible drawbacks of this approach. Thus PSA has become an important complement to DSA, identifying the weaknesses of safety systems and mitigation measures, improving design and operation procedures of complex and hazardous facilities, such as nuclear power or petrochemical plants.

A simple case study, adapted from the literature examples, is presented to illustrate some important issues on DSA and PSA. The conservative features of DSA, based on worst-case scenarios, as well as the more PSA realistic analysis, selecting alternative scenarios, have been highlighted. Even without additional preventive and mitigating measures, the consequences of the analyzed PSA alternative scenario were restricted to the area of company ownership and could be treated by local emergency planning. Many apparently unnecessary design and administrative modification alternatives could result from the analysis if only the DSA conservative outcomes are adopted to comply with the acceptance criteria.

The reader will also find in this chapter a very useful compilation and description of the available techniques used in DSA and PSA, their similarities and differences, as well as their advantages and disadvantages in applications required in all steps of qualitative and quantitative analyses.

ACKNOWLEDGMENTS

The authors thank the following Brazilian institutions: Centro de Desenvolvimento da Tecnologia Nuclear (CDTN), Comissão Nacional de Energia Nuclear (CNEN), Financiadora de Estudos e Projetos (FINEP), and Conselho Nacional de Desenvolvimento Científico e Tecnológico (CNPq) for supporting the elaboration of this chapter.

REFERENCES

[1] International Atomic Energy Agency (IAEA), Safety Glossary Terminology Used in Nuclear Safety and Radiation Protection, IAEA, Vienna, 2016.
[2] International Organization for Standardization (ISO), Risk Management: Principles and Guidelines, first ed., ISO, Geneva, 2009 (ISO 31000:2009(E)).

[3] International Atomic Energy Agency (IAEA), Deterministic Safety Analysis for Nuclear Power Plants, IAEA, Vienna, 2009 (Specific Safety Guide No. SSG-2).
[4] International Atomic Energy Agency (IAEA), Development and Application of Level 1 Probabilistic Safety Assessment for Nuclear Power Plants, IAEA, Vienna, 2010 (Specific Safety Guide No. SSG-3).
[5] NOPSEMA, National Offshore Petroleum Safety and Environmental Management Authority (ALARP Guidance Note. N-04300-GN0166 Revision 6), NOPSEMA, Melbourne, 2015.
[6] U.S. Nuclear Regulatory Commission (USNRC), Glossary of Risk-Related Terms in Support of Risk-Informed Decision-Making (NUREG 2122), USNRC, Washington, DC, 2013.
[7] A. Franks, Lines of Defence/Layers of Protection Analysis in the COMAH Context, Amey VECTRA Limited, Westbrook, Warrington, 2017.
[8] American Nuclear Society (ANS), Glossary of Definitions and Terminology, ANS, La Grange Park, IL, 2016.
[9] U.S. Environmental Protection Agency (USEPA), Risk Management Program Guidance for Offsite Consequence Analysis (EPA 550B99009), Office of Solid Waste and Emergency Response, Washington, DC, 2009.
[10] U.S. Nuclear Regulatory Commission (USNRC), WASH-1400: Reactor Safety Study (NUREG-75/014), USNRC, Washington, DC, 1975.
[11] F.M. Christensen, O. Andersen, N.J. Duijm, P. Harremoës, Risk terminology—a platform for common understanding and better communication, J. Hazard. Mater. 103 (2003) 181–203.
[12] International Atomic Energy Agency (IAEA), Needs for Safety Assessment & the Safety Assessment Process, http://www-ns.iaea.org/downloads/ni/training/specific_expert_knowledge/safety%20assessment/II%202%20and%20II%203_2%20-%20Needs%20for%20Safety%20Assessment%20&%20the%20Safety%20.pdf, 2009. Accessed 6 February 2018.
[13] International Atomic Energy Agency (IAEA), Safety Assessment for Facilities and Activities, IAEA, Vienna, 2009 (General Safety Requirements No. GSR Part 4).
[14] V. Vasconcelos, W.A. Soares, R.O. Marques, Integrated engineering approach to safety, reliability, risk management and human factors, in: F. Felice, A. Petrillo (Eds.), Human Factors and Reliability Engineering for Safety and Security in Critical Infrastructures: Decision Making, Theory, and Practice, Springer International Publishing AG, Cham, 2017, pp. 77–107.
[15] U.S. Nuclear Regulatory Commission (USNRC), Probabilistic Safety Analysis Procedures Guide (NUREG/CR-2815), USNRC, Washington, DC, 1983.
[16] C. Kirchsteiger, On the use of probabilistic and deterministic methods in risk analysis, J. Loss Prev. Process Ind. 12 (1999) 399–419.
[17] E. Calixto, Gas and Oil Reliability Engineering. Modeling and Analysis, Elsevier, Amsterdan, 2013.
[18] International Atomic Energy Agency (IAEA), Procedures for Conducting Probabilistic Safety Assessment for Non-reactor Nuclear Facilities, IAEA, Vienna, 2002 (TECDOC-1267).
[19] ReliaSoft, System Analysis Reference: Reliability, Availability and Optimization, ReliaSoft Corporation, Tucson, AZ, 2015.
[20] International Atomic Energy Agency (IAEA), Risk-Informed Regulation of Nuclear Facilities: Overview of the Current Status, IAEA, Vienna, 2005 (TECDOC-1436).
[21] International Atomic Energy Agency (IAEA), Guidelines for Integrated Risk Assessment and Management in Large Industrial Areas, IAEA, Vienna, 1998 (TECDOC-994).
[22] F.P. Lees, Loss Prevention in the Process Industries: Hazard Identification, Assessment and Control, fourth ed., Butterworth-Heinemann, Oxford, 2012.

[23] International Atomic Energy Agency (IAEA), Periodic Safety Review for Nuclear Power Plants, IAEA, Vienna, 2013 (Specific Safety Guide No. SSG-25).
[24] I.S. Sutton, Process Hazards Analysis, first ed., Southwestern Books, Ashland, VA, 2001.
[25] M. Stamatelatos, Probabilistic Risk Assessment Procedures Guide for NASA Managers and Practitioners: Version 1.1, Office of Safety and Mission Assurance: NASA Headquarters, Washington, DC, 2002.
[26] R. Fullwood, Probabilistic Safety Assessment in the Chemical and Nuclear Industries, first ed., Butterworth-Heinemann, Oxford, 1999.
[27] N. Iqbal, M.H. Salley, Fire Dynamic Tools (FDTs): Quantitative Fire Hazard Analysis Methods for U.S. Nuclear Regulatory Commission Fire Protection Inspection Program, Division of System Safety and Analysis Office of Nuclear Reactor Regulation, USNRC, Washington, DC, 2004 (NUREG-1805).
[28] G. Apostolakis, S. Kaplan, B.J. Garrick, R.J. Duphily, Data specialization for plant-specific risk studies, Nucl. Eng. Des. 56 (1980) 321–329.
[29] ReliaSoft, Accelerated Life Testing Reference, ReliaSoft Corporation, Tucson, AZ, 2015.
[30] K.N. Fleming, Markov models for evaluating risk-informed in-service inspection strategies for nuclear power plant piping systems, Reliab. Eng. Syst. Saf. 83 (2004) 27–45.
[31] U.S. Nuclear Regulatory Commission (USNRC), Integrated Safety Analysis: Guidance Document (NUREG-1513), Office of Nuclear Material Safety and Safeguards, Washington, DC, 2001.
[32] F. Di Maio, E. Zio, C. Curtis Smith, V. Rychkov, Integrated deterministic and probabilistic safety analysis for safety assessment of nuclear power plants, in: Science and Technology of Nuclear Installations, Hindawi Publishing Corporation, New York, NY, 2015, pp. 1–2.
[33] International Atomic Energy Agency (IAEA), Best Estimate Safety Analysis for Nuclear Power Plants: Uncertainty Evaluation, IAEA, Vienna, 2008 (Safety Report Series No. 52).
[34] American Institute of Chemical Engineers (AIChE), Guidelines for Chemical Process Quantitative Risk Analysis, second ed., Center for Chemical Process Safety (CCPS), AIChE, New York, NY, 2000.
[35] H. Ozog, Hazard identification, analysis and control, Chem. Eng. 18 (1985) 161–170.
[36] U.S. Nuclear Regulatory Commission (USNRC), Evaluations of Explosions Postulated to Occur at Nearby Facilities and on Transportation Routes near Nuclear Power Plants (Regulatory Guide 1.91, Revision 2), USNRC, Washington, DC, 2011.
[37] U.S. Nuclear Regulatory Commission (USNRC), Fire Dynamics Tools (FDTs): Quantitative Fire Hazard Analysis Methods for the U.S. Nuclear Regulatory Commission Fire Protection Inspection Program (NUREG 1805), USNRC, Washington, DC, 2004.
[38] W.E. Vesely, F.F. Goldberg, N.H. Roberts, D.F. Haasl, Fault Tree Handbook (NUREG-0492), Office of Nuclear Regulatory Research, USNRC, Washington, DC, 1981.

FURTHER READING

[39] A. Weele, Deterministic or probabilistic analysis? in: RISKWorld, Risktec, Warrington, 2002. p. 2 (issue 1 spring 2002).

CHAPTER 3

Reliability-Game Theory

Lirong Cui, Chen Fang
School of Management & Economics, Beijing Institute of Technology, Beijing, China

Abstract

Game theory is applicable to situations where decision-makers are competing for a limited resource. Game theory offers rational or limited rational justification for selecting strategies that lead to specific outcomes. Game theory has been used very successfully in economics, and it has also been applied in the social sciences, computer science, and biological sciences, among others.

Game theory can be classified into two types: *cooperative* and *noncooperative*. In the field of reliability, the players in noncooperative games could be consumers, factories, retail networks, regulatory agencies, etc. Each player is assumed to act rationally to maximize its profit or minimize its effort.

Ramamurthy's book (1990, Kluwer Academic Publishers) on cooperative game theory, titled *Coherent Structures and Simple Games*, was among the first to bridge the gap between reliability and game theory. Ramamurthy notes that a number of concepts, as well as the applications of game theory, were rediscovered only fairly recently by researchers in reliability.

In this chapter, we formulate problems in the reliability field by defining them as games, and we define the players, strategies, payoff functions, and other elements that constitute game models. The major advantages of the method are that it can provide new ways to examine reliability problems and can often lead to novel solutions to these problems.

The chapter discusses different types of games, such as static games, dynamic games, etc. Information on rules, kinds of strategies, and equilibrium solutions are also covered. Some of the classical game theory models are shown to have equivalent applications in the reliability field. In addition, numerical examples are presented in detail to demonstrate specific applications of game theory in reliability.

Keywords: Reliability, Game theory, Nash equilibrium, Static game, Dynamic game

Game theory [1] deals with some conflicting problems among interacting decision-makers. Differing from the theory of decision-making, which has only one decision-maker, game theory deals with multiple players or decision-makers. Therefore we can treat game theory as a generalization of the theory of decision-making.

Games have existed in our daily life for a long time, such as chess and playing cards. The studies of game theory are to deal with the corresponding

https://doi.org/10.1016/B978-0-12-815906-4.00003-8

outcomes and rational behaviors for selecting strategies of players whose benefits are dependent upon each other. In game theory, there are lots of different definitions for various games, but the common point is the strategic intercourse or strategic dependence, that is, the strategies taken by players are not necessarily independent of each other.

Although with a long history, but it was not until 1928 that van Neumann published the general theory for solving zero-sum games. In 1944, John Von Neumann and Oskar Morgenstern published the famous book *Game Theory and Economic Behavior*, which is a sign of birth of game theory. From 1950 to 1960, many important developments took place; the best known among them is the concept of the Nash equilibrium, which sets up a foundation of a noncooperative game. During the period of 1960–80, game theory has grown greatly; specifically, both cooperative noncooperative games have rapidly developed. Since the 1980s, game theory has formed into a complete system, and various applications have been developed.

Game theory can be classified into two areas: *cooperative* and *noncooperative*. This chapter will provide some applications of game theory in reliability. We formulate the reliability problems by defining them as games, including players, strategies, payoff functions, and other elements that have been used in the frames of game theory. The major advantages of using the concept of gaming are:

(1) Game theory leads to solutions to the some reliability problems.
(2) Game theory provides some new views for the proposed reliability problems.
(3) There are some reliability problems that can be interpreted by game theory; in this chapter, we deal with them formally.
(4) Some relationships between reliability and game theory can be addressed through the discussions.

The rest of the chapter is organized as follows. The chapter is composed of six sections. In Section 1, problems and basic concepts in reliability and game theory are presented, including a brief introduction to game theory, general reliability problems related to games, and reliability-game theory. Section 2 analyzes some reliability problems (in a narrow sense) related to games, for which several game theory models are built and the solutions to these problems are presented. Sections 3 and 4 follow the similar contents for viewing problems in maintainability, testability, logistics, and safety in terms of game theory. Section 5 discusses the problems in reliability tests and other reliability-related problems with game theory. Finally, further developments are presented in Section 6.

1 PROBLEMS AND BASIC CONCEPTS

It is well known that the elements for a game are players, strategies, actions, information, payoff functions, outcomes, and equilibrium. They are briefly addressed as follows:

(1) *Players*: The parties or individuals (at least two) in a game, who are the main bodies of the game, who may or may not be in conflict with their benefits (at least the conflicts exist in two different parties or individuals). For example, in reliability, the producers hope to pay less effort to produce the product and to get higher benefits, whereas users want to pay less money to obtain a higher quality of product, which is an obvious conflict. The producers and the users are called players in the game if we balance their benefits in some sense. We denote players by the set $N=\{1,2,\ldots,n\}$, the element i represents player i.

(2) *Strategies*: A set of rules of actions for players when some information is available. There are two kinds of strategies in game theory. One is called pure strategy, which is a fixed action when each player chooses it. The set of pure strategies is denoted by S_i for player i. Another one is called a mixed strategy, which is a probability distribution over pure strategies; namely, for player i, the set of its pure strategies is $S_i=\{s_1^{(i)}, s_2^{(i)},\ldots,s_{N_i}{}^{(i)}\}$, then its mixed strategy is S_i with probability distribution $P\left\{s_j^{(i)} \text{ is taken}\right\}=p_j$, $\sum_{j=1}^{N_i}p_j=1$, $p_j\geq 0$, N_i is cardinality of set S_i.

(3) *Actions or moves*: The order of play determined in games, which is an important aspect in game theory. If all players make their actions simultaneously, or if they do not at the same time, but latter players don't know anything from the former players' actions in a game, it is called a static game; otherwise, if all players make their actions within the specified order, and the latter players know all actions the former players did, it is called a dynamic game.

(4) *Information*: There are usually two kinds of knowledge within a game: common knowledge and private knowledge. The information contains the strategies of the opponents, payoffs for all players, types and actions of all players, etc. It usually falls within one of the following classes of information:

- **(A)** *Complete information*: All players know the other players' payoffs.
- **(B)** *Incomplete information*: When some players do not know the payoffs of the others, the game is said to have incomplete information. Many games of interest have incomplete information at least of some extent.

   (C) *Perfect information*: Players move one at a time, and each player knows all previous actions when making his decision.
   (D) *Other classes of information*: Imperfect information, symmetric information, asymmetric information, etc. They are omitted here because we shall not discuss them in this chapter.

(5) *Payoff*: It gives player *i*'s utility $u_i(s)$ for each profile $s=(s_1, s_2, \ldots, s_n)$ of strategies, which are either discrete or continuous.

(6) *Equilibrium*: All players have their optimal strategies, in which no one can get a better payoff when one changes its strategy and others keep their strategies.

(7) *Outcome*: This are what all players are interested in, including a combination of equilibrium strategies, combination of equilibrium actions, and combination of equilibrium payoffs, etc.

As previously mentioned, game theory can be classified into two areas: cooperative and noncooperative. The former one is the main topic discussed in this chapter, and the latter one is mentioned simply in terms of the work of Ramamurthy in Section 6. In noncooperative games, the payoff (benefits/reward/gain) for a given player depends not only on the strategy chosen but also on the strategies chosen by the other players, and each player is assumed to act rationally to maximize profit or minimize loss. The Nash equilibrium in game theory plays a very important role, which involves each player choosing some strategies, even though they are not the strategies that maximize the payoff for a player. A Nash equilibrium exists if, for a given set of strategies chosen by other players, each player's strategy is an optimal response to those strategies. Thus, at a Nash equilibrium, a player's payoff decreases if it changes its strategy, assuming all other players' strategies remain the same. In our reliability-game problems, it is a main task to find the Nash equilibrium and its related strategy equilibrium to analyze the given problems in reliability by forming it as a game problem.

In this chapter, we mainly discuss the following games that are raised in reliability field.

(1) Static game with complete information, in which the Nash equilibrium needs to be found

(2) Static game with incomplete information, in which the Bayesian-Nash equilibrium needs to be found

(3) Dynamic game with complete information, in which the subgames complete the Nash equilibrium that needs to be found

Reliability is a broad subject, which contains five subbranches: reliability (here, in a narrow sense; hereinafter, reliability is for a wide sense without specifications), maintainability, testability, logistics, and safety. Many aspects need to be balanced in the reliability field, because they sometimes conflict. On the other hand, some problems in reliability are essentially game theory topics. Thus the extension of game theory to reliability field is a nature idea, which is one of main motivations for writing the chapter. The following problems listed are examples in reliability that may be analyzed by using game theory:

(1) Design for optimal reliability, such as analysis of optimal structure, optimization allocation in components, analysis of optimal usage for reliability test equipment, analysis of optimal maintenance, etc.
(2) Analysis of trade-off factors in reliability, such as analysis of reliability indexes trade-off, analysis of trade-off for reliability and maintainability, etc.
(3) Problems in reliability sampling, such as establishment of reliability sampling plan, analysis of balance of risk for customer and producer, analysis for choosing sampling plans, etc.
(4) Reliability growth test, such as analysis of strategies for reliability growth test, game problems in various testers in system reliability growth test, etc.
(5) System reliability management, such as game problems in various bodies in reliability management, design of reliability management mechanism, analysis of costs for reliability management, etc.
(6) Stress-strength analysis, such as analysis of stress and strength, analysis of strategies for improving reliability for stress-strength systems, etc.
(7) Problems in reliability accelerated tests, such as analysis of ways for reliability accelerated tests, analysis for various factors in reliability accelerated tests, etc.

In fact, many reliability problems can be analyzed from the point of view of game theory. In this chapter, we only provide some examples to show our ideas by cases. We do not think that the reliability-game theory mentioned in the chapter is a new one, although it is another way to view reliability-related problems and applications of game theory in reliability. However, it indeed brings some new results for us, which sometimes are different from these we knew in reliability. In reliability-game theory, the elements in games, such as players, strategy, payoff, and rules of playing, are built in terms of the combination of reliability-related problems and game theory.

For completeness in this chapter, we will state the methods of finding the Nash equilibrium for static and dynamic games.

(I) Finite static games with complete information
A finite game is assumed that every player strategy space S_i has finite elements. For a finite static game, the Nash equilibrium $(\overset{*}{s_1}, \ldots, \overset{*}{s_n})$ satisfies the following equations,

$$u_i(s_i^*, s_{-i}^*) \geq u_i(s, s_{-i}^*), \quad \text{for } \forall s_i \in S_i, \quad \text{and } \forall i,$$

or in other words, $s_i^* = \arg\max_{s_i} \left\{ u_i\left(s_1^*, \ldots, s_{i-1}^*, s_i, s_{i+1}^*, \ldots, s_n^*\right), s_i \in S_i \right\}$, $i = 1, \ldots, n$,

where $\overset{*}{s_{-i}} = (\overset{*}{s_1}, \ldots, \overset{*}{s_{i-1}}, \overset{*}{s_{i+1}}, \ldots, \overset{*}{s_n})$.

The procedure for finding the Nash equilibrium will be given through an example: A two-person game in which each player has two possible strategies, that is, $S_1 = \{\theta_1, \theta_2\}$ and $S_2 = \{\psi_1, \psi_2\}$, because the similar procedures can be done for other finite static game. The payoff functions are listed in Table 1.

(1) To find the maximum $u_2(i, \bullet)$, $i = 1, 2$, denoted by $\overline{u}_2(i, \bullet)$ in each line of Table 1, without loss of generality, we assume that

$$\overline{u}_2(1, \bullet) = u_2(1, 2), \text{ i.e.}, u_2(1, 2) \geq u_2(1, 1);$$
$$\overline{u}_2(2, \bullet) = u_2(2, 1), \text{ i.e.}, u_2(2, 1) \geq u_2(2, 2).$$

(2) Similarly, in each column of Table 1, to find the maximum $u_1(\bullet, j)$, $j = 1, 2$, denoted by $\overline{u}_1(\bullet, j)$ without loss of generality, we also assume that

$$\overline{u}_1(\bullet, 1) = u_1(1, 1), \text{ i.e.}, u_1(1, 1) \geq u_1(2, 1);$$
$$\overline{u}_1(\bullet, 2) = u_1(1, 2), \text{ i.e.}, u_1(1, 2) \geq u_1(2, 2).$$

Luckily, in position (1, 2) of Table 1, we have two maximum values $\{u_1(1,2), u_2(1,2)\}$, otherwise, no pure Nash equilibrium exists, so that the Nash equilibrium is $\{u_1(1,2), u_2(1,2)\}$. Note that if there are some inequalities hold in nonstrict, then we may have more Nash equilibrium; otherwise,

Table 1 The payoff functions for a finite game

		Player 2	
	Strategies	ψ_1	ψ_2
Player 1	θ_1	$(u_1(1,1), u_2(1,1))$	$(u_1(1,2), u_2(1,2))$
	θ_2	$(u_1(2,1), u_2(2,1))$	$(u_1(2,2), u_2(2,2))$

if all inequalities hold in strict, then only one Nash equilibrium may exist. On the other hand, the procedure mentioned is valid only when the payoff functions are better if they are larger. When the payoff functions are worse if they are large, then it is better to find minimal values rather than to find maximal values.

The Nash equilibria found in the previously discussed procedure are called pure Nash equilibria. Unfortunately, not all games have pure Nash equilibria, which may be found in the prior procedure. However, Nash (1950) gave the following theorem:

Theorem 1 [2]
Every finite strategic form game has a mixed-strategy equilibrium.

The so-called mixed-strategy and mixed-strategy Nash equilibrium are defined as follows. Suppose that player i has a strategy space $S_i = \{s_{i1}, s_{i2}, \ldots, s_{ik}\}$ and a corresponding probability distribution $\pi_i = (\pi_{i1}, \pi_{i2}, \ldots, \pi_{ik})$, that is, player i takes strategy s_{ij} with probability π_{ij}, $j = 1, 2, \ldots, k$, where $\sum_{j=1}^{k} \pi_{ij} = 1$, $\pi_{ij} \geq 0$, which is called a mixed strategy, denoted by

$$\sigma_i = \left\{ \begin{array}{c} (s_{i1}, s_{i2}, \ldots, s_{ik}) \\ (\pi_{i1}, \pi_{i2}, \ldots, \pi_{ik}) \end{array} \right\}.$$

A mixed-strategy Nash equilibrium $(\sigma_1^*, \sigma_2^*, \ldots, \sigma_n^*)$ satisfies the following inequality,

$$u_i(\sigma_i^*, \sigma_{-i}^*) \geq u_i(\sigma_i, \sigma_{-i}^*), \quad \text{for } \forall \sigma_i \text{ and } \forall i,$$

where $u_i(\bullet, \bullet)$ is the payoff function for player i, $\sigma_{-i}^* = (\sigma_1^*, \ldots, \sigma_{i-1}^*, \sigma_{i+1}^*, \ldots, \sigma_n^*)$.

(II) Dynamic games with complete information
In a dynamic game, players take their feasible actions within a specified order, that is, some players take their actions first, then other players perform secondly, until all players have taken their actions; the latter players know the actions taken by the former players, then the dynamic game is ended. An extensive form representation is typically used to describe and analyze the dynamic games, which consists of the set of players, the order of actions (moves, i.e., who moves when), the space of actions (i.e., what the players' choices are when they move), the information set for players (i.e., what each player knows when it makes its choices), the players' payoff as a function of the actions that were made, and the probability distributions over any exogenous events. A finite dynamic game assumes that the number of stages is finite, and

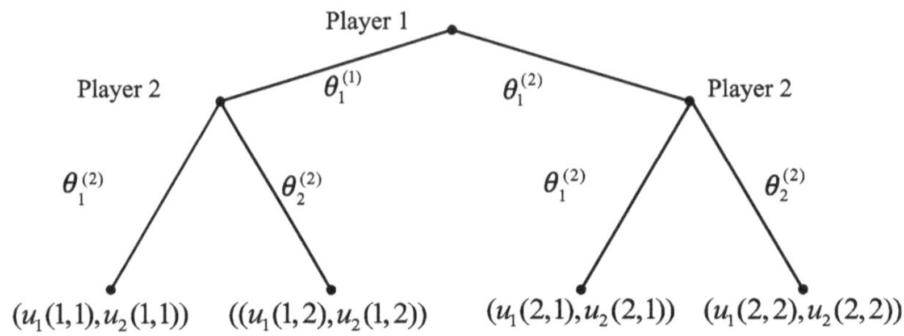

Fig. 1 A game tree.

the number of feasible actions at any stage is finite as well. An extensive form representation for a finite dynamic game can be depicted by a so-called game tree. For example, the set of players is $\Gamma=\{1,2\}$; the order of play is that player 1 takes its action first, then player 2, who knows the action taken by player 1, does the second. The space actions are $S_k=\{\theta_1^{(k)},\theta_2^{(k)}\}$, $k=1$, 2; assume that the information is complete, that is, all necessary information is common knowledge. The payoff functions are denoted as $u_k(i,j)$, $k=1$, 2, $i=1$, 2, $j=1$, 2, and an exogenous event is one type, that is, the probability distribution at one point with probability 1, where $u_k(i,j)$ denotes player k's payoff when players 1 and 2 take their strategies $\theta_i^{(1)}$ and $\theta_j^{(2)}$, respectively. The game tree (similar to a decision tree) for the extensive form representation of the dynamic game is shown in Fig. 1.

A pure-strategy Nash equilibrium $(\overset{*}{s}_1, \overset{*}{s}_2, \ldots, \overset{*}{s}_n)$ in an extensive form representation is similar to that in a static game, that is, it maximizes the payoff functions for strategies. Similarly, the detailed procedure for finding the pure-strategy Nash equilibria in a dynamic complete information game will be given through the previous example by the following procedure:

(1) To find the maximum $u_2(i,\bullet)$, $i=1$, 2, denoted by $\overline{u}_2(i,\bullet)$ in each subbranch of Fig. 1 (first two left ramifications and last right two ramifications), without loss of generality, we assume that

$$\overline{u}_2(1,\bullet)=u_2(1,1),\text{i.e.,}\,u_2(1,2)\le u_2(1,1);$$
$$\overline{u}_2(2,\bullet)=u_2(2,2),\text{i.e.,}\,u_2(2,1)\le u_2(2,2).$$

(2) To find the corresponding maximum $\overline{u}_1(*,\bullet)$ among two positions in which $\overline{u}_2(1,\bullet)$ and $\overline{u}_2(2,\bullet)$ are located. In our example,

$$\overline{u}_1(*,\bullet)=\max\{u_1(1,1),u_1(2,2)\},$$

without loss of generality, we assume that $\overline{u}_1(*, \bullet) = u_1(2,2)$, i.e., $u_1(2,2) \geq u_1(1,1)$.

Thus the pure-strategy Nash equilibrium is $\{u_1(2,2), u_2(2,2)\}$. Note the similar arguments for inequality hold, just like those in the static game previously mentioned. The pure-strategy Nash equilibria exist for a given finite dynamic game with perfect information, which results from the following theorem.

Theorem 2 [2]
A finite game of perfect information has a pure-strategy Nash equilibrium.

(III) Other games
The methods for finding the strategy Nash equilibria will be simply depicted in the following corresponding sections, or readers can refer to some game theory books.

2 ANALYSIS FOR RELIABILITY PROBLEMS

In Section 1, we discussed generally some problems in game theory, which are related to reliability in a broad sense. In this section, we put the focus specifically on the classical definition of reliability in the context of reliability-game theory. In the following, some examples are used to illustrate the models of problems of reliability-game theory, and analysis and solutions are given in detail. Also some numerical examples are presented as well, which can help readers to understand reliability as it relates to game theory.

2.1 Model 1

Type of problem: Reliability in a narrow sense [3, 4].

Description of problem: Two suppliers, A and B, work together to produce a reliability system that is a series system with two subsystems. Each supplier produces one subsystem; there are different reliabilities for both subsystems due to different designs. Supplier A has four possible alternatives, that is, plans $P_1^{(1)}$, $P_2^{(1)}$, $P_3^{(1)}$, $P_4^{(1)}$, which result in subsystem 1 with reliabilities p_{11}, p_{12}, p_{13}, p_{14}, respectively. Supplier B has three possible alternatives, that is, plans $P_1^{(2)}$, $P_2^{(2)}$, $P_3^{(2)}$, which result in subsystem 2 with reliabilities p_{21}, p_{22}, p_{23}, respectively. Three cases will be considered in the following. For Case I, the plan to be adopted for both suppliers must be fixed before the plan can be carried out further. What plan should suppliers A and

B adopt to maximize both suppliers' benefits? The problem forms the following game.

2.1.1 Case I: Type of Game: A Static Complete Information Game

(1) *Players*: Suppliers A and B.
(2) *Strategy sets*: $S_1=\{P_1^{(1)}, P_2^{(1)}, P_3^{(1)}, P_4^{(1)}\}$, $S_2=\{P_1^{(2)}, P_2^{(2)}, P_3^{(2)}\}$.
(3) *Order of actions*: Suppliers A and B make their actions simultaneously;
(4) *Payoff functions*: The payoff function for each supplier is the benefit function, that is,

$$u_1\left(P_i^{(1)}, P_j^{(2)}\right)=f\left(p_{1i}, p_{2j}\right)p_{1i}-c_1p_{1i},$$

$$u_2\left(P_i^{(1)}, P_j^{(2)}\right)=f\left(p_{1i}, p_{2j}\right)p_{2j}-c_2p_{2j},$$

where the function $f(p_{1i}, p_{2j})$ represents total income for system, for example, $f(p_{1i}, p_{2j})=2C-C\exp(-\lambda p_{1i}p_{2j})$, $\lambda>0$, that is, the total income function is just related to system reliability (under independence assumption); if we assume that $c_1=c_2=C$, we have the payoff functions as follows, $u_k(p_i^{(1)}, p_j^{(2)})=(C-C\exp(-\lambda p_i^{(1)}p_j^{(2)}))p_{iI\{k=1\}+jI\{k=2\}}^{(k)}$, where $I\{\bullet\}$ is an indicator function, and $k=1, 2, i=1, 2, 3, 4, j=1, 2, 3$.
(5) *Procedure of the game solution*: A numerical example is given in the following with $C=20$, $p_{11}=0.7$, $p_{12}=0.8$, $p_{13}=0.9$, $p_{14}=0.95$, $p_{21}=0.75$, $p_{22}=0.85$, $p_{23}=0.94$. We developed the following game table (Table 2).
(6) *Equilibrium*: The Nash equilibrium is $(P_4^{(1)}, P_3^{(2)})$ (marked in Table 2), and equilibrium benefits for suppliers A and B are 6.84, and 6.77, respectively. In fact, the benefits and payoff functions are increasing in each $p_1^{(i)}$ and $p_2^{(j)}$ respectively.

If the payoff functions are changed as described in Case II, then we have the following detailed contents.

Table 2 The payoffs for Case I of Model 1

		Supplier B		
	Strategies	$P_1^{(2)}$	$P_2^{(2)}$	$P_3^{(2)}$
Supplier A	$P_1^{(1)}$	(3.23, 3.46)	(3.60, 4.37)	(3.92, 5.27)
	$P_2^{(1)}$	(4.15, 3.89)	(4.61, 4.90)	(5.01, 5.89)
	$P_3^{(1)}$	(5.16, 4.30)	(5.72, 5.40)	(6.21, 6.48)
	$P_4^{(1)}$	(5.69, 4.50)	(6.31, 5.65)	**(6.84, 6.77)**

2.1.2 Case II: Type of Game: A Stat

(1) *Players*: Suppliers A and B.

(2) *Strategy sets*: $S_1=\{P_1^{(1)}, P_2^{(1)}, P_3^{(1)}, P_4^{(1)}\}$, $S_2=\{P_1^{(2)}, P_2^{(2)}, P_3^{(2)}\}$.

(3) *Order of actions*: Suppliers A and B make their actions simultaneously.

(4) *Payoff functions*: The payoff function for each supplier is the benefit function, respectively,

$$u_1\left(p_i^{(1)}, p_j^{(2)}\right)=1000\sqrt{p_i^{(1)}p_j^{(2)}}\frac{p_i^{(1)}}{p_i^{(1)}+p_j^{(2)}}-c_i^{(1)}\frac{p_i^{(1)}}{1-p_i^{(1)}},$$

$$u_2\left(p_i^{(1)}, p_j^{(2)}\right)=1000\sqrt{p_i^{(1)}p_j^{(2)}}\cdot\frac{p_i^{(2)}}{p_i^{(1)}+p_j^{(2)}}-c_i^{(2)}\frac{p_i^{(2)}}{1-p_i^{(2)}}.$$

The reasons for taking the previously described payoff functions are as follows. The benefit functions are $u_k\left(p_i^{(1)}, p_j^{(2)}\right)=1000\sqrt{p_i^{(1)}p_j^{(2)}}\frac{p_i^{(k)}}{p_i^{(1)}+p_j^{(2)}}$, $k=1, 2$, in which the term $1000\sqrt{p_i^{(1)}p_j^{(2)}}$ is for the whole working system benefit, which depends on the series system reliability $p_i^{(1)}p_j^{(2)}$, 1000 is a factor coefficient The term $\frac{p_i^{(k)}}{p_i^{(1)}+p_j^{(2)}}$ results from the contribution percentage of subsystem k for the system reliability; and $c_i^{(k)}\frac{p_i^{(k)}}{1-p_i^{(k)}}$ represents cost for subsystem k with reliability $p_i^{(k)}$.

(5) *Procedure of the game solution*: Given a numerical example in the following: $p_1^{(1)}=0.4$, $p_2^{(1)}=0.6$, $p_3^{(1)}=0.7$, $p_4^{(1)}=0.9$; $p_1^{(2)}=0.6$, $p_2^{(2)}=0.7$, $p_3^{(2)}=0.85$; $c_1^{(1)}=1$, $c_2^{(1)}=5$, $c_3^{(1)}=25$, $c_4^{(1)}=40$; $c_1^{(2)}=15$, $c_2^{(2)}=20$, $c_3^{(2)}=30$, we developed the following game table.

(6) *Equilibrium*: Based on Table 3, we can get the Nash equilibrium $(P_3^{(1)}, P_2^{(2)})$ (marked in Table 3), namely, the equilibrium benefits

Table 3 The payoffs for Case II of Model 1

		Supplier B		
	Strategies	$P_1^{(2)}$	$P_2^{(2)}$	$P_3^{(2)}$
Supplier A	$P_1^{(1)}$	(195.29, 271.43)	(191.75, 290.07)	(185.92, 226.50)
	$P_2^{(1)}$	(**292.50**, 277.50)	(291.61, 302.29)	(288.00, 248.69)
	$P_3^{(1)}$	(290.63, 276.61)	(**291.67, 303.33**)	(290.02, 253.00)
	$P_4^{(1)}$	(80.91, 271.43)	(86.47, 300.59)	(89.82, 254.82)

for suppliers A and B are 291.67 and 303.33, respectively. It is a unique Nash equilibrium strategy for this problem. However, when suppliers A and B take their strategies $(P_2^{(1)}, P_1^{(2)})$ (marked in Table 3) to make supplier A get its maximum benefit 292.50, which differs from the benefit supplier A previously received based on the Nash equilibrium.

If we consider a dynamic game, Case III, with complete information, that is, supplier A takes its action first, then supplier B can too, and both have complete information. The game tree is as follows; the payoff functions and parameters are the same as that in Case I of Model 1.

2.1.3 Case III: Type of Game: A Dynamic Complete Information Game

(1) *Players*: Suppliers A and B.
(2) *Strategy sets*: $S_1=\{P_1^{(1)}, P_2^{(1)}, P_3^{(1)}, P_4^{(1)}\}$, $S_2=\{P_1^{(2)}, P_2^{(2)}, P_3^{(2)}\}$.
(3) *Order of actions*: Supplier A does its action first, then supplier B does it.
(4) *Payoff functions*: The payoff function for each supplier is the same as that in Case I of Model 1.
(5) *Procedure of the game solution*: The parameters are the same as those in Case I of Model 1.
(6) *Equilibrium*: The subgame complete Nash equilibrium is $(P_3^{(1)}, P_2^{(2)})$, which is the same as the solution in Case I of Model 1, and the equilibrium benefits for both suppliers are 291.67 and 303.33, respectively. The game tree is shown in Fig. 2.

From the three results of the games described, we know that the Nash equilibria are different from each other due to different payoff functions and different games. In the first game, the payoff functions are increasing functions

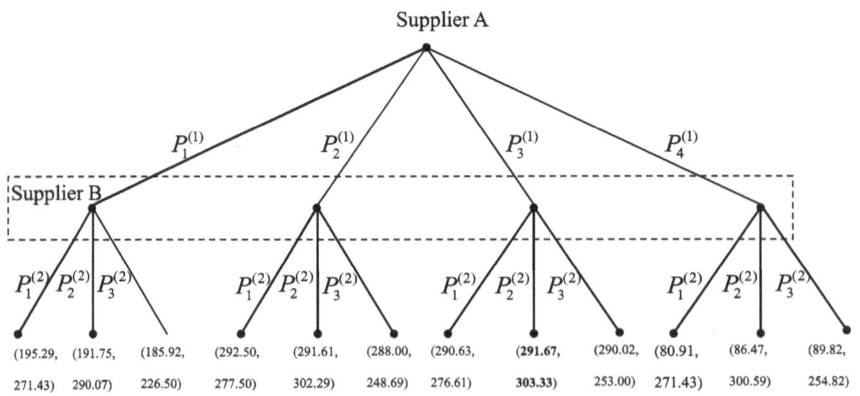

Fig. 2 A game tree for Case III of Model 1.

in $p_i^{(1)}$ and $p_i^{(2)}$, respectively, so that the maximal payoff is in the point $(p_4^{(1)}, p_3^{(2)})$. In Cases I and II, the different games have the same Nash equilibrium.

Of course, we may consider other cases under some restrictions, such as $c_1(p_1) + c_2(p_2) \leq D$, $i = 1$, 2, which are the cost functions of subsystems for suppliers A and B, respectively, but here they are omitted.

2.2 Model 2

Type of problem: Reliability in a narrow sense.

Description of problems: Similar to "the prisoner's dilemma" problem in game theory, we can discuss the following reliability problem. The system consists of two components, which are produced by suppliers 1 and 2; if the system fails, then both suppliers are punished, because it is difficult to know which component's failure results in the system failure. The punishments are represented by the payoffs, which are listed in Table 4. Each player knows the punishments, so how do the suppliers decide on which strategy to take to make their punishments minimal?

Type of game: A static complete information game.

(1) *Players*: Suppliers 1 and 2.
(2) *Strategy sets*: $S_1 = S_2 = \{\text{lower reliability, high reliability}\}$.
(3) *Order of actions*: Suppliers 1 and 2 make their actions simultaneously.
(4) *Payoff functions*: The payoffs for each supplier are listed in Table 4.
(5) *Procedure of the game solution*: The parameters are listed in Table 4.
(6) *Equilibrium*: The Nash equilibrium is (high reliability, high reliability) (marked in Table 4), and both equilibrium benefits for suppliers 1 and 2 are all 2.

In this complete information static game, the players hope payoffs will be minimal, thus the Nash equilibrium satisfies the following inequality,

$$u_i(s_i^*, s_{-i}^*) \leq u_i(s_i, s_{-i}^*), \quad \forall s_i \neq s_i^*, \quad s_{-i}^* = \left(s_1^*, \ldots, s_{i-1}^*, s_{i+1}^*, \ldots, s_n^*\right).$$

If we take the payoffs in Table 5, we can discuss the general situation.

Table 4 The game for Model 2

		Supplier 2	
	Strategies	**Lower reliability**	**High reliability**
Supplier 1	Lower reliability	(5, 5)	(7, 1)
	High reliability	(1, 7)	**(2, 2)**

Table 5 The game for general reliability problem 2

		Supplier 2	
	Strategies	**L**	**H**
Supplier 1	L	(x, x)	(w, y)
	H	(y, w)	(z, z)

Table 6 The game for general reliability problem 2

		Supplier 2	
	Strategy	**L**	**H**
Supplier 1	L	(10, 7)	(15, 5)
	H	(5, 8)	(3,1)

If $x, y, w, z > 0$, $x > z$, $y < z$, $w > z$. If we follow the "reasonable punishment" rules, then the Nash equilibrium is (H,H) (Table 5). In the previous example, if payoff functions are given in Table 6, then we have the following results.

The Nash equilibrium is (H,H) whose corresponding payoffs are (3,1). In fact, we can discuss the 8 numbers, when 6 of them are fixed. The key point in this kind of problem is the payoffs, because the different payoffs result in the different results. For example, supplier 1 should be punished more, because it has advanced technical method and good equipment conditions, however, it provides lower reliability product, which is used in the series system.

2.3 Model 3

Type of problem: Reliability in a narrow sense.

Description of problems: A two-component parallel system is made by producers A and B; each producer has two strategies for its production plan. The benefits containing the incomes, costs, and punishments for them depend on the system reliability. If both producers make their production plan strategies by considering both strategies taken by them, what are the optimal strategies to make their benefits maximal?

Type of game: A static complete information game.

(1) *Players*: Producers 1 and 2.
(2) *Strategy sets*: $S_1 = \{\sigma_{11}, \sigma_{12}\}$, $S_2 = \{\pi_{21}, \pi_{22}\}$.
(3) *Order of actions*: Producers 1 and 2 make their actions simultaneously.
(4) *Payoff functions*: The payoffs for each producer are listed in Table 7.
(5) *Procedure of the game solution*: The parameters are listed in Table 7.

Table 7 The game payoffs for Model 3

		Producer B	
	Strategies	π_{21}	π_{22}
Producer A	σ_{11}	(4, 2)	(−2, 5)
	σ_{12}	(−2, 3)	(−1, 1)

The problem does not have the pure Nash equilibrium solution, but we can consider the mixed-strategy Nash equilibrium solution. We assume that producer A uses a mixed-strategy $\sigma_1 = (\theta, 1-\theta)$, that is, the producer A takes the strategies σ_{11} and σ_{12} with probabilities θ and $1-\theta$, respectively. Similarly, the producer B takes the strategies π_{21} and π_{22} with probabilities λ and $1-\lambda$, respectively. Then the expected utility function for producer A is,

$$\begin{aligned} u_A(\sigma,\pi) &= 4\theta\lambda - 2\lambda(1-\theta) - 2\theta(1-\lambda) - (1-\theta)(1-\lambda) \\ &= 7\theta\lambda - \theta - \lambda - 1 = \theta(7\lambda - 1) - (1+\lambda). \end{aligned}$$

Form the prior equation, we know that

(1) when $7\lambda - 1 < 0$, i.e., $\lambda < 1/7$, $\theta = 0$, which makes $u_A(\sigma,\pi)$ maximal, that is, when $\lambda < 1/7$, the producer A takes strategy σ_{12};

(2) when $7\lambda - 1 > 0$, i.e., $\lambda > 1/7$, $\theta = 1$, which makes $u_A(\sigma,\pi)$ maximal, that is, when $\lambda > 1/7$, the producer A takes strategy σ_{11};

(3) when $7\lambda - 1 = 0$, i.e., $\lambda = 1/7$, $\theta \in [0,1]$, which makes $u_A(\sigma,\pi)$ maximal, that is, when $\lambda = 1/7$, the producer A can take any strategy.

Similarly, we have the expected utility function for producer B as,

$$\begin{aligned} u_B(\sigma,\pi) &= 2\theta\lambda + 3\lambda(1-\theta) + 5\theta(1-\lambda) + (1-\theta)(1-\lambda) \\ &= -5\theta\lambda + 4\theta + 2\lambda + 1 = \lambda(2 - 5\theta) + 4\theta + 1. \end{aligned}$$

We know that

(4) when $2 - 5\theta < 0$, i.e., $\theta > 2/5$, $\lambda = 0$, which makes $u_B(\sigma,\pi)$ maximal, that is, when $\theta > 2/5$, the producer B takes strategy π_{22};

(5) when $2 - 5\theta > 0$, i.e., $\theta < 2/5$, $\lambda = 1$, which makes $u_B(\sigma,\pi)$ maximal, that is, when $\theta < 2/5$, the producer B takes strategy π_{21};

(6) when $2 - 5\theta = 0$, i.e., $\theta = 2/5$, $\lambda \in [0,1]$, which makes $u_B(\sigma,\pi)$ maximal, that is, when $\theta = 2/5$, the producer B can take any strategy.

(6) Equilibrium: The mixed-strategy Nash equilibrium has been given as follows, that is, for both producers are,

$$\sigma = \begin{pmatrix} \sigma_{11}, & \sigma_{12} \\ \theta, & 1-\theta \end{pmatrix}, \quad \pi = \begin{pmatrix} \pi_{21}, & \pi_{22} \\ \lambda, & 1-\lambda \end{pmatrix}.$$

Fig. 3 depicts the mixed-strategy Nash equilibrium in detail.

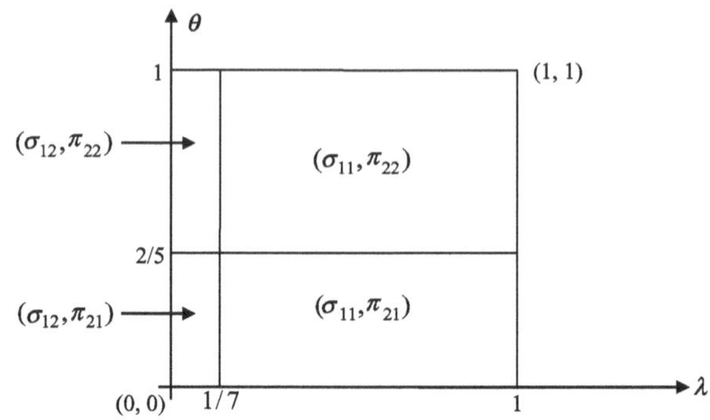

Fig. 3 The mixed-strategy Nash equilibrium areas.

3 ANALYSIS FOR MAINTAINABILITY PROBLEMS

In Section 2, we discussed some reliability problems in a narrow sense in reliability-game theory. In this section, similar analyses can be performed for maintainability problems [5–9].

3.1 Model 4

Type of problem: Maintainability.

Description of problems: For a two-stage maintenance problem, assume the system lifetime is distributed with $F(x)=1-e^{-x^2}$, $(x\geq 0)$. When the system fails for the first time, the repair action can be taken by company 1 with repair degree $\theta=0$ or 1, representing perfect and minimal repairs, respectively. When the second failure occurs, then company 2 will perform the repair action with repair degree $\theta=0$ or 1, representing perfect and minimal repairs, respectively. Repairs can only be taken twice for the failed system. If we also assume that a new system is put into operation at time $t=0$ and all failures are instantly detected, repair time is negligible. The repair decision made by each company is independent of each other. How do we take repair strategies to maximize the company's net profits? This is a two-stage maintenance problem, which can be described by a dynamic game with complete information; that is to say, each company knows all previous situations when making its decision.

Type of game: A dynamic complete information game.

(1) *Players*: Companies 1 and 2

(2) *Strategy set*: $S_1=S_2=\{\theta_1=0,\theta_2=1\}$

(3) *Order of actions*: Company 1 takes its action first, then company 2 does second.

(4) *Payoff functions:*

$$\begin{cases} u_1(1,1)=10\sqrt{D}-11, \\ u_2(1,1)=12\sqrt{D}-11, \end{cases} \quad \begin{cases} u_1(1,2)=10\sqrt{D}-11, \\ u_2(1,2)=11\sqrt{D}-10, \end{cases}$$

$$\begin{cases} u_1(2,1)=10\sqrt{D}-9, \\ u_2(2,1)=12\sqrt{D}-11, \end{cases} \quad \begin{cases} u_1(2,2)=9\sqrt{D}-9, \\ u_2(2,2)=10\sqrt{D}-10, \end{cases}$$

where

$D=$ expected lifetime of the system $=E(X_1)+E(X_2)+E(X_3), X_i, \quad i=1,2,3,$

where $E(X_i)$ is lifetime of the system at stages i, $i=1, 2, 3$. $u_k(i,j)$, $k, i, j=1, 2$; k denotes company k, and i and j denote companies 1 and 2's strategies, respectively. Namely, the strategy (i,j) represents $(\theta_i^{(1)}, \theta_j^{(2)})=(i-1, j-1)$, where $\theta_i^{(1)}$ and $\theta_j^{(2)}$ are the repair degrees for companies 1 and 2 adopted, respectively. Thus there are four possible cases shown in Fig. 4.

We know that

$$E(X_1)=\int_0^\infty [1-F(x)]dx, E(X_2)=\int_0^\infty [\overline{V}_2(x)-\overline{V}_1(x)]dx,$$

$$E(X_3)=\int_0^\infty [\overline{V}_3(x)-\overline{V}_2(x)]dx,$$

where $\overline{V}_1(x)=1-F(x)$, $\overline{V}_{n+1}(x)=\overline{V}_n(x)+\int_0^x \frac{1-F(x)}{1-F(y)} dV_n(y)]$, $\overline{V}_n(x)=1-V_n(x), \quad n\geq 1.$

For details, see Chi et al. [4]. Thus we have,

$$\overline{V}_2(x)=e^{-x^2}+x^2e^{-x^2}, \quad \overline{V}_3(x)=e^{-x^2}+x^2e^{-x^2}+0.5x^4e^{-x^2},$$

$$E(X_1)=0.8862, \quad E(X_2)=0.4431, \quad E(X_3)=0.3323.$$

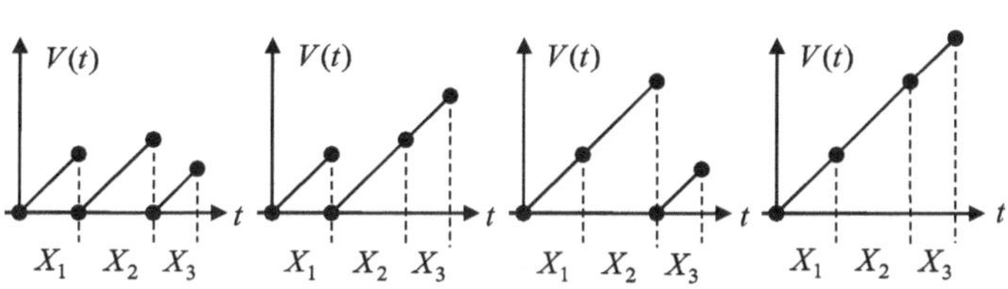

Fig. 4 Four possible sample paths of virtual age $V(t)$.

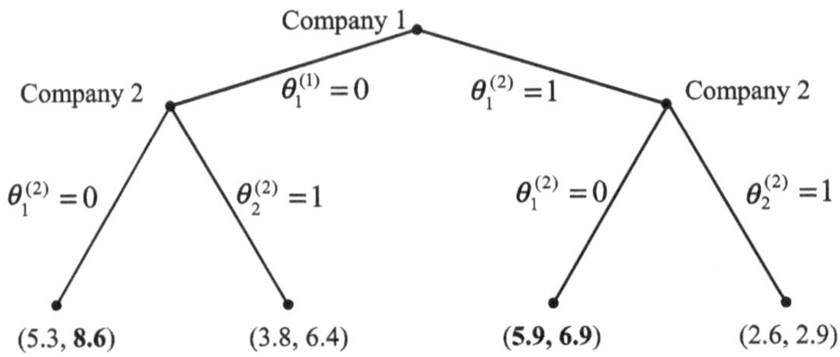

Fig. 5 A game tree for Model 4.

(5) *Procedure of the game solution*: The game tree is shown in Fig. 5.
(6) *Equilibrium*: The subgame complete Nash equilibrium is $(\theta_1=1, \theta_2=0)$, and companies 1 and 2 have 5.9 and 6.9 benefits, respectively. However, from Fig. 5 we know that company 2 can get its maximum benefit 8.6 if company 1 takes strategy $\theta_1=0$, but company 1 considers its benefit, it does not take strategy $\theta_1=0$.

3.2 Model 5

There are two cases to be considered as follows.

3.2.1 Case I: Type of Problem: Maintainability

Description of problems: For a single-unit repairable Markov system, the lifetime and repair time of the system are both exponentials with parameters λ, μ, respectively. It is well known that the steady–state availability $A=\frac{\mu}{\mu+\lambda}$. The system is produced by a company who has some strategies in which the device failure rate can be chosen $\lambda \in (0,\infty)$. The producer makes its strategy first, then the repair action is carried out by another company when a failure occurs. Each company can get some profits due to the operation of the system, which are functions of the steady-state A, of course, their inputs are different due to different strategies. The second company knows the first company's strategy after its action. What is the best strategy for both companies to make their profits maximal in the steady-state, if the second company has strategies $\mu \in (0,\infty)$

Type of game: A dynamic perfect information game.

(1) *Players*: Producer and Repairman, named as 1 and 2, respectively.
(2) *Strategy sets*: $S_1=\{\lambda:\lambda\in(0,\infty)\}$, $S_2=\{\mu:\mu\in(0,\infty)\}$.
(3) *Order of actions*: Player 1 makes its action first, then player 2 does second.

(4) *Payoff functions*:

$$u_1(\lambda, \mu) = g(A) - c_1\lambda^{-1},$$
$$u_2(\lambda, \mu) = g(A) - c_2\mu.$$

(5) *Procedure of the game solution*: The producer gets its maximal profit if it thinks the repairman uses the strategy μ, then we have

$$\frac{\partial u_1(\lambda, \mu)}{\partial \lambda} = 0 \Rightarrow \frac{\mu}{(\mu+\lambda)^2}\frac{\partial g(x)}{\partial x} = \frac{c_1}{\lambda^2}.$$

Also the second company gets its maximal profit if it thinks the first company's strategy, then we have,

$$\frac{\partial u_2(\lambda, \mu)}{\partial \mu} = 0 \Rightarrow \frac{\lambda}{(\mu+\lambda)^2}\frac{\partial g(x)}{\partial x} = c_2.$$

The Nash equilibrium is obtained from the solution of the set of equations above, that is, we have,

$$\begin{cases} \dfrac{\mu}{(\mu+\lambda)^2}\dfrac{\partial g(A)}{\partial A} = \dfrac{c_1}{\lambda^2}, \\ \dfrac{\lambda}{(\mu+\lambda)^2}\dfrac{\partial g(A)}{\partial A} = c_2. \end{cases} \Rightarrow \begin{cases} \lambda = \lambda^*, \\ \mu = \mu^*. \end{cases} \Rightarrow \mu^* = \frac{c_1}{\lambda^* c_2},$$

where the solution (λ^*, μ^*) is the Nash equilibrium.

(6) *Equilibrium*: The subgame complete Nash equilibrium is (λ^*, μ^*).

3.2.2 Case II: Type of Problem: Maintainability

Description of problems: Most of the situation is the same as that in Case I, and the producer uses strategies λ_1, λ_2 with probabilities p_1, $1-p_1$, respectively, then the repairman uses the strategies μ_1, μ_2 with probabilities p_2, $1-p_2$, respectively. What are the results for both producer and repairman to have their maximal profits?

Type of game: A dynamic perfect information game.

(1) *Players*: Producer and Repairman, named as 1 and 2, respectively.

(2) *Strategy sets*: $S_1 = \{\lambda_1, \lambda_2\}$, $S_2 = \{\mu_1, \mu_2\}$.

(3) *Order of actions*: Player 1 makes its action first, then player 2 does second.

(4) *Payoff functions:*

$$u_1(\lambda,\mu)=p_1\left[g\left(\frac{\mu}{\mu+\lambda_1}\right)-c_1\lambda_1^{-1}\right]+(1-p_1)\left[g\left(\frac{\mu}{\mu+\lambda_2}\right)-c_1\lambda_2^{-1}\right],$$

$$u_2(\lambda_1,\mu)=p_1p_2\left[g\left(\frac{\mu_1}{\mu_1+\lambda_1}\right)-c_2\mu_1\right]+p_1(1-p_2)\left[g\left(\frac{\mu_2}{\mu_2+\lambda_1}\right)-c_2\mu_2\right],$$

$$u_2(\lambda_2,\mu)=(1-p_1)p_2\left[g\left(\frac{\mu_1}{\mu_1+\lambda_2}\right)-c_2\mu_1\right]+(1-p_1)(1-p_2)\left[g\left(\frac{\mu_2}{\mu_2+\lambda_2}\right)-c_2\mu_2\right].$$

(5) *Procedure of the game solution*: To use the calculus knowledge, we can easily get the Nash equilibrium strategy.

(6) *Equilibrium*: The subgame complete Nash equilibrium is (λ^*,μ^*), where $\lambda^*=\arg\max\limits_{\lambda_1,\lambda_2}\{u_1(\lambda,\mu_1),u_1(\lambda,\mu_2)\}$, which is the equilibrium benefit for the first producer;

$$\mu^*=\begin{cases}\arg\max\limits_{\mu_1,\mu_2}\{u_2(\lambda_1,\mu)\}, & \text{if the first company takes strategy } \lambda_1,\\ \arg\max\limits_{\mu_1,\mu_2}\{u_2(\lambda_1,\mu)\}, & \text{if the first company takes strategy } \lambda_2,\end{cases}$$

which is the equilibrium strategy for the repairman.

(7) *Discussion*: If we consider the continuous case, we assume that the first player has the strategy set $\{\lambda:\text{with proability } F(\lambda)\}$, and the second player has the strategy set $\{\mu:\text{with proability } G(\mu)\}$, where the $F(\lambda)$ and $G(\mu)$ are probability distribution functions of nonnegative random variables.

The payoff functions are

$$u_1(\lambda,\mu)=\left[g\left(\frac{\mu}{\mu+\lambda}\right)-\frac{c_1}{\lambda}\right]F(\lambda)G(\mu),$$

$$u_2(\lambda,\mu)=\left[g\left(\frac{\mu}{\mu+\lambda}\right)-c_1\mu\right]F(\lambda)G(\mu).$$

The Nash equilibrium strategy is

$$(\lambda^*,\mu^*)=\arg\left\{\max_{\lambda,\mu}u_1(\lambda,\mu),\ \max_{\lambda,\mu}u_2(\lambda,\mu)\right\}.$$

4 ANALYSIS FOR TESTABILITY, LOGISTICS, AND SAFETY PROBLEMS

4.1 Model 6

Type of problem: Testability.

Description of problems: The tested system consists of two components with series structure. Two testers, A and B, do their testing work for components 1 and 2, respectively. The testers A and B have three and two strategies, respectively.

Type of game: A static complete information game.

(1) *Players*: Testers A and B.

(2) *Strategy sets*: $S_1=\{\sigma_{11},\sigma_{12},\sigma_{13}\}$, $S_2=\{\pi_{21},\pi_{22}\}$.

(3) *Order of actions*: Testers 1 and 2 make their actions simultaneously.

(4) *Payoff functions*:

$$u_A(\sigma(p_1),\pi(p_2))=100p_1p_2-6(1-p_1p_2)-\textit{test cost of } A,$$
$$u_B(\sigma(p_1),\pi(p_2))=120p_1p_2-12(1-p_1p_2)-\textit{test cost of } B.$$

The principle of utility function: If the state of the system can be judged correctly, then both testers obtain benefits, respectively. If the state of the system is judged wrongly, the punishments are taken for both testers. Thus we give the utility functions for both testers.

(5) *Procedure of the game solution*: All parameters are shown in Table 8. Using the game theory, we get Table 9.

Table 8 Parameters of Model 6

	Judgment correction probability based on testing/test cost
Strategies for tester A	**(1)** $\sigma_{11}=0.8/1$
	(2) $\sigma_{12}=0.9/2$
	(3) $\sigma_{13}=0.95/3$
Strategies for tester B	**(1)** $\pi_{21}=0.9/3$
	(2) $\pi_{22}=0.96/5$

Table 9 Payoffs for the tested system game

		Tester B	
	Strategies	π_{21}	π_{22}
Tester A	σ_{11}	(69.32, 80.04)	(74.408, 84.376)
	σ_{12}	(77.86, 91.92)	(83.584, 97.408)
	σ_{13}	(81.630, 97.860)	(**87.672, 103.384**)

(6) *Equilibrium*: The Nash equilibrium is (θ_{13}, π_{22}) (marked in Table 9), and testers 1 and 2 have 87.672 and 103.384 benefits, respectively.

4.2 Model 7

Type of problem: Logistics.

Description of problems: A reliability system needs two-kind standby subsystems. However, because of randomness, we do not know how many of the standby subsystems are needed; using the related information, we just know that the amounts of subsystems needed are Poisson random variables, say X and Y, with parameters λ and μ $(\lambda, \mu > 0)$, that is,

$$P\{X=s\} = \frac{\lambda^s e^{-\lambda}}{s!}, \quad s=0,1,2,\ldots, P\{Y=s\} = \frac{\mu^s e^{-\mu}}{s!}, \quad s=0,1,2,\ldots.$$

Now suppliers A and B can provide both subsystems. Assume that n_1 and n_2 are the amounts of subsystems 1 and 2, respectively, supplier A provides, and their unit costs are c_1 and c_2, respectively. The similar symbols for supplier B are m_1 and m_2 with unit costs d_1 and d_2. Like the Cournot model in game theory, we have the following analytic results. We consider the case of perfect information static game, that is, the suppliers A and B give their decisions simultaneously, and assume that the demand function is linear, which is similar to the Cournot model.

Type of game: A static complete information game.

(1) *Players*: Suppliers A and B.

(2) *Strategy sets*: $S_1 = \{(n_1, n_2), n_j = 0, 1, 2, \ldots\}$, $S_2 = \{(m_1, m_2), m_j = 0, 1, 2, \ldots\}$, $j = 1, 2$.

(3) *Order of actions*: Suppliers 1 and 2 make their actions simultaneously.

(4) *Payoff functions*:

The profit functions of suppliers A and B, which are random variables, are

$$\begin{cases} u_A(1, X) = n_1(X - n_1 - m_1 - c_1), \\ u_B(1, X) = m_1(X - n_1 - m_1 - d_1), \end{cases} \quad \begin{cases} u_A(2, Y) = n_2(Y - n_2 - m_2 - c_2), \\ u_B(2, Y) = m_2(Y - n_2 - m_2 - d_2). \end{cases}$$

Then the expected benefits for suppliers A and B are, respectively,

$$\begin{cases} u_A = E[u_A(1, X)] + E[u_A(2, Y)] = n_1(1/\lambda - n_1 - m_1 - c_1) + n_2(1/\mu - n_2 - m_2 - c_2), \\ u_B = E[u_B(1, X)] + E[u_B(2, Y)] = m_1(1/\lambda - n_1 - m_1 - d_1) + m_2(1/\mu - n_2 - m_2 - d_2). \end{cases}$$

(5) *Procedure of the game solution*: All parameters are not specified numerically; they are algebraic.

(6) *Equilibrium*: To maximize the both supplier's benefits, if we do a continuousness extension, then we have,

$$\begin{cases}\dfrac{\partial u_A}{\partial n_1}=0,\\[2ex] \dfrac{\partial u_A}{\partial n_2}=0,\\[2ex] \dfrac{\partial u_B}{\partial m_1}=0,\\[2ex] \dfrac{\partial u_A}{\partial m_2}=0.\end{cases} \Rightarrow \begin{cases}\dfrac{\partial u_A}{\partial n_1}=\dfrac{1}{\lambda}-2n_1-m_1-c_1=0,\\[2ex] \dfrac{\partial u_A}{\partial n_2}=\dfrac{1}{\mu}-2n_2-m_2-c_2=0,\\[2ex] \dfrac{\partial u_B}{\partial m_1}=\dfrac{1}{\lambda}-2m_1-n_1-d_1=0,\\[2ex] \dfrac{\partial u_B}{\partial m_2}=\dfrac{1}{\mu}-2m_2-n_2-d_2=0.\end{cases} \Rightarrow \begin{cases}n_1=\dfrac{1+d_1\lambda-2c_1\lambda}{3\lambda},\\[2ex] m_1=\dfrac{1-2d_1\lambda+c_1\lambda}{3\lambda},\\[2ex] n_2=\dfrac{1+d_2\mu-2c_2\mu}{3\mu},\\[2ex] m_2=\dfrac{1-2d_2\mu+c_2\mu}{3\mu}.\end{cases}$$

Then the Nash equilibrium is $\{(n_1^*, n_2^*), (m_1^*, m_2^*)\}$, which are integer solutions that can be obtained easily from the solutions of the previous equations to maximize both benefits of suppliers A and B. In fact, in similar situations, a dynamic game can be considered, like Stackelberg model, which is omitted here.

4.3 Model 8

Type of problem: Safety.

Description of problems: Consider a system consisting of an operating subsystem and a protector. The operating subsystem performs its function for the system operation; the protector is to prevent a dangerous failure from occurring. The system has two kinds of failure states: fail-safe state and fail-dangerous state. Once a failure occurs, the failed system goes into one of the two failure states, and for the fail-safe state and fail-dangerous state with probabilities p and $1-p$, respectively. If the failed system reaches fail-dangerous state and the protector works normally, then the system is a safe state, otherwise, a dangerous event will occur. The subsystem and protector are provided by suppliers A and B, respectively. If the system operates normally, then both suppliers can benefit, and the system fails, then supplier A will be punished; if a dangerous event occurs, then both suppliers will be punished and supplier B have done more.

Type of game: A static complete information game.

(1) *Players*: Suppliers A and B.

(2) *Strategy sets*: $S_1=\{R_1, R_2, R_3\}, \quad S_2=\{p_1, p_2\}$.

(3) *Order of actions*: Suppliers 1 and 2 make their actions simultaneously.

(4) *Payoff functions*: The game strategies and payoffs are shown in Table 10.

Table 10 The payoffs of Model 8

	Strategies	Supplier B ($p_1 < p_2$) Protector reliability p_1	Protector reliability p_2
Supplier A ($R_1 < R_2 < R_3$) R_i is the reliability of the subsystem.	Reliability R_1	(7, 10)	(4, 12)
	Reliability R_2	(6, 8)	(5, 7)
	Reliability R_3	(5, 9)	(4, 6)

(5) *Procedure of the game solution*: Because the Nash equilibrium does not exist, we need to find a mixed-strategy Nash equilibrium. Suppose that supplier A has a mixed-strategy $\sigma_A = (\theta_1, \theta_2, 1-\theta_1-\theta_2)$, the supplier B has $\sigma_B = (\pi, 1-\pi)$.

Then supplier A has the following expected benefit,

$$\begin{aligned} u_A(\sigma_A, \sigma_B) &= \theta_1[7\pi + 4(1-\pi)] + \theta_2[6\pi + 5(1-\pi)] \\ &\quad + (1-\theta_1-\theta_2)[5\pi + 4(1-\pi)] \\ &= \theta_1(3\pi + 4) + \theta_2(\pi + 5) + (1-\theta_1-\theta_2)(\pi + 4) \\ &= 2\theta_1\pi + \theta_2 + \pi + 4. \end{aligned}$$

Thus the following equations are established,

$$\begin{cases} \max \quad u_A(\sigma_A, \sigma_B) = 2\theta_1\pi + \theta_2 + \pi + 4, \\ \text{s.t.} \quad \theta_1 + \theta_2 \le 1, \\ \qquad 0 \le \theta_1, \theta_2 \le 1, \\ \qquad 0 \le \pi \le 1. \end{cases}$$

We get that, when $\pi = 0$, $\theta_1 = 0$, $\theta_2 = 1$; $\pi = 1$, $\theta_1 = 1$, $\theta_2 = 0$. That is, when supplier B takes the strategy p_2, supplier A should take the strategy R_2; when supplier B takes the strategy p_1, supplier A should take the strategy R_1, which also can be known in terms of Table 10.

Similarly, we have the expected benefit for supplier B as follows,

$$u_B(\sigma_A, \sigma_B) = \pi(3 - 7\theta_1 - 2\theta_2) + 6\theta_1 + \theta_2 + 6,$$

We have that, when supplier A takes the strategies $\sigma_A = \{(\theta_1, \theta_2, 1-\theta_1-\theta_2) : 3 - 7\theta_1 - 2\theta_2 > 0\}$, supplier B should take the strategy p_1; when supplier A takes the strategies $\sigma_A = \{(\theta_1, \theta_2, 1-\theta_1-\theta_2) : 3 - 7\theta_1 - 2\theta_2 < 0\}$, supplier B should take the strategy p_2; and when supplier A takes the strategies $\sigma_A = \{(\theta_1, \theta_2, 1-\theta_1-\theta_2) : 3 - 7\theta_1 - 2\theta_2 = 0\}$, supplier B can take any strategies.

(6) *Equilibrium*: The game strategies have been previously mentioned. The strategy for supplier B can be indicated in Fig. 6.

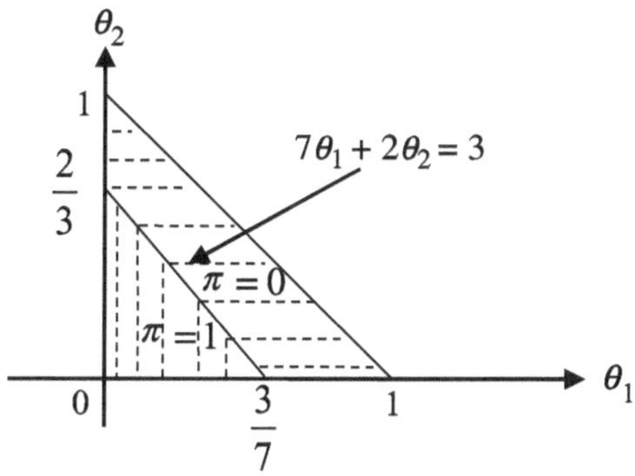

Fig. 6 The strategy representation for supplier B.

4.4 Model 9

In the following three complete information cases are considered, Case I for static, Case II for dynamic with supplier A takes action first, Case III for dynamic with supplier B takes action first.

4.4.1 Case I: Type of Problem: Safety

Description of problems: A system has reliability R, and to prevent a dangerous event from occurring when the system fails, a protector with working reliability p is needed. If the system operates normally, no dangerous event occurs; if the system fails and the protector works, then the dangerous event still cannot occur, otherwise, the dangerous event occurs once the system fails. The system and protector are provided by suppliers A and B, and both suppliers can benefit when no dangerous event occurs. The benefits for both suppliers are allocated in terms of the proportion of reliability provided by each supplier.

Type of game: A static complete information game.

(1) *Players*: Suppliers A and B.
(2) *Strategy sets*: $S_1=\{R_1=0.7, R_2=0.72\}$, $S_2=\{p_1=0.8, p_2=0.9\}$.
(3) *Order of actions*: Suppliers 1 and 2 make their actions simultaneously.
(4) *Payoff functions*: The benefits functions or payoff functions in the game are as follows:

$$\begin{cases} u_1(R,p)=\left[100\sqrt{R}-c_1R+50\sqrt{(1-R)p}-c_2p\right]\dfrac{R}{R+p}, \\ u_2(R,p)=\left[100\sqrt{R}-c_1R+50\sqrt{(1-R)p}-c_2p\right]\dfrac{p}{R+p}, \end{cases}$$

where $c_1=2$, $c_2=4$ are parameters of costs for the system and protector.

Table 11 Payoffs for Case I of Model 9

	Strategies	Supplier B $p_1=0.8$	Supplier B $p_2=0.9$
Supplier A	$R_1=0.7$	(48.3, 55.3)	**(46.8, 58.9)**
	$R_2=0.72$	(49.2, 54.7)	(46.6, 58.3)

The reasons for defining the payoff functions are, (1) the benefit for the system increases as its reliability increases, and the increment becomes smaller and smaller as its reliability increases; (2) the benefit for the protector has the same property, and when the system fails, its function can be useful, that is, the allocation of total benefits in terms of the proportion of reliability is provided by each supplier.

(5) *Procedure of the game solution*: The payoffs are obtained in Table 11.

(6) *Equilibrium*: The Nash equilibrium is $(R_1=0.7, p_2=0.9)$ (marked in Table 11).

4.4.2 Case II: Type of Problem: Safety

Description of problems: It is the same as that in Case I of Model 9.

Type of game: A dynamic complete information game.

(1) *Players*: Suppliers A and B.

(2) *Strategy sets*: $S_1=\{R_1=0.7, R_2=0.72\}$, $S_2=\{p_1=0.8, p_2=0.9\}$.

(3) *Order of actions*: Supplier A takes its actions first, then supplier B does second.

(4) *Payoff functions*: The payoff values are the same as that listed in Table 11.

(5) *Procedure of the game solution*: The game tree is shown in Fig. 7.

(6) *Equilibrium*: The subgame complete Nash equilibrium is $(R_1=0.7, p_2=0.9)$, which is the same as that in Case I of Model 9.

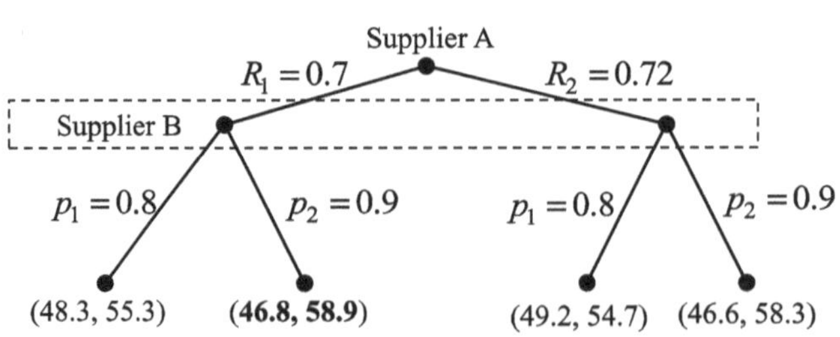

Fig. 7 The game tree for Case II of Model 9.

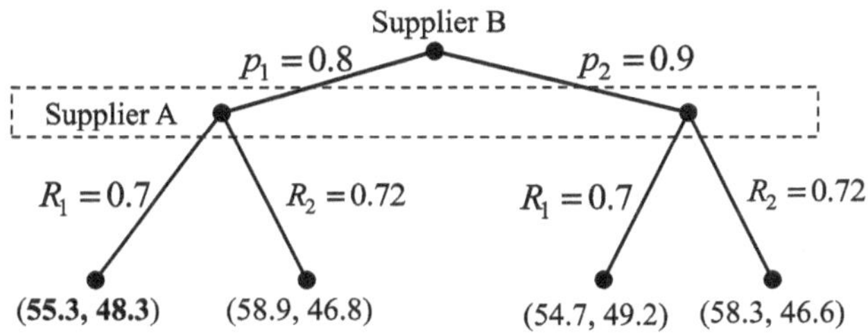

Fig. 8 The game tree for Case III of Model 9.

4.4.3 Case III: Type of Problem: Safety

Description of problems: It is the same as that in Case I of Model 8.

Type of game: A dynamic complete information game.

(1) *Players*: Suppliers A and B.
(2) *Strategy sets*: $S_1 = \{p_1 = 0.8, p_2 = 0.9\}$, $S_2 = \{R_1 = 0.7, R_2 = 0.72\}$.
(3) *Order of actions*: Supplier B takes its actions first, then supplier A does second.
(4) *Payoff functions*: The payoff values are the same as that listed in Table 11.
(5) *Procedure of the game solution*: The game tree is shown in Fig. 8.
(6) *Equilibrium*: The subgame complete Nash equilibrium is $(p_1 = 0.9, R_1 = 0.7)$, which is different from that in Case II of Model 9.

Of course, we can consider to take the payoff allocation functions as follows, but here it is omitted.

$$\rho_1 = \frac{\omega_1 R}{\omega_1 R + \omega_2 p}, \quad \rho_2 = \frac{\omega_2 p}{\omega_1 R + \omega_2 p},$$

where the weights ω_1 and ω_2 satisfy the relation $\omega_1 + \omega_2 = 1$, $\omega_1, \omega_2 > 0$.

The different payoff allocation functions will result in different Nash equilibria. In Cases I and II, the suppliers A and B get the benefits 46.8 and 58.9, respectively, but in Case III, they get the benefits 48.3 and 55.3 in the Nash equilibria, respectively.

4.5 Model 10

Type of problem: Logistics.

Description of problems: Two suppliers 1 and 2 transfer some devices from place A to place B via place C. The suppliers 1 and 2 have two choices, respectively. Each choice can transfer X or Y devices by suppliers 1 and 2 with successful probabilities $p_1^{(i)}$ or $p_2^{(i)}$ $(i = 1, 2)$, respectively. The transportation diagram is shown in Fig. 9.

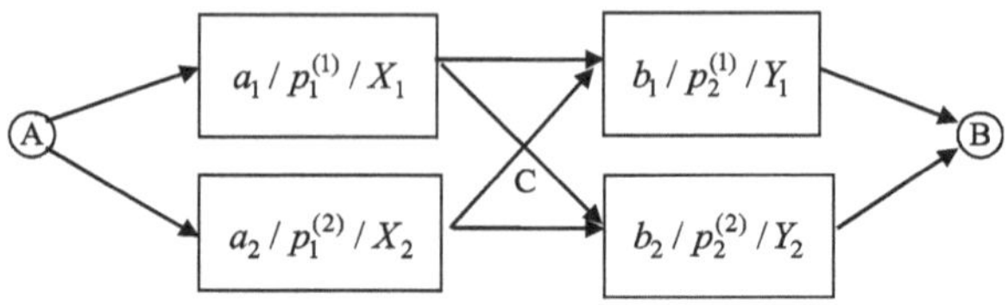

Fig. 9 The transportation diagram.

In road a_i, $(i=1,2)$ with successful probabilities $p_1^{(i)}$, the transport amount of devices is X_i with probability distributions as follows,

$$\begin{cases} X_1=5, \text{ with } P\{X_1=5\}=0.2, \\ X_1=6, \text{ with } P\{X_1=6\}=0.5, \\ X_1=7, \text{ with } P\{X_1=7\}=0.3, \end{cases} \quad \begin{cases} X_2=6, & \text{with } P\{X_2=6\}=0.3, \\ X_2=7, & \text{with } P\{X_2=7\}=0.3, \\ X_2=8, & \text{with } P\{X_2=8\}=0.2, \\ X_2=9, & \text{with } P\{X_2=9\}=0.2. \end{cases}$$

Similarly, for road b_i with successful probabilities $p_2^{(i)}$, $(i=1,2)$, the transport amount of devices is Y_i with probability distributions as follows,

$$Y_1 = \begin{cases} X-2, & \text{with } P\{Y_1=X-2\}=0.3, \\ X-1, & \text{with } P\{Y_1=X-1\}=0.3, \\ X, & \text{with } P\{Y_1=X\}=0.4, \end{cases}$$

$$Y_2 = \begin{cases} X-2, & \text{with } P\{Y_2=X-2\}=0.2, \\ X-1, & \text{with } P\{Y_2=X-1\}=0.6, \\ X, & \text{with } P\{Y_2=X\}=0.2, \end{cases}$$

where the random variable X is amount of devices transferred from place A to place C by the supplier 1. How can the suppliers make their actions to maximize their benefits?

Type of game: A dynamic complete information game.

(1) *Players*: Suppliers 1 and 2.

(2) *Strategy sets*: $S_1=\{a_1,a_2\}$, $S_2=\{b_1,b_2\}$.

(3) *Order of actions*: Supplier 1 makes its action first, then supplier 2 does second.

(4) *Payoff functions*:

$$u_1(X_i, Y_j) = p_1^{(i)} p_2^{(j)} E(Y_j | i) - c_i E(X_i), \quad i=1,2; \ j=1,2;$$

$$u_2(X_i, Y_j) = d_j p_1^{(i)} p_2^{(j)} E(Y_j | i), \quad i=1,2; \ j=1,2;$$

where $E(Y_j | i)$ is the expectation of Y_j given the supplier 1 chose road a_i.

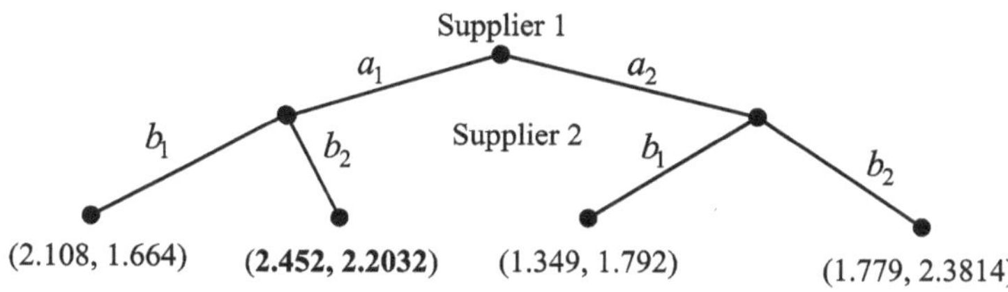

Fig. 10 The game tree for complete information dynamic game.

(5) *Procedure of the game solution*: We have

$$E(X_1) = 6.1,\quad E(X_2) = 7.3,\quad E(Y_1|\,1) = 5.2,\quad E(Y_1|\,2) = 6.4,$$
$$E(Y_2|\,1) = 5.1,\quad E(Y_2|\,2) = 6.3.$$

If other parameters have the following values,

$$p_1^{(1)} = 0.8,\quad p_1^{(2)} = 0.7,\quad p_2^{(1)} = 0.8,\quad p_2^{(2)} = 0.9,\quad c_1 = 0.2,\quad c_2 = 0.3,$$
$$d_1 = 0.5,\quad d_2 = 0.6.$$

We get the game tree as shown in Fig. 10.

(6) *Equilibrium*: The subgame complete Nash equilibrium is (a_1, b_2), and suppliers 1 and 2 have 2.452 and 2.2032 benefits. Of course, as discussed before, we can consider other cases with games; here they are omitted.

5 ANALYSIS FOR RELIABILITY TESTING AND OTHER RELIABILITY-RELATED PROBLEMS

5.1 Model 11

Type of problem: Reliability growth.

Description of problems: The system consists of two components with series structure. The total resource for improving the system is 4; first supplier chooses its improvement action first, the alternative action set is $A = \{0, 1, 2, 3\}$, where 0, 1, 2, 3 represents the corresponding action's resource consumption, respectively. The alternative action set for the second supplier is $B = \{0, 1, 2, 3\}$ if it is available. The utility functions are,

$$u_1 = C\left(R_{sys}\right) - c_j^{(1)} a_i,\quad a_i \in A,$$

Table 12 The values for resource input and increment reliability of components

a_i	0	1	2	3	b_j	0	1	2	3
$\Delta R_1(a_i)$	0	0.1	0.15	0.2	$\Delta R_2(b_j)$	0	0.15	0.25	0.3

$$u_2 = C\left(R_{sys}\right) - c_j^{(2)} b_j, \quad b_j \in B, \text{ and subject to}$$

$$a_i + b_j \leq 4.$$

The original component reliabilities are $R_1^{(0)} = 0.6$, $R_2^{(0)} = 0.7$ for components 1 (supplier A produces) and 2 (supplier B produces), respectively. The relation between the resource input and increment reliability of components are listed in Table 12.

What are the strategies for suppliers A and B to maximize their benefits under the constraints?

Type of game: A dynamic complete information game.

(1) *Players*: Suppliers A and B.

(2) *Strategy sets*: $S_1 = \{a_1, a_2, a_3, a_4\}$, $S_2 = \{b_1, b_2, b_3, b_4\}$ and $a_i + b_j \leq 4$.

(3) *Order of actions*: Supplier A makes its action first, then supplier B does second.

(4) Payoff functions:

$$u_1\left(a_j, b_j\right) = \rho_1\left[R_1^{(0)} + \Delta R_1\left(a_j\right)\right]\left[R_2^{(0)} + \Delta R_2\left(b_j\right)\right] - 0.5a_j,$$

$$u_2\left(a_j, b_j\right) = \rho_2\left[R_1^{(0)} + \Delta R_1\left(a_j\right)\right]\left[R_2^{(0)} + \Delta R_2\left(b_j\right)\right] - 0.5b_j.$$

(5) *Procedure of the game solution*: The game tree is given in Fig. 11, if $\rho_1 = 7$, $\rho_2 = 6$.

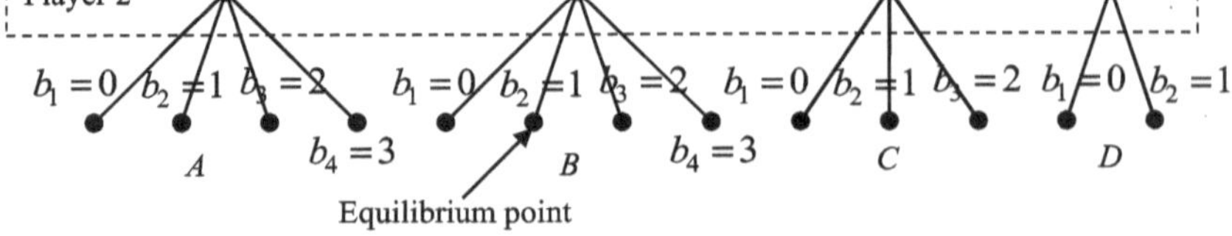

Fig. 11 The game tree for Model 11.

$$A:\begin{cases} u_1(0,0)=2.94, u_2(0,0)=2.52;\\ u_1(0,1)=3.57, u_2(0,1)=2.56;\\ u_1(0,2)=3.99, u_2(0,2)=2.42;\\ u_1(0,3)=4.20, u_2(0,3)=2.1;\end{cases}$$

$$B:\begin{cases} u_1(1,0)=2.93, u_2(1,0)=2.94;\\ u_1(1,1)=\mathbf{3.665}, u_2(1,1)=\mathbf{3.07};\\ u_1(1,2)=4.155, u_2(1,2)=2.99;\\ u_1(1,3)=4.4, u_2(1,3)=2.7;\end{cases}$$

$$C:\begin{cases} u_1(2,0)=2.675, u_2(2,0)=3.15;\\ u_1(2,1)=3.463, u_2(2,1)=3.325;\\ u_1(2,2)=3.988, u_2(2,2)=3.275;\end{cases}$$

$$D:\begin{cases} u_1(3,0)=2.42, u_2(3,0)=3.36;\\ u_1(3,1)=3.26, u_2(3,1)=3.58.\end{cases}$$

(6) *Equilibrium*: The subgame complete Nash equilibrium is $(a_2=1, b_2=1)$. Suppliers A and B have 3.665 and 3.07 benefits, respectively, which are not the maximum benefits for every supplier.

5.2 Model 12

Type of problem: Network reliability.

Description of problems: Consider a reliability system with a bridge structure shown in Fig. 12.

The reliability for component i, $(1 \le i \le 6)$ is p_i and $p_2=p_1$, $p_3=p_4=p$. Components 1 and 5 can be treated as two separate players in the game, with two possible reliability values. The payoff function is the Birnbaum importance. Assume the possible reliability values for components 1 and 5 are

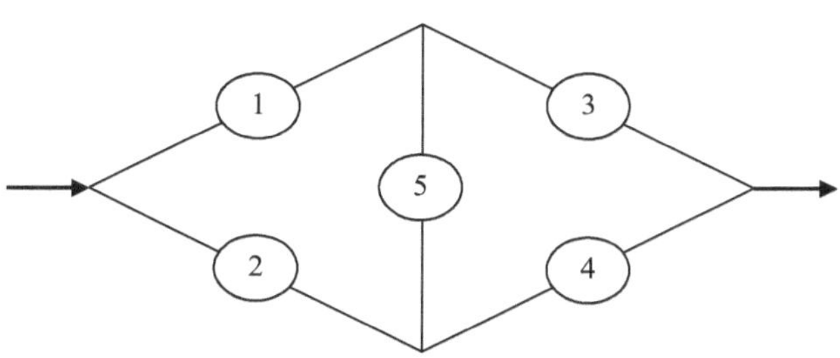

Fig. 12 The structure of a bridge reliability system.

given as $p_1^{(1)}=0.6$, $p_1^{(2)}=0.7$, $p_5^{(1)}=0.9$, $p_5^{(2)}=0.95$. What are the optimum strategies for components 1 and 5 to maximize their Birnbaum importance? To answer this question, we need to give the game equilibrium in various information conditions. For simplicity, we just consider the static complete information game.

Type of game: A static complete information game.

(1) *Players*: Components 1 and 5.

(2) *Strategy sets*: $S_1=\{p_1^{(1)}=0.6, p_2^{(1)}=.7\}$, $S_2=\{p_1^{(5)}=0.9, p_2^{(5)}=0.95\}$.

(3) *Order of actions*: Components 1 and 5 make their actions simultaneously.

(4) *Payoff functions*: The payoff functions are Birnbaum importance for components 1 and 5. It is well known that the Birnbaum importance for component 1 [1] is

$$\begin{aligned} I_1(p) &= \frac{\partial h(p)}{\partial p_1} = h(1, p_2, \ldots, p_5) - h(0, p_2, \ldots, p_5) \\ &= 1 - [1 - (1 - (1 - p_5)(1 - p_2))p_4][1 - p_3] - [p_2(1 - (1 - p_3 p_5)(1 - p_4))]. \end{aligned}$$

Similarly, the Birnbaum importance for component 5 is

$$\begin{aligned} I_5(p) = {} & [1 - (1 - p_1)(1 - p_2)][1 - (1 - p_3)(1 - p_4)] \\ & - [1 - (1 - p_1 p_3)(1 - p_2 p_4)]. \end{aligned}$$

Thus we have the payoff functions as follows,

$$\begin{aligned} u_1\left(p_1^{(1)}, p_5^{(1)}\right) &= 0.82p - 0.42p^2, \\ u_1\left(p_1^{(1)}, p_5^{(2)}\right) &= 0.81p - 0.41p^2, \\ u_1\left(p_1^{(2)}, p_5^{(1)}\right) &= 0.64p - 0.34p^2, \\ u_1\left(p_1^{(2)}, p_5^{(2)}\right) &= 0.62p - 0.32p^2, \\ u_2\left(p_1^{(1)}, p_5^{(1)}\right) = u_2\left(p_1^{(1)}, p_5^{(2)}\right) &= 0.48p - 0.48p^2, \\ u_2\left(p_1^{(2)}, p_5^{(1)}\right) = u_2\left(p_1^{(2)}, p_5^{(2)}\right) &= 0.42p - 0.42p^2. \end{aligned}$$

(5) *Procedure of the game solution*: Because the payoff functions are a polynomial function of p, it is easy to get the Nash equilibrium.

(6) *Equilibrium*: The Nash equilibrium is $(\sigma_1^*, \sigma_5^*) = (p_1^{(1)}, p_5^{(1)})$. The result is coincident with our intuition, because component 1 has lower reliability, which indicates to its importance to the system reliability.

5.3 Model 13

Type of problem: Problem for reliability trade-off.

Description of problems: In the reliability subject, it is well known that system 1 has more reliability than system 2 does, which are shown in Fig. 13.

In both systems 1 and 2, components 1 and 3 are identical, and components 2 and 4 are identical. In system 1, components 1 and 3, components 2 and 4, are packaged into, respectively, two-box with volumes $V_1(1,3)$ and $V_1(2,4)$; similarly, in system 2, for components 1 and 2, components 3 and 4, we have $V_2(1,2)$ and $V_2(3,4)$, respectively. Because of the same and different components, we have

$$V_1(1,3) + V_1(2,4) > V_2(1,2) + V_2(3,4),$$

that is, system 1 has bigger volume than system 2 does. However, it is well known that system 1 has higher reliability than system 2 does.

If two suppliers A and B do the package and reliability of the systems, respectively, then the benefits can be obtained in terms of the following payoff functions, which are determined according to the reasons: if there is a bigger space, the higher reliability can be obtained; the lower reliability results in a smaller space required, which raises a balance problem between the two factors: volume and reliability. What are the optimal strategies for suppliers A and B to maximize their benefits?

Type of game: A dynamic complete information game.

(1) *Players*: Suppliers A and B.

(2) *Strategy sets*:

$$\{(V,R): 5(V-2)+3(R-0.5) \leq 1.5, 5(V-2) + 3(R-0.5) \geq 1, 0.5 \leq R \leq 1, V \geq 2\},$$

which is the possible strategy solution space.

(3) *Order of actions*: Supplier B makes its action first, then supplier A does second.

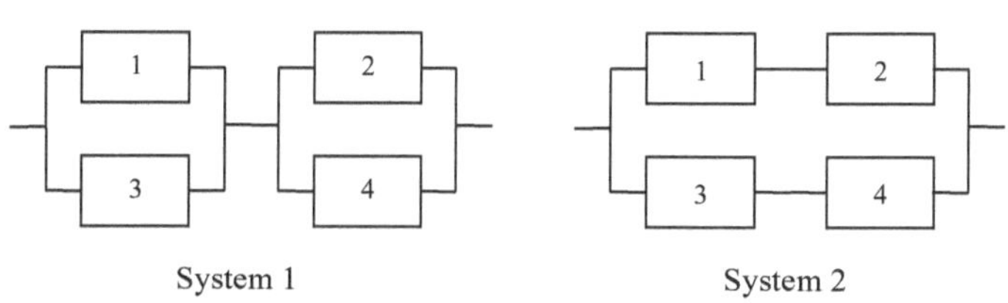

Fig. 13 The diagrams of two systems.

(4) Payoff functions:

$$\begin{cases} u_A(\sigma_V, \sigma_R) = 3V^2 + 8R, \\ u_B(\sigma_V, \sigma_R) = 8V^2 + 3R. \end{cases}$$

(5) *Procedure of the game solution*: Thus we can get the optimal strategy for supplier A given the supplier B takes its strategy R, that is,

$$\begin{cases} \underset{V}{\text{Max}} & u_A(\sigma_V, \sigma_R) = 3V^2 + 8R, \\ \text{s.t.} & 5(V-2) + 3(R-0.5) \leq 1.5, \\ & 5(V-2) + 3(R-0.5) \geq 1, \\ & V \geq 2,\ 0.5 \leq R \leq 1. \end{cases}$$

We get the supplier's optimal strategy is $V^* = 2.3 - 0.6(R - 0.5)$.

Supplier B thinks that if supplier A takes its optimal strategy V^*, then its maximal benefit is the solution of the following equations.

$$\begin{cases} \underset{R}{\text{Max}} & u_B(\sigma_{V^*}, \sigma_R) = 8V^{*2} + 3R, \\ \text{s.t.} & V^* = 2.3 - 0.6(R - 0.5), \\ & 0.5 \leq R \leq 1. \end{cases}$$

We get the subgame complete Nash equilibrium (σ_V^*, σ_R^*).

(6) *Equilibrium*: The subgame complete Nash equilibrium $(\sigma_V^*, \sigma_R^*) = (V^* = 2.3, R^* = 0.5)$ based on the equations previously provided.

5.4 Model 14

Type of problem: Reliability test.

Description of problems: This model is very similar to the Cournot game model, the differences are in the following points:

(1) Payoff functions are different.

(2) In this model, there is a constraint between two strategy variables.

(3) It is a discrete case rather than a continuous.

The model is as follows. There are two testers, one is for doing destructive test, another for nondestructive test, and they are named testers 1 and 2, respectively. All possible tested components (or units) are at most m, and the costs for each tester at every test are c_1 and c_2, respectively. Let n_1 and n_2 be the numbers of tests for the two testers. The two kinds of tests have a constraint $|n_1 - n_2| \leq \Delta$. As our common knowledge, the information received from the test is increasing at first and then decreasing as the number of tests increase. In a mathematical point of view, the following functions have this property, that is, it is increasing first then decreasing in some interval.

$$f_1(x) = x(a - 2x); \; f_2(x) = x(a - x), \quad a > 0.$$

Based on the prior discussion, we define the payoff functions as follows,

$$u_1(n_1, n_2) = n_1(m - 2n_1 - n_2 - c_1), \quad u_2(n_1, n_2) = n_2(m - 2n_1 - n_2 - c_2),$$

which tells us that the destructive test has less information than the nondestructive test does, and the benefits for both testers are related to each other. The further assumptions are that testers 1 and 2 know the cost of the destructive test each time, but the cost of nondestructive test each time is known just for tester 2, not for tester 1, as tester 1 just knows the cost of the nondestructive test each time may be c_2^L or c_2^H ($c_1^L < c_2^H$) with probability μ and $1 - \mu$, and μ is known by both testers.

Type of game: A static incomplete information game.

(1) *Players*: Testers 1 and 2.

(2) *Strategy sets*: $S_i = \{n_i, \text{each } n_i \text{ satisfies the following set}\}, \quad i = 1, 2,$

$$\{(n_1, n_2) : n_1 + n_2 \le m, |n_1 - n_2| \le \Delta, n_i \ge 0, \; i = 1, 2\}.$$

(3) *Order of actions*: Testers 1 and 2 makes their actions simultaneously, but with different information.

(4) Payoff functions:

$$u_1(n_1, n_2) = n_1(m - 2n_1 - n_2 - c_1), \quad u_2(n_1, n_2) = n_2(m - 2n_1 - n_2 - c_2).$$

(5) *Procedure of the game solution*: We will discuss it with the following parameters, $m = 20, \quad c_1 = 2, \quad c_2^L = 1, \quad c_2^H = 3, \quad \Delta = 5, \quad \mu = 0.5$. Thus the possible strategies for both testers are the following shadowed area shown in Fig. 14.

Because tester 2 knows its own test cost for each test and the test cost for tester 1, tester 2 will choose n_2^* to maximize $u_2(n_1^*, n_2^*)$, that is, the n_2^* satisfies the following inequalities,

$$u_2(n_1^*, n_2^* - 1) \le u_2(n_1^*, n_2^*) \ge u_2(n_1^*, n_2^* + 1),$$

We get $\frac{19 - 2n_1^* - c_2}{2} \le n_2^* \le \frac{21 - 2n_1^* - c_2}{2}$, that is, $n_2^* = g(n_1^*, c_2)$, n_2^* is a function of n_1^* and c_2. On the other hand, tester 1 knows the possible costs for tester 2 are

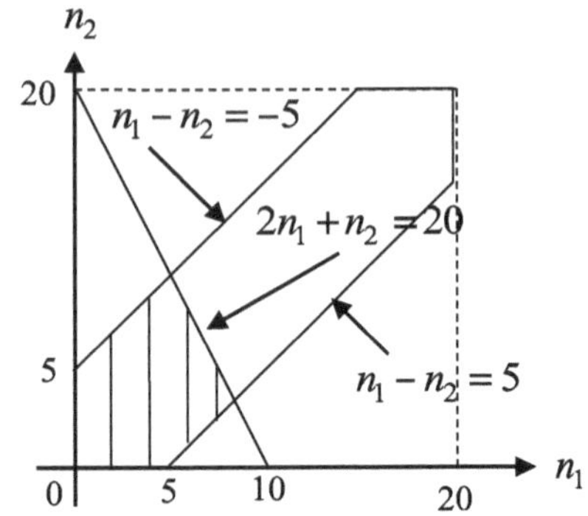

Fig. 14 The possible strategy area.

c_2^L and c_2^H with probabilities μ and $1-\mu$. Tester 1 wants to maximize its expected cost, that is,

$$\max\left\{\frac{1}{2}n_1^*\left(20-2n_1^*-g\left(n_1^*, c_2^L\right)-c_1\right)+\frac{1}{2}n_1^*\left(20-2n_1^*-g\left(n_1^*, c_2^H\right)-c_1\right)\right\}$$
$$=\max\left\{n_1^*\left(20-2n_1^*-c_1\right)-\frac{g\left(n_1^*, c_2^L\right)+g\left(n_1^*, c_2^H\right)}{2}\right\}.$$

However,

$$g\left(n_1^*, c_2^L\right)=\left[\frac{21-2n_1^*-c_2^L}{2}\right]=9-n_1^*$$

or

$$g\left(n_1^*, c_2^L\right)=\left[\frac{21-2n_1-c_2^L}{2}\right]=10-n_1^*,$$

$$g\left(n_1^*, c_2^H\right)=\left[\frac{19-2n_1^*-c_2^H}{2}\right]=8-n_1^*$$

or

$$g\left(n_1^*, c_2^H\right)=\left[\frac{21-2n_1-c_2^H}{2}\right]=9-n_1^*$$

. For four possible combinations, we get three possible values as follows,

$$\frac{g\left(n_1^*, c_2^L\right)+g\left(n_1^*, c_2^H\right)}{2}=\begin{cases}8.5-n_1^*,\\9-n_1^*,\\9.5-n_1^*.\end{cases}$$

By using the three possible values, we get four Bayesian-Nash equilibria as follows:

$$\begin{cases} n_1^* = 5, \\ n_2^{L^*} = 3, \\ n_2^{H^*} = 4, \end{cases} \quad \begin{cases} n_1^* = 5, \\ n_2^{L^*} = 4, \\ n_2^{H^*} = 4, \end{cases} \quad \begin{cases} n_1^* = 5, \\ n_2^{L^*} = 3, \\ n_2^{H^*} = 5, \end{cases} \quad \begin{cases} n_1^* = 5, \\ n_2^{L^*} = 4, \\ n_2^{H^*} = 5. \end{cases}$$

(6) *Equilibrium*: The Bayesian-Nash equilibria $(n_1^*, n_2^{L^*}, n_2^{H^*})$ are given based on the equations previously provided. We are lucky that all the Bayesian-Nash equilibria meet the constraint $|n_1 - n_2| \leq \Delta$. If it were not, we need more computation to get the Bayesian-Nash equilibria, but it is tedious rather than difficult.

6 FUTURE DEVELOPMENT ON RELIABILITY-GAME THEORY

Further developments on reliability-game theory are definitely needed on both theory and practical applications. The authors believe that the developments have some necessary conditions such as good knowledge on reliability and game theory, the promotion of practical reliability problems, and a good technique to bridge both subjects. The development may take two directions: one is to extend the game theory applications on reliability field, which is the main and easy one; another is to build a bridge between game theory and reliability, then to develop some reliability theory to promote game theory, which is more difficult than the first way, but if it is done, it would be a great contribution.

Reliability problems are transferred into problems of game theory, in which how to construct the elements of game theory is the key step. Based on the elements of a game, first the players should be constructed, such as components of reliability systems, elements considered in the reliability problems, suppliers, producers, users, etc. Of course, more than two players should be considered in further developments on reliability-game theory. Second, the rules of a reliability-game need to be established, such as sequential order on individual on group, rule of information share, etc. Third, the payoff functions must be determined. Usually the work is more difficult, because the payoff functions have greatly affected the solutions, but how to determine the payoff function properly for a real reliability problem needs to consider more factors. It is well known that the same situation is faced in the game theory. The possible payoff functions may be used such as reliability function, availability function, safety function, testability function, effectiveness, and other reliability measures. Fourth, alternative strategies for each player in reliability-game theory can be reliability executive plans; reliability test ways; reliability input costs and system reliability structure adopted etc. Fifth, other factors in game problems are also studied, such as procedure of finding solutions, equilibrium, etc. to form and present some reasonable solutions on reliability problems. To consider the characteristics of reliability, maintainability, testability, logistics, and safety are important points to build a reliability-game problem. On the other hand, we also need some situation to impetus the development of reliability-game theory, in this point, while we can solve the related reliability problems in different point of view. Once we form a game for a reliability problem, some modified games such as related dynamic and static games can be discussed, which can extend the reliability-game problems. The discrete and continuous reliability games have been studied in this chapter, but there still is a big room to develop them, especially on continuous reliability-game problems in the future.

Although many kinds of games have been considered in the chapter, we need to consider other classes of games, such as zero-sum games, nonzero-sum games, sequential games, more-player games, repeated games, truncated games, multistage games, Bayesian games etc., which definitely can enrich the contents of reliability-game theory.

Here, we want to emphasize that the differences existed between convention optimal problems and reliability-game theory, in both directions, the form of problems is usually different, and the factors concerned to solutions of problems are different, and sometimes their solutions are coincident, but most times are not.

For cooperative games, as pointed by Ramamurthy (1990), the similarity between reliability and simple games results in a number of concepts as well as results that were well known in game theory were rediscovered much later by researchers in reliability. He pointed out that semicoherent structures and simple games are equivalent, and he also presented a table for correspondence between the terminology of reliability and game theory; on this direction, readers can refer his book published in 1990.

By ending the chapter, the authors list some examples in the following for the future developments on reliability-game theory that may need to be done, which are stress-strength problems in reliability, reliability sampling problems, reliability test problems, shock reliability problems, reliability accelerated test problems, and n-component systems. The authors wish the combination of reliability and game theory will bring more fruitful results and new ideas on both subjects.

ACKNOWLEDGEMENT

This work was supported by the National Natural Science Foundation of China under grant 71631001.

REFERENCES

[1] D. Fudenberg, J. Tirole, Game Theory, MIT Press, Cambridge, MA, 1991.
[2] K.G. Ramamurthy, Coherent Structures and Simple Games, Kluwer Academic Publishers, Dordrecht, Netherlands, 1990.
[3] G.J. Chang, L.R. Cui, F.K. Hwang, Reliabilities of Consecutive-k Systems, Kluwer Academic Publishers, Dordrecht, Netherlands, 2000.
[4] L.R. Cui, W. Kuo, J.L. Li, M. Xie, On the dual reliability systems of (n,f,k) and $<n,f,k>$, Statist. Probab. Lett. 76 (2006) 1081–1088.
[5] L.R. Cui, W. Kuo, H.T. Loh, M. Xie, Optimal allocation of minimal & perfect repairs under resource constraints, IEEE Trans. Reliab. 53 (2) (2004) 193–199.
[6] L.R. Cui, H.J. Li, Opportunistic maintenance for multi-component shock model, Math. Meth. Oper. Res. 63 (2006) 493–511.
[7] L.R. Cui, M. Xie, H.T. Loh, Inspection schemes for general system, IIE Trans. 36 (9) (2004) 817–825.
[8] W. Kuo, V.R. Prasad, F.A. Tillman, C.L. Hwang, Optimal Reliability Design Fundamentals and Applications, Cambridge University Press, Cambridge, United Kingdom, 2001.
[9] W. Kuo, R. Wan, Recent advances in optimal reliability allocation, IEEE Trans. Syst. Man Cybern. 37 (2) (2007) 143–156.

CHAPTER 4

Integrating Reliability Models and Adaptive Algorithms for Wind Power Forecasting

F. De Caro*, A. Vaccaro*, D. Villacci*
*Department of Engineering, University of Sannio, Benevento, Italy

Abstract

The high proliferation of wind generators used in modern electrical grids determines several critical issues pushing power system operators to improve critical operation functions, such as security analysis and spinning reserve assessment, by taking into account the effects of intermittent and nonprogrammable power profiles. To address this challenging issue, a large number of frameworks for wind power forecasting have been proposed in the literature. Although these tools reliably allow prediction of wind speed and theoretical generated power, more complex phenomena need to be investigated to comprehensively model wind power uncertainty and its effect on power system operation. To address this issue, this chapter proposes a probabilistic model based on Markov chains, which predicts the injected power profiles considering wind speed forecasting uncertainty and generator operation states. Experimental results obtained on a real case study are presented and discussed to assess the performance of the proposed method.

Keywords: Markov chain, Wind speed forecasting, ECMWF, SCADA, Model fusion, Transition matrix

1 INTRODUCTION

The decarbonization process passes through renewable energies deployment, and wind power stands as one of the most efficient technologies.

Unfortunately, due to the high proliferation of wind generators in modern electrical grids, the complexity in the control, protection, and prediction of power system operation now grows as a result of several technical and economical side-effects, which are mainly caused by nonprogrammable and intermittent wind generation profiles [1].

In the current scenario, the role of wind power forecasting tools has been well explored in the literature, outlining their importance in supporting both energy producers and power system operators in mitigating the effects of wind uncertainty, reducing imbalance charges, and obtaining strategic information in day-ahead and real-time energy market trading [2, 3]. Several methodologies, based on both physical and statistical models, have been proposed in the literature for wind power forecasting. Many of them exhibit adaptive features to properly describe wind dynamics, dealing with the intrinsic time-varying phenomena affecting them, and are characterized by low computational resources to satisfy the time constraints of energy management systems [2].

Although the integration of these tools in several power system operation functions allows the enhancement of their performance, primarily by reducing the effects of wind data uncertainty, more complex phenomena need to be investigated to reliably predict real wind generation profiles, assessing the effect of predicted weather variables on the generator operation state.

In the light of this need, modern research trends are oriented toward the conceptualization of more sophisticated methods that describe wind generators as stochastic multistate systems, where the state probabilities model the levels of available wind energy [4–6].

In this context the application of the Markov chains has been revealed as one of the most promising research direction [7], and has been adopted in assessing wind generators reliability indices [8]. The main idea is to estimate the effects of the wind speed profiles on the generators' operational states, which can be roughly determined as "in service" or "faulted." This process requires the discretization of the wind speed in a certain number of classes and the classification of the generator states for each speed class.

Unfortunately, in this kind of approach, the granularity level adopted in discretizing the state variables affects the model accuracy and its complexity. Hence, the identification of a proper level of discretization is an open problem requiring further investigation.

To deal with this problem, several approaches have been developed. Wu et al. [9] have proposed a linear rounding method for reducing the number of operational states on the basis of the states' sequence concept, whereas on the basis of the same concept, a probabilistic model for wind farm modeling in a time-based reliability analysis has been proposed [10]. These approaches are useful for assessing several strategic reliability indices such as the Loss of Energy Expectation and Loss of Load Expectation.

The described models have been improved by Jiang et al. [11], who describe the effects of the derated working points by using a Monte Carlo method.

In accordance with previous, the adoption of a Markov chain base has been only restricted to effectively address the planning problems, whereas their integration in power system operation frameworks are still in the embryonic state and need further investigation.

In this context a very promising research approach deals with the fusion of information coming from wind forecasting tools for reliable power profiles assessment for each available wind generator.

To address the described problem, this chapter proposes the integration of an adaptive wind forecasting model, which has been proposed by Vaccaro et al. [12], and a probabilistic reliability model based on Markov chains, whose parameters are continuously adjoined by processing real-time operation data.

The objective is to compute the wind speed forecasting for the next day on each generator, compute the corresponding forecasting error bounds, and asses the system state probabilities in the function of each wind speed span, by solving a probabilistic multistate system.

The described process allows a power system operator to estimate, for each hour of the next day, the expected generated power considering the cumulative effect of all the generation operational states (i.e., alarm, fault, derating operation) and the forecasting errors.

2 PROPOSED METHOD

The objective of this chapter is to predict the produced power and generator operation state over time, for both the next day and for each wind generator.

To this aim, the fusion of a hybrid forecasting algorithm, which fuses and processes the statistical and physical models' outputs, and of an adaptable reliability model that is continuously adjoined by real-time operation data, has been developed.

The main idea is based on the estimation of the probability density function of wind forecasting error, computing the wind speed tolerance bound, followed by the estimation of the corresponding probability to find the generator in a defined operation state. This is obtained by scientifically by solving a reliability model.

Hence, the proposed framework's main features will be described in the next sections.

2.1 Wind Speed Forecasting

The wind forecasting model applied here has been deployed by Vaccaro et al. [12], in which the prediction supplied by synoptic and local forecasting models are amalgamated by a supervised learning system.

In that work the considered synoptic model is a primitive equation for the atmospheric general circulation model, which has been developed by the European Center for Medium-range Weather Forecasts (ECMWF) [13].

Several physical interactions between each of these systems, such as ocean, atmosphere, soil wetness, and snow covering, are covered in that work.

Hence, the synoptic model output is corrected by employing an adaptive learning algorithm to improve the forecast accuracy of the wind speed by processing experimental data. The adaptive feature means the opportunity to update the model allows us to consider the effects of "new" operating conditions.

In light of the produced results by several case studies, the effectiveness of this forecasting algorithm has been confirmed, proving its adoption in the proposed work.

2.2 Generator Reliability Model

In this section, the following generator operation states have been defined to model generator reliability:

1. Alarm: operation in the presence of anomalous working conditions.
2. Faulted: operation inhibited due to a failure condition.
3. Derated: operation in the presence of an external reduction of generated power.
4. Run: normal operation.

The wind speed, which is one of the most influential variables ruling the generation state transitions, is classified in the following classes:

1. $w < w_1$
2. $w_1 \leq w < w_2$
3. $w_2 \leq w < w_3$
4. $w \geq w_3$

where w_1, w_2, and w_3 are the cut-in, rated, and cut-off speeds, respectively.

Hence, the state transition diagram in Fig. 1 allows us to describe the generator reliability.

The computation of the operation data stored in the Supervisory Control And Data Acquisition (SCADA) event register allows us to determine the

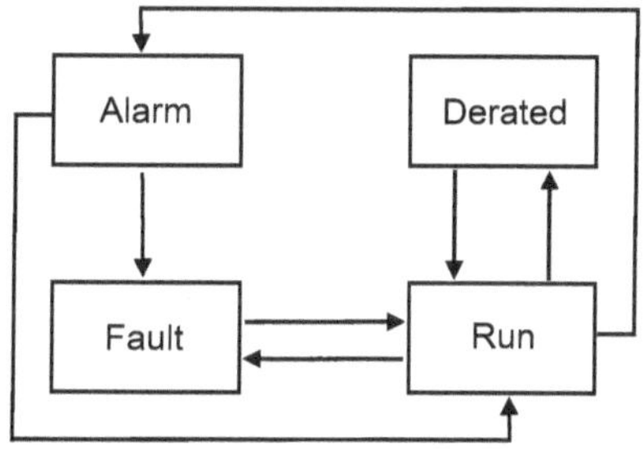

Fig. 1 Markov model states diagram, preliminary study.

model parameters. Data are organized in a double-column matrix $\mathbf{R}_z$, whose dimensions are $[M(z), 2]$, with $z \in [1, N_{wg}]$, where $m \in [1, M(z)]$.

For each row of this matrix correspond the mth transition to a new state: the two columns mean the transition between two different states' recorded time and the corresponding arrival state codes, respectively.

Hence, for each measurement set of $\mathbf{S}_z$, a label has been assigned with a code number by employing Algorithm 1.

Then, by adopting Algorithm 2, a square matrix $\mathbf{MAT}_z$ is obtained, which contains for each element the number of transitions from the state ith to jth one.

Algorithm 1 Labeling: $\forall z \in [1, N_{wg}]$

for $m = 1{:}(M(z) - 1)$ **do**
  for $h = 1{:}H$ **do**
    if $\mathbf{S}_z(h, 1) \geq \mathbf{R}_z(m, 1)$ **and** $\mathbf{S}_z(h, 1) < \mathbf{R}_z(m + 1, 1)$ **then**
      $\mathbf{S}_z(h, V + 1) = \mathbf{R}_z(m, 2)$

Algorithm 2 Transitions Counting: $\forall z \in [1, N_{wg}]$

$\mathbf{MAT}_z = zeros\,(N, N)$
for $h = 1{:}(H - 1)$ **do**
  from $= \mathbf{S}_z(h, V + 1)$
  to $= \mathbf{S}_z(h + 1, V + 1)$
  $\mathbf{MAT}_z(\text{from, to}) = \mathbf{MAT}_z(\text{from, to}) + 1$
1

Where the number of total states is N.

Matrix $\mathbf{MAT}_z$ allows us to compute the transition probabilities as follows:

$$P(j|i) = \frac{\mathbf{MAT}_z(i,j)}{\sum_{j=1}^{N} \mathbf{MAT}_z(i,j)} \tag{1}$$

Then, the generation state probabilities can be computed at each time class t by iterating the following set of linear equations:

$$\mathbf{p}^t = \mathbf{p}^{t-1}\mathbf{P} \tag{2}$$

The latter equation allows us to calculate the steady-state probabilities $\mathbf{x}$ as follows:

$$\begin{cases} x_1(P_{11}-1) + x_2P_{12} + \cdots + x_NP_{1N} & = 0 \\ x_1P_{21} + x_2(P_{21}-1) + \cdots + x_NP_{2N} & = 0 \\ \cdots & = 0 \\ x_1 + x_2 + \cdots + x_N & = 1 \end{cases} \tag{3}$$

Or in a matrix-based formalism, as:

$$\mathbf{xP}^* = \mathbf{b} \tag{4}$$

where $\mathbf{I}$ and $\mathbf{P}^*$ are the identity and the modified transition matrix, respectively.

2.3 Model Fusion

Wind forecasting and generator reliability models have been combined with the purpose of predicting the generated power for each turbine, considering the corresponding expected operation states.

Then, to calculate the state probabilities, the formula of total probability, which considers the estimation of the probability density function of the wind speed forecasting error, has been applied:

$$P(x_z) = \sum_{k=1}^{N} P(x(t)_{k_z} | w_k(t) \le w_z(t) \le w_{k+1}) \times P(w_k \le w_z(t) + e(w_z(t)) \le w_{k+1}) \tag{5}$$

In this equation, $x(t)_k$ represents the probability at time t that zth wind generator is in a fault or alarm state, whereas the true wind speed value $w_{k_z}(t)$

is included in the kth cluster (w_k and w_{k+1}) at the same time. N is the total number of classes, and $P(w_k \leq w_z(t) + e(t) \leq w_{k+1})$ is the corresponding probability that the forecasted wind value $w_z(t)$, which is affected by the forecasting error e, is included in the k class at the same time.

3 CASE STUDY

The proposed method has been applied in the task of estimating the effect of wind forecasting error on the generators' operation states, considering also the derating states, which are related to the congestion level of the network. For this case study, which is composed of 19 wind generators with 2 MVA rated power, the derating conditions are dictated by the Transmission System Operator to mitigate the effects of power systems' congestion induced by large wind energy generation.

The generator model shown in Fig. 1 and characterized by the transition matrix reported in Table 1 has been deployed. The transition probabilities have been calculated for each wind generator by processing 1 year of first- and second-level SCADA data, with 10 minutes of sampling time.

Applying the equation system in Eq. (2) on the latter calculated transition matrix, the following steady-state probabilities have been computed (Table 2) for each wind generator. Their summarization is shown in Fig. 2, with particular reference to the turbine number 3.

As shown by the previous results, the estimated average value of derated operational state probability is about the 5% in a 1-year time-window. Unfortunately the described system is not still able to supply information related to the effect of the wind speed on the described generator operation states. Hence, to estimate how the wind speed influences the probability of power curtailments, the Markov model in Fig. 3 has been employed

Table 1 WTG 3: Transition matrix

	To (j)			
From (j)	**Alarm**	**Derated**	**Faulted**	**Run**
Alarm	0.9091	0.0021	0.0021	0.0867
Derated	0.0003	0.0971	0.0009	0.0274
Faulted	0.0043	0.0043	0.8369	0.1545
Run	0.0008	0.0019	0.0006	0.9966

Table 2 Steady-state probabilities

WTG	Alarm	Derated	Faulted	Run
1	0.0139	0.0586	0.0137	0.9136
2	0.0105	0.0663	0.0043	0.9188
3	0.0090	0.0611	0.0044	0.9253
4	0.0162	0.0621	0.0035	0.9181
5	0.0091	0.0595	0.0064	0.9249
6	0.0186	0.0640	0.0028	0.9145
7	0.0152	0.0522	0.0065	0.9259
8	0.4436	0.0000	0.0000	0.5563
9	0.0048	0.0000	0.0000	0.9951
10	0.0062	0.0000	0.0000	0.9937
11	0.0189	0.0598	0.0055	0.9157
12	0.0189	0.0622	0.0063	0.9125
13	0.0175	0.0566	0.0069	0.9190
14	0.0099	0.0496	0.0031	0.9372
15	0.0199	0.0491	0.0026	0.9283
16	0.0199	0.0564	0.0077	0.9159
17	0.0126	0.0470	0.0013	0.9390
18	0.0149	0.0529	0.0014	0.9306
19	0.0197	0.0501	0.0013	0.9287

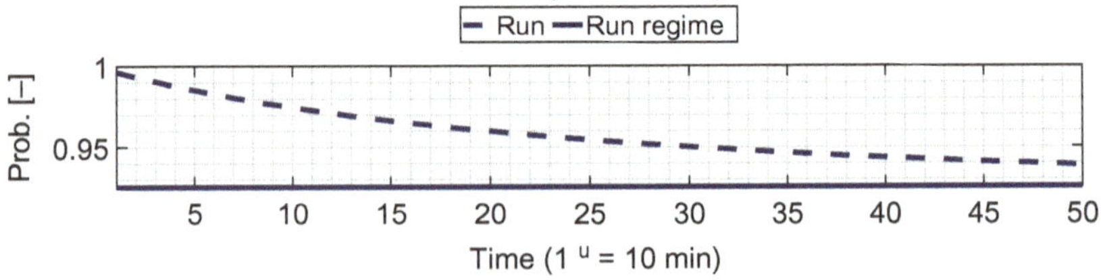

Fig. 2 Markov's transient probability evolution.

considering the wind speed clusters in Table 3. Algorithm 1 allows us to calculate the following Transition Matrix, which is reported in Table 4.

The computed transition matrix is indispensable to calculate the generators' operation probabilities by applying Eq. (2), whose results are shown in

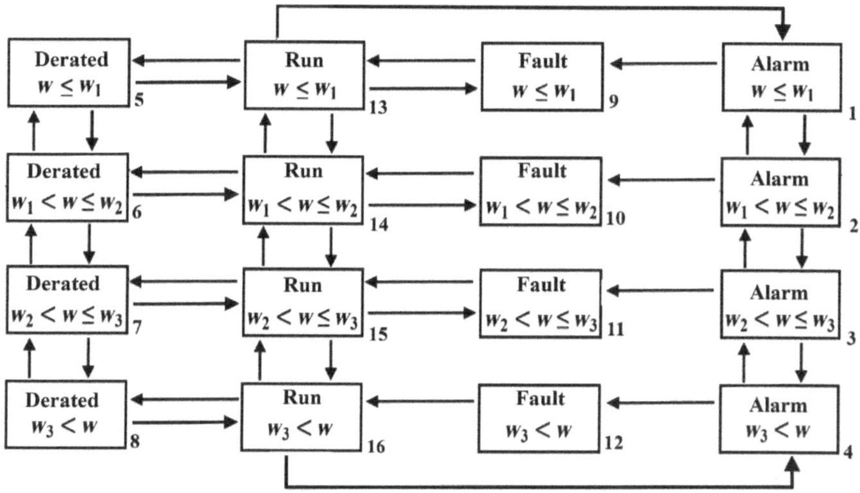

Fig. 3 Considered Markov model states diagram.

Table 3 Bins codes

	$w < w_1$	$w > w_1$ $w < w_2$	$w > w_2$ $w < w_3$	$w > w_3$
Alarm	1	2	3	4
Derated	5	6	7	8
Faulted	9	10	11	12
Run	13	14	15	16

Table 2. Unfortunately this analysis reveals lack of homogeneity with reference to the considered number of wind speed classes. In fact, a further subdivision of wind speed cluster in greater class numbers allows us to estimate the probabilistic profile of the wind effect on the generator operation state (Table 5).

Hence, to quantify this effect, Eq. (5) has been employed allowing the estimation of the derating generated power profiles for each turbine, which are summarized in Fig. 4. It is remarkable how the wind turbines show different levels of derated generated power in a period of 1 year that the different derating probabilities derive. This should suggest curtailment sequence policies in the function of potential wind resource availability.

Then, as a natural consequence of the described data analysis, the conclusive study of this chapter is the statistical assessment on the impact level of wind

Table 4 Transition matrix with wind speed clusters: WTG 3

			To (*j*)															
			Alarm				Derated				Faulted				Run			
			1	2	3	4	5	6	7	8	9	10	11	12	13	14	15	16
From (*i*)	Alarm	1	0.8038	0.0696	0.0000	0.0000	0.0000	0.0000	0.0000	0.0000	0.0000	0.0000	0.0000	0.0000	0.1013	0.0253	0.0000	0.0000
		2	0.0461	0.8582	0.0142	0.0000	0.0000	0.0035	0.0000	0.0000	0.0000	0.0035	0.0000	0.0000	0.0035	0.0709	0.0000	0.0000
		3	0.0000	0.1212	0.8788	0.0000	0.0000	0.0000	0.0000	0.0000	0.0000	0.0000	0.0000	0.0000	0.0000	0.0000	0.0000	0.0000
		4	0.0000	0.0000	0.0000	0.0000	0.0000	0.0000	0.0000	0.0000	0.0000	0.0000	0.0000	0.0000	0.0000	0.0000	0.0000	0.0000
	Derated	5	0.0000	0.0000	0.0000	0.0000	0.7967	0.0488	0.0000	0.0000	0.0000	0.0000	0.0000	0.0000	0.0732	0.0813	0.0000	0.0000
		6	0.0000	0.0005	0.0000	0.0000	0.0049	0.9043	0.0584	0.0000	0.0000	0.0010	0.0000	0.0000	0.0025	0.0265	0.0020	0.0000
		7	0.0000	0.0000	0.0000	0.0000	0.0000	0.1149	0.8778	0.0000	0.0000	0.0000	0.0009	0.0000	0.0000	0.0000	0.0055	0.0009
		8	0.0000	0.0000	0.0000	0.0000	0.0000	0.0000	0.0000	0.0000	0.0000	0.0000	0.0000	0.0000	0.0000	0.0000	0.0000	0.0000
	Faulted	9	0.0000	0.0000	0.0000	0.0000	0.0000	0.0000	0.0000	0.0000	0.6600	0.2000	0.0000	0.0000	0.1400	0.0000	0.0000	0.0000
		10	0.0000	0.0083	0.0000	0.0000	0.0000	0.0000	0.0000	0.0000	0.0744	0.7107	0.0248	0.0000	0.0083	0.1736	0.0000	0.0000
		11	0.0000	0.0000	0.0000	0.0000	0.0000	0.0000	0.0164	0.0000	0.0000	0.0492	0.8197	0.0000	0.0000	0.0000	0.1148	0.0000
		12	0.0000	0.0000	0.0000	0.0000	0.0000	0.0000	0.0000	0.0000	0.0000	0.0000	1.000	0.0000	0.0000	0.0000	0.0000	0.0000
	Run	13	0.0009	0.0000	0.0000	0.0000	0.0003	0.0000	0.0000	0.0000	0.0004	0.0001	0.0000	0.0000	0.9026	0.0957	0.0000	0.0000
		14	0.0001	0.0008	0.0000	0.0000	0.0003	0.0020	0.0000	0.0000	0.0000	0.0006	0.0000	0.0000	0.0569	0.9342	0.0051	0.0000
		15	0.0000	0.0000	0.0000	0.0000	0.0000	0.0048	0.0133	0.0000	0.0000	0.0000	0.0072	0.0012	0.0000	0.1812	0.7899	0.0024
		16	0.0000	0.0000	0.0000	0.0000	0.0000	0.0000	0.0000	0.0000	0.0000	0.0000	0.0000	0.0000	0.0000	0.0000	0.5000	0.5000

Table 5 Steady-state probabilities with wind speed clusters

	Alarm				Derated				Faulted				Run			
WTG	1	2	3	4	5	6	7	8	9	10	11	12	13	14	15	16
1	0.0029	0.0091	0.0019	0.0000	0.0016	0.0277	0.0294	0.0000	0.0019	0.0105	0.0012	0.0001	0.2527	0.6284	0.0321	0.0005
2	0.0039	0.0051	0.0015	0.0000	0.0025	0.0333	0.0304	0.0000	0.0016	0.0002	0.0007	0.0000	0.3342	0.5613	0.0232	0.0002
3	0.0030	0.0054	0.0006	0.0000	0.0023	0.0383	0.0203	0.0000	0.0010	0.0023	0.0012	0.0000	0.3403	0.5694	0.0157	0.0001
4	0.0098	0.0058	0.0007	0.0000	0.0217	0.0276	0.0127	0.0000	0.0016	0.0017	0.0002	0.0000	0.4039	0.4920	0.0225	0.0001
5	0.0037	0.0049	0.0005	0.0000	0.0036	0.0392	0.0166	0.0000	0.0019	0.0032	0.0013	0.0000	0.3603	0.5496	0.0151	0.0001
6	0.0103	0.0075	0.0008	0.0000	0.0054	0.0270	0.0314	0.0000	0.0005	0.0019	0.0004	0.0000	0.3995	0.4925	0.0227	0.0000
7	0.0068	0.0073	0.0011	0.0000	0.0015	0.0264	0.0242	0.0000	0.0024	0.0026	0.0013	0.0002	0.3463	0.5452	0.0344	0.0002
8	0.2751	0.1510	0.0177	0.0000	0.0000	0.0000	0.0000	0.0000	0.0000	0.0000	0.0000	0.0000	0.2029	0.3192	0.0340	0.0001
9	0.0038	0.0011	0.0000	0.0000	0.0000	0.0000	0.0000	0.0000	0.0000	0.0000	0.0000	0.0000	0.6732	0.2758	0.0458	0.0003
10	0.0046	0.0017	0.0000	0.0000	0.0000	0.0000	0.0000	0.0000	0.0000	0.0000	0.0000	0.0000	0.4596	0.4797	0.0540	0.0004
11	0.0129	0.0058	0.0002	0.0000	0.0026	0.0283	0.0288	0.0000	0.0029	0.0022	0.0003	0.0001	0.3781	0.5105	0.0271	0.0002
12	0.0124	0.0061	0.0005	0.0000	0.0051	0.0302	0.0267	0.0000	0.0017	0.0037	0.0010	0.0000	0.4111	0.4739	0.0275	0.0002
13	0.0137	0.0036	0.0003	0.0000	0.0050	0.0283	0.0231	0.0000	0.0016	0.0037	0.0016	0.0000	0.4551	0.4452	0.0187	0.0002
14	0.0071	0.0028	0.0001	0.0000	0.0048	0.0216	0.0230	0.0000	0.0005	0.0017	0.0010	0.0000	0.4855	0.4278	0.0238	0.0004
15	0.0157	0.0042	0.0000	0.0000	0.0051	0.0252	0.0186	0.0000	0.0015	0.0006	0.0004	0.0000	0.4691	0.4332	0.0262	0.0001
16	0.0164	0.0036	0.0000	0.0000	0.0072	0.0306	0.0184	0.0000	0.0058	0.0016	0.0002	0.0000	0.4977	0.4002	0.0181	0.0000
17	0.0096	0.0031	0.0000	0.0000	0.0083	0.0321	0.0066	0.0000	0.0006	0.0003	0.0004	0.0000	0.5514	0.3763	0.0115	0.0000
18	0.0084	0.0049	0.0016	0.0000	0.0051	0.0256	0.0225	0.0000	0.0005	0.0008	0.0001	0.0000	0.3936	0.5017	0.0351	0.0000
19	0.0141	0.0048	0.0008	0.0000	0.0050	0.0234	0.0216	0.0000	0.0004	0.0007	0.0002	0.0000	0.4438	0.4629	0.0223	0.0000

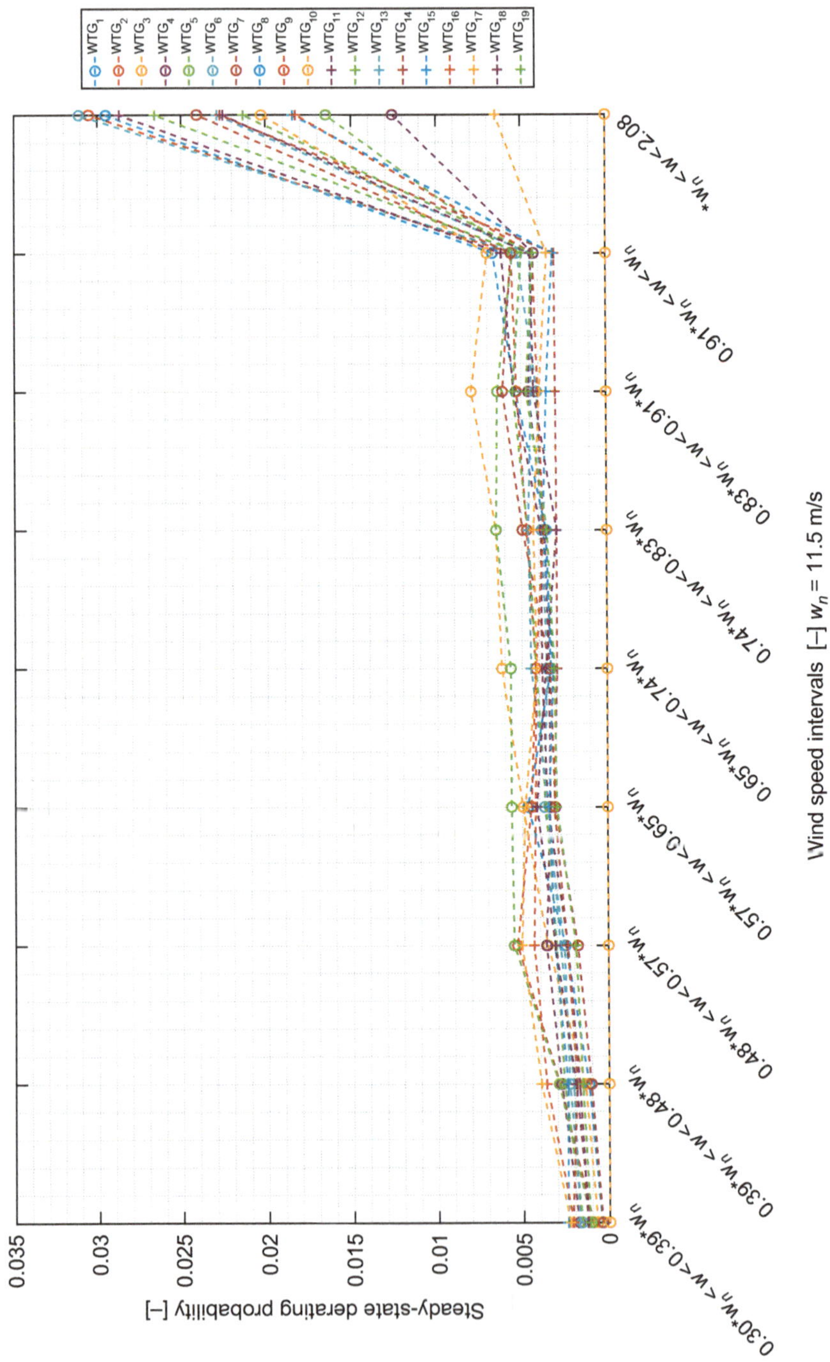

Fig. 4 Derating probability.

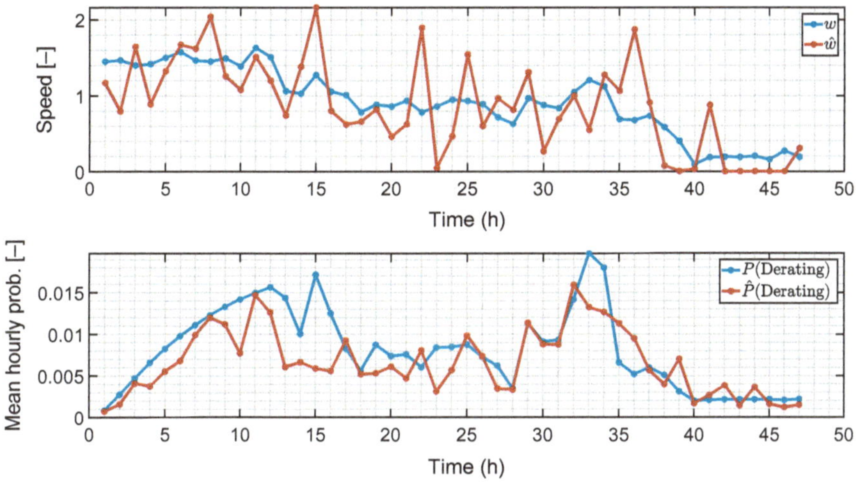

Fig. 5 WTG 3 derating forecasting on next 48 hours.

forecasting uncertainty on wind generator derating power probabilities by calculating them for each turbine. To estimate the time evolution of wind forecasting errors, the application of Eq. (5) becomes crucial. The deviation of the expected profile from the real one allows us to quantify the effect of the actual wind speed figure on the generation operation, as shown in Fig. 5.

Thus this analysis reveals it is not worth claiming the relevant effect of the wind forecasting uncertainty in the perturbation of generator operation estimation profiles. In fact the induced deviations on the estimated profile could determine conservative curtailments by the system operator to prevent a risk of possible network congestion. Moreover, it is evident that there is a proportional correlation between the wind speed magnitude and the derating probabilities, namely the increase of wind speed corresponds with a rise of curtailment risks.

4 CONCLUSION

The massive proliferation of wind generators in modern power systems pushes the system operator to equip themselves with advanced forecasting tools for the optimal management of the generator assets, dealing with the crucial issue of mitigating data uncertainties effects. To face this issue, this chapter proposes a forecasting framework aimed at both predicting wind power profiles and the effects of weather variables on the generator operation state. A probabilistic model, based on Markov chains, has been adopted to predict the reduced generated power profiles to prevent congestion on the network, considering the effects of wind speed forecasting uncertainty and the expected generator operation state. To prove the effectiveness of the proposed method, several case studies have been presented, and the most relevant results have been analyzed.

REFERENCES

[1] J. Lerner, M. Grundmeyer, M. Garvert, The role of wind forecasting in the successful integration and management of an intermittent energy source, Energy Cent. Wind Power 3 (8) (2009) 1–6.
[2] I. González-Aparicio, A. Zucker, Impact of wind power uncertainty forecasting on the market integration of wind energy in Spain, Appl. Energy 159 (2015) 334–349.
[3] Y.V. Makarov, P.V. Etingov, J. Ma, Z. Huang, K. Subbarao, Incorporating uncertainty of wind power generation forecast into power system operation, dispatch, and unit commitment procedures, IEEE Trans. Sustain. Energy 2 (4) (2011) 433–442.
[4] R. Billinton, R. Karki, Y. Gao, D. Huang, P. Hu, W. Wangdee, Adequacy assessment considerations in wind integrated power systems, IEEE Trans. Power Syst. 27 (4) (2012) 2297–2305.
[5] A. Ghaedi, A. Abbaspour, M. Fotuhi-Firuzabad, M. Moeini-Aghtaie, Toward a comprehensive model of large-scale DFIG-based wind farms in adequacy assessment of power systems, IEEE Trans. Sustain. Energy 5 (1) (2014) 55–63.
[6] F. Chen, F. Li, Z. Wei, G. Sun, J. Li, Reliability models of wind farms considering wind speed correlation and WTG outage, Electr. Power Syst. Res. 119 (2015) 385–392.
[7] S. Sulaeman, M. Benidris, J. Mitra, C. Singh, A wind farm reliability model considering both wind variability and turbine forced outages, IEEE Trans. Sustain. Energy 8 (2) (2017) 629–637.
[8] A.P. Leite, C.L.T. Borges, D.M. Falcao, Probabilistic wind farms generation model for reliability studies applied to Brazilian sites, IEEE Trans. Power Syst. 21 (4) (2006) 1493–1501.
[9] L. Wu, J. Park, J. Choi, A.A. El-Keib, M. Shahidehpour, R. Billinton, Probabilistic reliability evaluation of power systems including wind turbine generators using a simplified multi-state model: a case study, in: Power and Energy Society General Meeting, 2009. PES'09, IEEE, 2009, pp. 1–6.
[10] S. Kahrobaee, S. Asgarpoor, Short and long-term reliability assessment of wind farms, North American Power Symposium (NAPS), 2010, IEEE, 2010, pp. 1–6.
[11] C. Jiang, J. Zhang, L. Yu, Reliability probability evaluation of wind turbine based on Monte-Carlo simulation, in: 2012 China International Conference on Electricity Distribution (CICED), IEEE, 2012, pp. 1–4.
[12] A. Vaccaro, P. Mercogliano, P. Schiano, D. Villacci, An adaptive framework based on multi-model data fusion for one-day-ahead wind power forecasting, Electr. Power Syst. Res. 81 (3) (2011) 775–782.
[13] ECMWF Description, Available from: http://www.ecmwf.int/en/research/modelling-and-prediction/atmospheric-dynamics (Accessed 13 February 2013).

CHAPTER 5

Time-Dependent Analysis of Series-Parallel Multistate Systems Using Structure Function and Markov Processes

Miroslav Kvassay, Patrik Rusnak, Jan Rabcan
Department of Informatics, University of Zilina, Zilina, Slovakia

Abstract

Multistate systems (MSSs) represent one of the key concepts of reliability engineering. They are used to represent systems that can work at several performance levels. Such systems are usually composed of many various components, making their analysis much harder. Therefore one of the key issues of reliability engineering is development of methods for their efficient analysis. A prospective way to solve this problem is an application of a modular decomposition that allows developing fast algorithms for quantitative analysis of such systems. The quantitative analysis deals with computation of global reliability characteristics of an MSS, such as system-state probabilities or system availabilities. These characteristics can be static (if the investigated system is in a steady-state and time is not taken into account) or dynamic (if the analysis is performed with respect to time). If these characteristics are dynamic, then they become functions of time. In this chapter, we present how these characteristics can be computed in both a static and dynamic case if we know the system topology defined by structure function. We also show how these computations can be performed efficiently using modular decomposition. More precisely, we combine methods for computation of global reliability characteristics based on structure function together with Markov processes to obtain stochastic model of the system, whose time-dependent reliability characteristics can be quickly obtained using modular decomposition. In this chapter, we primarily deal with series-parallel MSSs without maintenance.

Keywords: Multistate system, Reliability, Structure function, Modular decomposition, Stochastic process

1 INTRODUCTION

One of the key challenges in reliability engineering is analysis of complex systems. Such systems are composed of many components with similar or different behaviors. Examples of such systems include various types of

Advances in System Reliability Engineering
https://doi.org/10.1016/B978-0-12-815906-4.00005-1

complex digital circuits [1, 2], distribution networks [3, 4] or sociotechnical systems, such as healthcare systems [5, 6], whose components can usually be classified as hardware, software, human factor, and organizational elements [7]. Some of these systems can be viewed as binary-state systems (BSSs), that is, systems that are just working or failed [8, 9]. Typical examples are digital circuits, which are working if the output of the circuit agrees with expectations and failed if the output is incorrect [1, 2]. Other types of systems can work at several performance levels, from perfect functioning to complete failure. Modeling such systems such as BSSs is usually very complicated. Because of that, they are modeled as systems with more than two states. Such systems are known as multistate systems (MSSs) [10–12].

MSSs allow defining more than two states in modeling system/component's behavior. These states are usually interpreted as performance levels at which the system/component can operate. The dependency between states of the system components and state of the system is expressed by structure function [8–12]. This function defines system topology. Because the topology usually does not change over time, the structure function is time-independent, and it is usually used in static analysis, that is, in analysis without a time factor.

The static analysis can be topological or reliability time-independent. The former is based only on the knowledge of the structure function, whereas the latter assumes that the structure function of the system and the state probabilities of the system components are known. Topological analysis allows us to compare different topologies and identify those that are more reliable [13] or identify components with the greatest influence on system operation from a topological point of view [13–16]. Time-independent reliability analysis allows us to estimate basic reliability measures of a system [8, 12] and investigate how these measures depend on the state probabilities of the system components [16–21]. Such an analysis can be useful in optimization of system reliability.

One of the principal advantages of static analysis, or analysis based on structure function, is the relative simplicity of models of studied systems. However, its main disadvantage is an inability of investigating how the system evolves over time, that is, how its reliability or performance changes as time flows. These and similar tasks are addressed in dynamic (time-dependent) reliability analysis. In this case, the system and its components are modeled as stochastic processes [8–12, 22]. Analysis of these processes allows us to find how the state probabilities of the system or its components change over time, which allows us to estimate time when the system performance, reliability, or other measures decrease below a specific value. Results of such an analysis can be useful in planning system maintenance or in development of high reliability systems. The principal issue of this analysis is complexity of models that has to be analyzed because they usually grow exponentially with an increasing number of system components or states of the components [8–12]. In this chapter, we present how this disadvantage can be overcome if the structure function, which is typically used in static analysis, of the system is known, and if we assume that the components are independent. This approach allows us to develop relatively simple dynamic (time-dependent) models of MSSs.

This chapter has three core sections. Section 2 summarizes basic principles of static and dynamic modeling of MSSs. In the first part of this section, the concept of structure function is introduced, and the bases of stochastic modeling of a single component are summarized. In the second part of Section 2, a process of finding time-dependent reliability measures of an MSS based on knowledge of the structure function and stochastic processes modeling behavior of the system components over time is presented. Section 3 deals with computation of reliability characteristics of series-parallel MSSs. This section explains how a modular decomposition [12, 14] can be used in analysis of such systems and how the reliability characteristics can be computed efficiently using it. Finally, Section 4 illustrates application of the approach presented in Sections 2 and 3 in time-dependent reliability analysis of a distributed computing system.

2 MATHEMATICAL BACKGROUND

Let us consider an MSS composed of n components denoted by numbers 1, 2, ..., n, and let us denote the set of the system components as $\mathbf{N} = \{1, 2, \ldots, n\}$. Next, let us assume that the system is homogeneous, which means that the system and all its components can operate at one of m possible performance levels [10–12]. These levels can be denoted by numbers from set $\{0, 1, \ldots, m-1\}$. These numbers are known as system/components states. In this notation, a greater number (a better state) agrees with a better performance level. Specially, state 0 agrees with complete failure of the system/component whereas state $m-1$ denotes its perfect functioning, that is, the best performance level. The dependency of system state on states of the system components is defined by structure function [12, 23]:

$$\phi(z_1, z_2, \ldots, z_n) = \phi(\boldsymbol{z}) : \{0, 1, \ldots, m-1\}^n \rightarrow \{0, 1, \ldots, m-1\}, \qquad (1)$$

where z_i is a variable that defines state of the i-th system component for $i \in \mathbf{N}$, and $\boldsymbol{z} = (z_1, z_2, \ldots, z_n)$ is a vector of components states (state vector).

The structure function carries information about system topology. Based on its properties, systems can be categorized into two classes: coherent, in which a degradation of any system component cannot result in improvement of system performance, and noncoherent, which contain one or more components whose degradation can improve system performance [17, 24–26]. In this chapter, we will deal with the first of them, that is, with coherent systems.

Knowledge of the structure function allows us to investigate topological properties of the system. For example, we can use it to find the most reliable topology in a set of systems with different topologies [13] or evaluate the importance of the components of a system and find those with the greatest influence on system operation from topological point of view [13, 15, 17, 18, 21]. However, its knowledge is not sufficient in performing time-dependent reliability analysis, which deals with evaluation of reliability of the system over time. For this purpose, the state probabilities of the system components have to be known. These probabilities can be discovered using a methodology of stochastic processes.

2.1 Stochastic Process Modeling Behavior of System Component

A state of component i of an MSS changes over time. These changes can be captured by state function $z_i(t)$. In case of a nonrepairable component, this function can look like one depicted in Fig. 1. Depending on time when the state of the component changes and types of changes that occur (degradation by one state, degradation by two states, etc.), an uncountable number of such functions exist for the component (Fig. 2). A collection of all these state functions constitutes a stochastic process describing all possible changes of the component over time.

Every state function $z_i(t)$ represents one possible way of how states of the component changes over time. If we take into account all these state functions at fixed time t_f, then they have to take a value from set $\{0, 1, \ldots, m-1\}$. The proportion of the functions that take value s, for $s \in \{0, 1, \ldots, m-1\}$, at this time agrees with the probability that the component will be in state s. More formally, at fixed time t_f, we can define random variable Z_i representing all possible states of the component (Fig. 3). Using this notation, we can introduce the probability that component i is in state s as follows [12, 19, 21]:

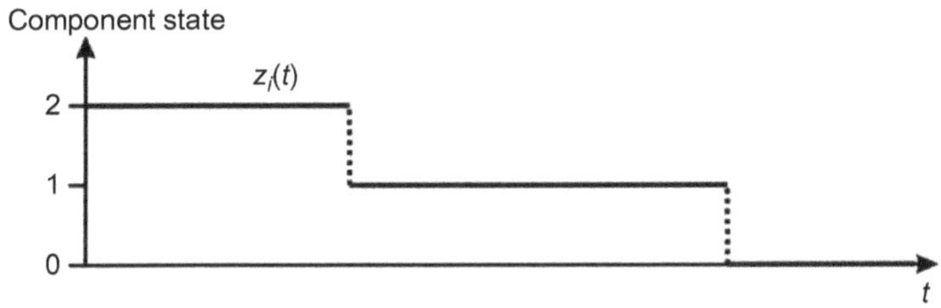

Fig. 1 Example of a state function of a nonrepairable component with three states.

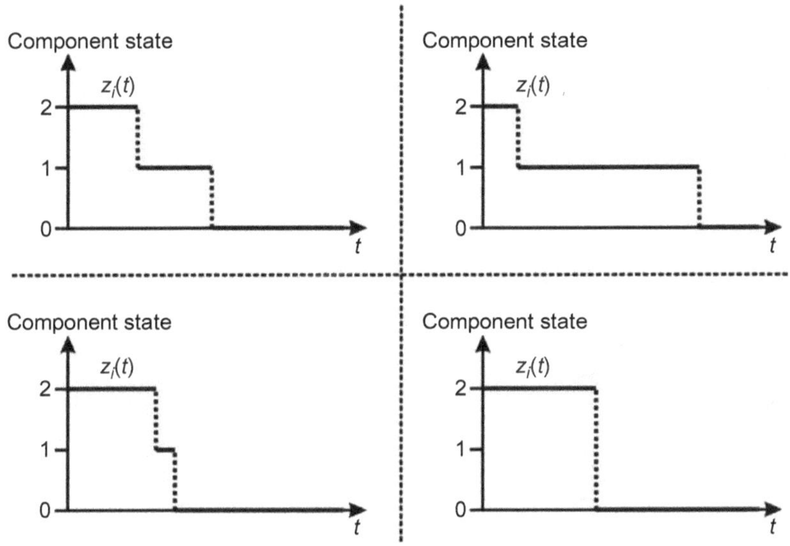

Fig. 2 Several state functions modeling behavior of a nonrepairable component with three states.

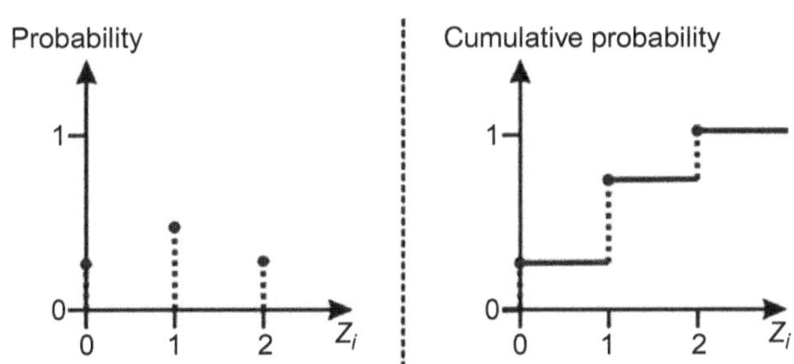

Fig. 3 Example of random variable Z_i modeling a component with three states at a fixed time.

$$p_{i,s} = \Pr\{Z_i = s\}, \quad s = 0, 1, \ldots, m-1,$$
$$\sum_{s=0}^{m-1} p_{i,s} = 1. \tag{2}$$

Random variable Z_i changes over time. This implies it can be interpreted as a function of time $Z_i(t)$. In this context, the stochastic process defining behavior of component i over time agrees with a collection of random variables $Z_i(t)$ [10, 22]:

$$\{Z_i(t); t \geq 0\}. \tag{3}$$

Because random variable Z_i depends on time, the probabilities defining possible states of component i have to change over time:

$$p_{i,s}(t) = \Pr\{Z_i(t) = s\}, \quad s = 0, 1, \ldots, m-1, \quad t \geq 0,$$
$$\sum_{s=0}^{m-1} p_{i,s}(t) = 1, \quad t \geq 0. \tag{4}$$

Behavior of the component over time can be very complex. A state of the component at time t_{w+1} can depend, for example, only on its state at time t_w, but also it can depend on the states in some other time points from interval $\langle 0, t_w)$. If the next state of the component depends just on its current state, then the stochastic process modeling behavior of the component is known as a Markov process. More formally, stochastic process $\{Z_i(t); t \geq 0\}$ is a Markov process if the next condition holds [10]:

$$\forall t_1, t_2, \ldots, t_w, t_{w+1}; 0 \leq t_1 < t_2 < \cdots < t_w < t_{w+1};$$
$$\forall s_1, s_2, \ldots, s_w, s_{w+1} \in \{0, 1, \cdots, m-1\}:$$
$$\Pr\{Z_i(t_{w+1}) = s_{w+1} | Z_i(t_w) = s_w, Z_i(t_{w-1}) = s_{w-1}, \ldots, Z_i(t_1) = s_1\}$$
$$= \Pr\{Z_i(t_{w+1}) = s_{w+1} | Z_i(t_w) = s_w\}. \tag{5}$$

In what follows, we will assume that stochastic processes modeling behavior of system components are Markov processes.

If we denote current time t_w as t, state of the component at that time as s, time t_{w+1} as $t + \Delta$, and state of the component at that time as r, the conditional probability $\Pr\{Z_i(t_{w+1}) = s_{w+1} \mid Z_i(t_w) = s_w\}$ can be denoted as follows [10]:

$$p_{i,s\to r}(t, \Delta) = \Pr\{Z_i(t+\Delta) = r | Z_i(t) = s\},$$
$$s, r \in \{0, 1, \ldots, m-1\}, t \geq 0, \Delta \geq 0. \tag{6}$$

This probability is known as the transition probability because it defines the probability that the state of the component will change from value s to r in time interval $(t, t+\Delta\rangle$ if we know it is in state s at time t [10, 22]. If we want to find how quickly the component moves between its states, then we can compute transition rates $a_{i,\ s \to r}(t)$. These rates are defined as follows [10, 22]:

$$a_{i,s\to r}(t) = \lim_{\Delta\to 0}\frac{p_{i,s\to r}(t,\Delta) - p_{i,s\to r}(t0)}{\Delta} = \begin{cases} \lim\limits_{\Delta\to 0}\dfrac{p_{i,s\to s}(t,\Delta)-1}{\Delta} & \text{if } r=s \\ \lim\limits_{\Delta\to 0}\dfrac{p_{i,s\to r}(t,\Delta)}{\Delta} & \text{if } r\neq s \end{cases},$$
$$s,r\in\{0,1,\ldots,m-1\},\ t\geq 0. \tag{7}$$

Clearly, the component has to be in a state from set $\{0,1,\ \ldots,\ m-1\}$ at time $t + \Delta$. This implies the following condition holds:

$$\sum_{r=0}^{m-1} p_{i,s\to r}(t,\Delta) = 1, \quad s\in\{0,1,\ldots,m-1\},\ t\geq 0,\ \Delta\geq 0. \tag{8}$$

This formula and definition (7) of transition rate result in the next formula:

$$\sum_{r=0}^{m-1} a_{i,s\to r}(t) = 0, \quad s\in\{0,1,\ldots,m-1\},\ t\geq 0. \tag{9}$$

Based on it, we can simply derive:

$$a_{i,s\to s}(t) = -\sum_{\substack{r=0\\ r\neq s}}^{m-1} a_{i,s\to r}(t), \quad s\in\{0,1,\ldots,m-1\},\ t\geq 0, \tag{10}$$

which implies just $m(m - 1)$ functions defining transition rates between component states have to be known to fully define how quickly the component moves between its states. This formula also implies if all transition rates at which the component moves from state s to a state different from s are nonnegative, then the transition rate at which the component remains in state s has to be nonpositive.

Transition rates can be arranged in transition matrix $\mathbf{A}_i(t)$ of m rows and m columns:

$$\mathbf{A}_i(t) = \begin{pmatrix} a_{i,0\to 0}(t) & a_{i,0\to 1}(t) & \cdots & a_{i,0\to m-1}(t) \\ a_{i,1\to 0}(t) & a_{i,1\to 1}(t) & \cdots & a_{i,1\to m-1}(t) \\ \vdots & \vdots & \ddots & \vdots \\ a_{i,m-1\to 0}(t) & a_{i,m-1\to 1}(t) & \cdots & a_{i,m-1\to m-1}(t) \end{pmatrix},\quad t\geq 0, \tag{11}$$

whose element in row s, for $s=0, 1, \ldots, m-1$, and column r, for $r=0, 1, \ldots, m-1$, defines transition rate at which component i moves from state s to r at time t. If we arrange the state probabilities of component i at time t into row vector $\boldsymbol{p}_i(t)$:

$$\boldsymbol{p}_i(t) = (p_{i,0}(t), p_{i,1}(t), \ldots, p_{i,m-1}(t)), \quad t \geq 0, \tag{12}$$

and if we assume that the component is perfect functioning at the beginning, then the transition matrix can be used to define the following system of differential equations [10, 22]:

$$\begin{aligned} &\frac{d\boldsymbol{p}_i(t)}{dt} = \boldsymbol{p}_i(t)\mathbf{A}_i(t), \quad t > 0, \\ &p_{i,0}(0) = p_{i,1}(0) = \cdots = p_{i,m-2}(0) = 0,\ p_{i,m-1}(0) = 1, \end{aligned} \tag{13}$$

which together with Formula (10) implies that the probability of the component being in state s at time t changes according to the following formula [10, 22]:

$$\frac{dp_{i,s}(t)}{dt} = \sum_{r=0}^{m-1} p_{i,r}(t) a_{i,r\to s}(t) = \sum_{\substack{r=0 \\ r\neq s}}^{m-1} p_{i,r}(t) a_{i,r\to s}(t) - p_{i,s}(t) \sum_{\substack{r=0 \\ r\neq s}}^{m-1} a_{i,s\to r}(t), \tag{14}$$

$$s \in \{0, 1, \ldots, m-1\}, \quad t > 0.$$

By solving this system of differential equations, we can find time-dependent functions describing evolution of the state probabilities of component i over time.

2.1.1 Binary-State Nonrepairable Component

Let us assume that we have a nonrepairable component i^* with just two states: state 1—the component is functioning, and state 0—the component fails. Because the component is nonrepairable, according to the previous formulae, the transition matrix has the following form:

$$\mathbf{A}_{i*}(t) = \begin{pmatrix} a_{i*,0\to 0}(t) & a_{i*,0\to 1}(t) \\ a_{i*,1\to 0}(t) & a_{i*,1\to 1}(t) \end{pmatrix} = \begin{pmatrix} 0 & 0 \\ a_{i*,1\to 0}(t) & -a_{i*,1\to 0}(t) \end{pmatrix}, \quad t \geq 0. \tag{15}$$

Based on Formula (7), the transition rate $a_{i^*,1\to 0}(t)$ is defined as follows:

$$a_{i*,1\to 0}(t) = \lim_{\Delta\to 0} \frac{p_{i*,1\to 0}(t, \Delta)}{\Delta}, \quad t \geq 0. \tag{16}$$

Transition probability $p_{i^*,1\to 0}(t, \Delta)$ agrees with the probability that component i^* will fail during time interval $(t, t+\Delta\rangle$ if we know it is functioning at time t. If we define random variable $T_{i^*,1\to 0}$ that agrees with time at which a failure of the component occurs, we can write:

$$p_{i*,1\to 0}(t,\Delta)=\Pr\left\{t<T_{i*,1\to 0}\le t+\Delta\middle|t<T_{i*,1\to 0}\right\}$$
$$=\frac{\Pr\left\{t<T_{i*,1\to 0}\le t+\Delta\right\}}{\Pr\left\{t<T_{i*,1\to 0}\right\}},\quad t\ge 0,\quad \Delta\ge 0. \tag{17}$$

By substituting this formula into Formula (16), we obtain:

$$a_{i*,1\to 0}(t)=\lim_{\Delta\to 0}\frac{\Pr\{t<T_{i^*,1\to 0}\le t+\Delta\}}{\Pr\left\{t<T_{i*,1\to 0}\right\}\Delta}=\frac{f_{i*,1\to 0}(t)}{\Pr\left\{t<T_{i*,1\to 0}\right\}}$$
$$=\frac{f_{i*,1\to 0}(t)}{p_{i*,1\to 0}(t,\infty)},\quad t\ge 0, \tag{18}$$

where $f_{i^*,1\to 0}(t)$ is a failure density function (also known as unconditional failure rate) of component i^* [8, 9]. This agrees with the definition of conditional failure rate (also known as hazard rate) $\lambda_{i^*}(t)$ of nonrepairable component i^* [8–10] and, therefore, transition matrix (15) for such a kind of component can be rewritten in the following form:

$$\mathbf{A}_{i*}(t)=\begin{pmatrix}0 & 0\\ \lambda_{i*}(t) & -\lambda_{i*}(t)\end{pmatrix},\quad t\ge 0. \tag{19}$$

Based on Formula (13) and assuming that the component is working at time $t=0$, we obtain the following system of differential equations:

$$\begin{aligned}&\frac{dp_{i*,0}(t)}{dt}=p_{i*,1}(t)\lambda_{i*}(t),\\ &\frac{dp_{i*,1}(t)}{dt}=-p_{i*,1}(t)\lambda_{i*}(t),\\ &p_{i*,0}(0)=0,\quad p_{i*,1}(0)=1,\end{aligned} \tag{20}$$

whose solution agrees with the following formulae [27]:

$$\left.\begin{aligned}&p_{i*,0}(t)=1-e^{-\int_0^t \lambda_{i*}(\tau)d\tau}\\ &p_{i*,1}(t)=e^{-\int_0^t \lambda_{i*}(\tau)d\tau}\end{aligned}\right\},\quad t\ge 0. \tag{21}$$

Finally, let us assume that conditional failure rate $\lambda_{i^*}(t)$ does not depend on time, that is:

$$\lambda_{i*}(t) = \lambda_{i*}, \quad t \geq 0. \tag{22}$$

In this case, the previous formulae are given the following form:

$$\begin{array}{l} p_{i*,0}(t) = 1 - e^{-\lambda_{i*} t} \\ p_{i*,1}(t) = e^{-\lambda_{i*} t} \end{array}, \quad t \geq 0. \tag{23}$$

These two equations define evolution of the state probabilities of a binary-state nonrepairable component with constant conditional failure rate λ_{i^*} over time assuming it is working at time $t = 0$.

2.2 Stochastic Process Modeling Behavior of Entire System

As in the case of individual system components, the behavior of an entire system can be expressed in a form of system-state function $z(t)$. If we know state functions $z_i(t)$ of all the system components, that is, for $i = 1, 2, \ldots, n$, then system-state function $z(t)$ can be obtained by combining them using system structure function (Fig. 4):

$$z(t) = \phi(z_1(t), z_2(t), \ldots, z_n(t)) = \phi(\boldsymbol{z}(t)) : < 0, \infty) \to \{0, 1, \ldots, m-1\}, \tag{24}$$

where $\boldsymbol{z}(t) = (z_1(t), z_2(t), \ldots, z_n(t))$ denotes a vector of state functions of the components. Although system-state function $z(t)$ is closely related to structure function $\phi(\boldsymbol{z})$, these two functions are very different in their nature because the former is a function of time, whereas the latter is a function defining system topology, which is independent of time.

As we previously mentioned, a system component can realize one of uncountable many state functions $z_i(t)$. Beacuse the system-state function can be viewed as a composition of system structure function and state functions of all the system components, system-state function $z(t)$ can also be viewed as one realization of uncountable many system-state functions. This implies that evolution of the system over time can be viewed as the following stochastic process:

$$\{Z(t); t \geq 0\}, \tag{25}$$

where $Z(t)$ is a random variable modeling behavior of the system at time t.

Let us evaluate function $Z(t)$ at fixed time t_f. In such a case, we obtain random variable Z that takes value j from set $\{0, 1, \ldots, m-1\}$ with probability P_j. The probabilities P_j, for $j = 0, 1, \ldots, m-1$, are known as system-state probabilities, and they represent one of the basic reliability

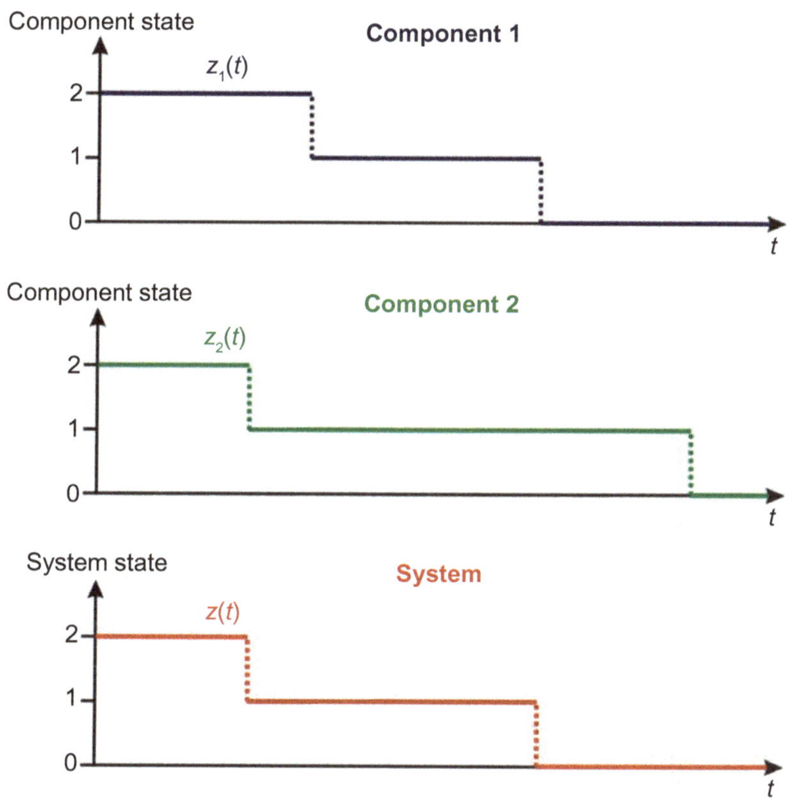

Fig. 4 State function of a system with structure function $\phi(z_1, z_2) = \min(z_1, z_2)$ for z_1, $z_2 \in \{0, 1, 2\}$ obtained by combining state functions of the components with the system structure function.

characteristic of an MSS. Clearly, these probabilities are defined as follows [12, 19, 21]:

$$P_j = \Pr\{Z = j\}, \quad \text{for } j \in \{0, 1, \ldots, m-1\},$$
$$\sum_{j=0}^{m-1} P_j = 1. \tag{26}$$

If we know random variable Z_i, which models behavior of component i at fixed time t_f, for each system component, that is, for $i = 1, 2, \ldots, n$, and if we assume the components are independent, then random variable Z can be obtained by combining random variables Z_i using the structure function. This allows us to compute the system-state probabilities using the following formula [21]:

$$P_j = \Pr\{\phi(\mathbf{Z}) = j\}, \quad \text{for } j \in \{0, 1, \ldots, m-1\}, \tag{27}$$

where $\mathbf{Z} = (Z_1, Z_2, \ldots, Z_n)$ is a vector of random variables modeling behavior of the system components at fixed time t_f. This definition implies that the system-state probabilities can be viewed as functions of component state probabilities (2) [12, 19, 21]:

$$\begin{aligned} &P_j = P_j(\mathbf{p}) = \Pr\{\phi(\mathbf{Z}) = j\}, \quad \text{for } j \in \{0, 1, \ldots, m-1\}, \\ &\sum_{j=0}^{m-1} P_j(\mathbf{p}) = 1, \end{aligned} \tag{28}$$

where $\mathbf{p} = (\boldsymbol{p}_1, \boldsymbol{p}_2, \ldots, \boldsymbol{p}_n)$ is a vector whose elements are vectors defining the state probabilities of individual system components, that is, $\boldsymbol{p}_i$ is a vector of the form of $(p_{i,0}, p_{i,1}, \ldots, p_{i,m-1})$, for $i = 1, 2, \ldots, n$.

Formula (28) allows us to find the system-state probabilities if we know the structure function of the system and the state probabilities of the components. It can be used to investigate how specific changes in state probabilities of one or more components influence the system-state probabilities or other reliability measures [12, 19, 21], but it does not allow us to perform dynamic (time-dependent) analysis of an MSS. For this task, random variable Z has to be replaced by function $Z(t)$, which defines how properties of random variable Z changes over time. In this case, the system-state probabilities become functions of time, that is:

$$\begin{aligned} &P_j(t) = P_j(\mathbf{p}(t)) = \Pr\{\phi(\mathbf{Z}(t)) = j\}, \quad \text{for } j \in \{0, 1, \ldots, m-1\}, \quad t \geq 0, \\ &\sum_{j=0}^{m-1} P_j(t) = 1, \quad t \geq 0, \end{aligned} \tag{29}$$

where $\mathbf{p}(t) = (\boldsymbol{p}_1(t), \boldsymbol{p}_2(t), \ldots, \boldsymbol{p}_n(t))$ is a vector whose elements are time-dependent vectors defining the state probabilities of individual system components over time (definition 12), and $\mathbf{Z}(t) = (Z_1(t), Z_2(t), \ldots, Z_n(t))$ is a vector of random variables modeling behavior of the system components over time. This function can be used to find how reliability of the system or importance of the components change as time flows.

The most important result of previous formulae is that the system-state probabilities can be viewed as a function of the components state probabilities (static analysis based on Formula 28) or as a composition of functions defining the state probabilities of the system components over time (time-dependent analysis based on Formula 29). This means if we know the structure function of the system and the state probabilities of the components, we are able to find the system-state probabilities in case of independent components without performing calculations similar to those in Section 2.1, for example, without necessity of finding and solving a system of differential equations describing behavior of the entire system in case of time-dependent analysis.

As one can see, an MSS can be analyzed with respect to time (dynamic analysis) or regardless of time (static analysis). This implies that reliability measures of an MSS can or might not depend on time. To avoid any confusion, we use the following notation in the rest of the chapter—symbols of reliability measures depending on time are followed by parentheses with t inside (e.g., $P_j(t)$ in Formula 29) whereas symbols of reliability measures that do not depend on time are not followed by t in parentheses (e.g., P_j in Formula 28).

2.3 Basic Reliability Measures

Reliability of MSSs is usually evaluated using three basic measures: system-state probabilities, system availability, and the expected performance of the system. The first of them was previously introduced. Now, we focus on the next two.

Firstly, let us assume static analysis. In this case, availability of an MSS is defined as the probability that the system operates at least in state j [12, 19, 21]:

$$A^{\geq j} = A^{\geq j}(\mathbf{p}) = \Pr\{\phi(\mathbf{Z}) \geq j\} = \sum_{h=j}^{m-1} P_h, \quad \text{for } j \in \{1, 2, \ldots, m-1\}. \tag{30}$$

The complement to this measure is known as system unavailability, and it is defined with respect to system state j as follows [12, 19, 21]:

$$\begin{aligned} &U^{\geq j} = U^{\geq j}(\mathbf{p}) = \Pr\{\phi(\mathbf{Z}) < j\} = \sum_{h=0}^{j-1} P_h, \quad \text{for } j \in \{1, 2, \ldots, m-1\}, \\ &A^{\geq j} + U^{\geq j} = 1, \quad \text{for } j \in \{1, 2, \ldots, m-1\}. \end{aligned} \tag{31}$$

For an illustration of the concept of system availability and unavailability, let us consider a communication network that can transfer data by 20, 10, or 0 Gbit/s (if it completely fails). These transfer rates can be viewed as performance levels at which the network can operate. These performance levels can be modeled by numbers from set {0,1,2}. The numbers denote states of the network, and each of them is linked with one transfer rate, that is, state 0 corresponds to 0 Gbit/s, state 1 to 10 Gbit/s, and state 2 to 20 Gbit/s. Now, let us assume that the network is used for communication between two processes running on two computers connected by the network, and that the correct execution of the processes requires transfer rate of at least 10 Gbit/s. In this context, availability of the network can be defined as the probability that the network is at least in state 1 (it is able to operate at transfer rates 10 or 20 Gbit/s), whereas its unavailability agrees with the probability that the network is in a state less than 1 (it failed).

The previous example illustrates that states of the system/components are just abstract numbers. If these numbers are linked with some performance levels, then we can compute another common characteristic of an MSS, which is known as the expected performance (output) of the system [12, 18, 19]:

$$O = O(\mathbf{p}) = \sum_{j=0}^{m-1} O_j P_j = O_0 + \sum_{j=1}^{m-1} (O_j - O_{j-1}) A^{\geq j}, \tag{32}$$

where O_j denotes performance linked with state j of the system. If we use j instead of O_j for all $j \in \{0, 1, \ldots, m-1\}$, then this formula allows us to obtain the expected state of the system.

The reliability measures previously introduced are based on the assumption that an MSS is analyzed at a fixed time. If we want to investigate how these characteristics change over time, then we have to replace vector $\boldsymbol{Z}$ of random variables modeling behavior of the system components by its time dependent version $\boldsymbol{Z}(t)$. After such a replacement, all static characteristics change into time-dependent functions. Thus system availability and unavailability with respect to system state j are transformed into availability and unavailability functions of the system:

$$\begin{aligned}
&A^{\geq j}(t) = A^{\geq j}(\mathbf{p}(t)) = \Pr\{\phi(\mathbf{Z}(t)) \geq j\} = \sum_{h=j}^{m-1} P_h(t), \quad \text{for } j \in \{1, 2, \ldots, m-1\}, \quad t \geq 0, \\
&U^{\geq j}(t) = U^{\geq j}(\mathbf{p}(t)) = \Pr\{\phi(\mathbf{Z}(t)) < j\} = \sum_{h=0}^{j-1} P_h(t), \quad \text{for } j \in \{1, 2, \ldots, m-1\}, \quad t \geq 0, \\
&A^{\geq j}(t) + U^{\geq j}(t) = 1, \quad \text{for } j \in \{1, 2, \ldots, m-1\}, \quad t \geq 0.
\end{aligned} \tag{33}$$

Similarly, the expected performance (32) is a function depending on time:

$$O(t) = \sum_{j=0}^{m-1} O_j P_j(t) = O_0 + \sum_{j=1}^{m-1} (O_j - O_{j-1}) A^{\geq j}(t), \quad t \geq 0, \tag{34}$$

and it describes how the expected performance of an MSS changes over time.

As in the case of the system-state probabilities, all these measures can be computed by combining the state probabilities of the components (or functions defining the state probabilities of the components) via the structure function. Furthermore, one can notice if the system-state probabilities are known, the remaining measures can be simply computed based on them. The same goes for the system availabilities. Therefore efficient computation of one of these two measures plays a key role in analysis of MSSs.

Another measure, which is typically used in analysis of nonrepairable systems, is system reliability. In case of BSSs, this measure is defined at time t as the probability that the system does not fail before or at time t, given it is working at time 0 [8, 9]. For MSSs, this measure can be defined with respect to system state j as the probability that the state of the MSS does not decrease below value j before or at time t, given the MSS is in a state greater than or equal to j at time 0, that is:

$$\begin{aligned}R^{\geq j}(t) &= \Pr\{\forall\tau\in(0,t>:\phi(\mathbf{Z}(\tau))\geq j|\phi(\mathbf{Z}(0))\geq j\}\\ &=\frac{\Pr\{\forall\tau\in<0,t>:\phi(\mathbf{Z}(\tau))\geq j\}}{\Pr\{\phi(\mathbf{Z}(0))\geq j\}}\\ &=\frac{\Pr\{\forall\tau\in<0,t>:\phi(\mathbf{Z}(\tau))\geq j\}}{A^{\geq j}(0)},\quad \text{for } j\in\{1,2,\ldots,m-1\},\quad t\geq 0.\end{aligned} \tag{35}$$

If we assume that the system is in state $m-1$ at time 0, then the following formulae are true:

$$\begin{aligned}P_0(0)=P_1(0)=\cdots=P_{m-2}(0)=0,\quad P_{m-1}(0)=1,\\ A^{\geq j}(0)=1,\quad \text{for } j\in\{1,2,\ldots,m-1\},\end{aligned} \tag{36}$$

what implies the system reliability can be computed as follows:

$$R^{\geq j}(t)=\Pr\{\forall\tau\in<0,t>:\phi(\mathbf{Z}(\tau))\geq j\},\quad \text{for } j\in\{1,2,\ldots,m-1\},\quad t\geq 0. \tag{37}$$

In the next stage, this formula can be transformed based on the conditional probability in the following manner:

$$\begin{aligned}R^{\geq j}(t)&=\Pr\{\phi(\mathbf{Z}(t))\geq j \text{ and } \forall\tau\in<0,t):\phi(\mathbf{Z}(\tau))\geq j\}\\ &=\Pr\{\{\forall\tau\in<0,t):\phi(\mathbf{Z}(\tau))\geq j\}|\{\phi(\mathbf{Z}(t))\geq j\}\}\Pr\{\phi(\mathbf{Z}(t))\geq j\},\\ &\quad \text{for } j\in\{1,2,\ldots,m-1\},\quad t\geq 0.\end{aligned} \tag{38}$$

Because we assume that the system is nonrepairable, event $\{\phi(\mathbf{Z}(t)) \geq j\}$ implies that event $\{\forall \tau \in \langle 0, t) : \phi(\mathbf{Z}(\tau)) \geq j\}$ certainly occurred and, therefore, the next condition holds:

$$\Pr\{\{\forall \tau \in < 0, t) : \phi(\mathbf{Z}(\tau)) \geq j\} | \{\phi(\mathbf{Z}(t)) \geq j\}\} = 1, \quad \text{for } j \in \{1, 2, \dots, m-1\}, \quad t \geq 0. \tag{39}$$

This implies that the system reliability can be computed as follows:

$$R^{\geq j}(t) = \Pr\{\phi(\mathbf{Z}(t)) \geq j\} = A^{\geq j}(t), \quad \text{for } j \in \{1, 2, \dots, m-1\}, \quad t \geq 0. \tag{40}$$

This result means that the reliabilities of an MSS agree with the corresponding availabilities if the MSS is nonrepairable and is in state $m-1$ at time 0. Because of that, it has no practical meaning to distinguish between these two measures in case of nonrepairable systems. In what follows, we will primarily deal with system availabilities.

3 RELIABILITY ANALYSIS OF SERIES-PARALLEL SYSTEMS

Two basic topologies of systems are series and parallel. They can be represented by reliability block diagrams depicted in Fig. 5. In case of MSSs, these topologies can be modeled by various structure functions depending on the properties of the system. In a study by Kolowrocki [28], a series MSS has been defined as a system whose structure function has the following form:

$$\phi_{ser}(\boldsymbol{z}) = \min(z_1, z_2, \dots, z_n), \tag{41}$$

whereas a parallel MSS as a system with the following structure function:

$$\phi_{par}(\boldsymbol{z}) = \max(z_1, z_2, \dots, z_n). \tag{42}$$

These definitions imply that the series connections in a reliability block diagram should be evaluated using min-function while the parallel as max-function.

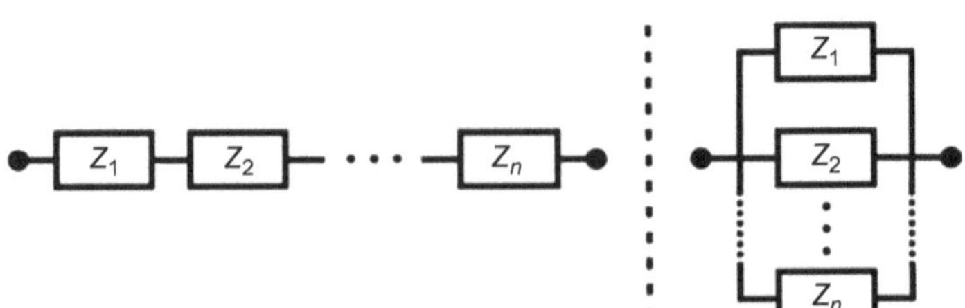

Fig. 5 Reliability block diagrams representing series and parallel MSSs.

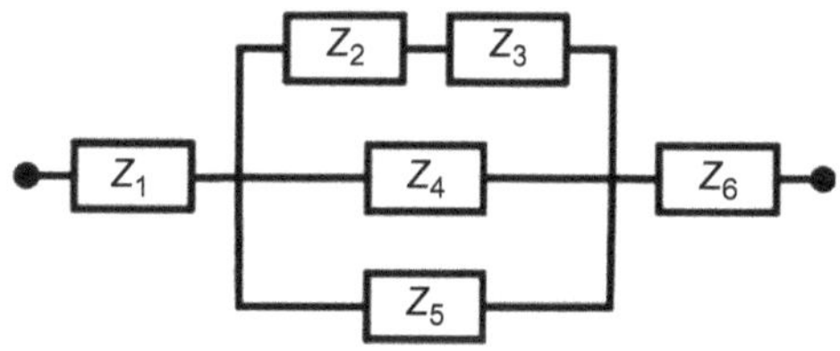

Fig. 6 Example of a series-parallel MSS.

A series-parallel system is a system composed of components from which some are arranged in series and others in parallel. Example of such a system is in Fig. 6. This system is composed of six components and, as one can see, components 2 and 3 are arranged in series, components 4 and 5 together with a group of components 2 and 3 in parallel, and components 1 and 6 together with a group of components 2, 3, 4, and 5 in series again. Based on previous definitions of evaluation of series and parallel connections using min- and max-functions, the structure function of this system can be expressed in the following form:

$$\phi(\boldsymbol{z}) = \min(z_1, \max(\min(z_2, z_3), z_4, z_5), z_6). \tag{43}$$

If we assume the components are independent and if we know their state probabilities (2), we can use this function to define random variable Z modeling behavior of the system depending on the state probabilities of the components:

$$Z = \min(Z_1, \max(\min(Z_2, Z_3), Z_4, Z_5), Z_6), \tag{44}$$

and then we can evaluate reliability of the system using characteristics (28), (30)–(32). However, if we want to evaluate reliability of the system over time using Formulae (29), (33)–(35), we have to know functions (4) defining the state probabilities of the components over time, and we have to transform function (44) into time-dependent function $Z(t)$. In this case, we obtain the following function stochastically modeling behavior of the system over time:

$$Z(t) = \min(Z_1(t), \max(\min(Z_2(t), Z_3(t)), Z_4(t), Z_5(t)), Z_6(t)). \tag{45}$$

If we are able to compute the probability that this function takes a specific value, we can find state probabilities functions (29) and availability function (33) of the system or evaluate how the expected performance of the system changes over time (34). The principal question is how this can be done in an efficient way. A possible solution to this problem is an application of a modular decomposition.

3.1 Modular Decomposition

A modular decomposition is one of the most useful methods for evaluation of systems composed of many components. It is a form of more general divide and conquer paradigm, which is used for design of efficient recursive algorithms. The modular decomposition is based on the idea of dividing the system into modules, their independent evaluation, and then combining the results of the evaluations to evaluate the entire system. The principal task in this method is division of the system into modules. According to Natvig [12], a module of an MSS is a pair ($\mathbf{B}$, $\phi^{\mathbf{B}}$) such that:

$$\phi(\boldsymbol{z}) = \psi\left(\phi^{\mathbf{B}}\left(\boldsymbol{z}^{\mathbf{B}}\right)\boldsymbol{z}^{\mathbf{B}^c}\right), \tag{46}$$

where $\mathbf{B}$ is a set of components that the module is composed of, $\mathbf{B}^c$ is a set of all the system components that are not present in the module (i.e., $\mathbf{B}^c = \mathbf{N} - \mathbf{B}$, where $\mathbf{N}$ is the set of all the system components), $\boldsymbol{z}^{\mathbf{B}}$ is a vector of state variables z_i such that $i \in \mathbf{B}$, $\boldsymbol{z}^{\mathbf{B}^c}$ is a vector of state variables z_i such that $i \in \mathbf{B}^c$, $\phi^{\mathbf{B}}(\boldsymbol{z}^{\mathbf{B}})$ is a function defining the structure of the module (it can be named as the structure function of the module or as a substructure function of the system), that is:

$$\phi^{\mathbf{B}}\left(\boldsymbol{z}^{\mathbf{B}}\right) : \{0, 1, \ldots, m-1\}^{|\mathbf{B}|} \rightarrow \{0, 1, \ldots, m-1\}, \tag{47}$$

$\psi(\phi^{\mathbf{B}}, \boldsymbol{z}^{\mathbf{B}^c})$ is a function combining substructure function $\phi^{\mathbf{B}}(\boldsymbol{z}^{\mathbf{B}})$ with the variables defining states of the components that are not present in the module ($\mathbf{B}$, $\phi^{\mathbf{B}}$) into the system structure function, and $|\cdot|$ denotes size (number of elements) of the argument interpreted as a set. Clearly, in this notation, an MSS can be viewed as a module ($\mathbf{N}$, $\phi^{\mathbf{N}}$), whose function $\phi^{\mathbf{N}}(\boldsymbol{z}^{\mathbf{N}})$ agrees with structure function $\phi(\boldsymbol{z})$ of the system.

Based on Natvig [12], a modular decomposition of an MSS can be viewed as a set of k disjoint modules denoted as ($\mathbf{N}_1$, $\phi^{\mathbf{N}_1}$), ($\mathbf{N}_2$, $\phi^{\mathbf{N}_2}$), …, and ($\mathbf{N}_k$, $\phi^{\mathbf{N}_k}$) together with function $\psi(\phi^{\mathbf{N}_1}, \phi^{\mathbf{N}_2}, \ldots, \phi^{\mathbf{N}_k}, \boldsymbol{z}^{\mathbf{N}_{k+1}})$ such that:

$$\begin{gathered}\phi(\boldsymbol{z}) = \psi\left(\phi^{\mathbf{N}_1}\left(\boldsymbol{z}^{\mathbf{N}_1}\right), \phi^{\mathbf{N}_2}\left(\boldsymbol{z}^{\mathbf{N}_2}\right), \ldots, \phi^{\mathbf{N}_k}\left(\boldsymbol{z}^{\mathbf{N}_k}\right), \boldsymbol{z}^{\mathbf{N}_{k+1}}\right), \\ \text{where } \mathbf{N} = \bigcup_{l=1}^{k+1} \mathbf{N}_l \text{ and } \mathbf{N}_{l_1} \cap \mathbf{N}_{l_2} = \{\} \text{ for any } l_1 \neq l_2.\end{gathered} \tag{48}$$

Clearly, a modular decomposition can be applied recursively to modules of an MSS, then submodules, etc. For this purpose, we will use the following notation in the rest of the chapter: ($\mathbf{N}_l$, $\phi^{\mathbf{N}_l}$)—the l-th module of an MSS, ($\mathbf{N}_{l1,l2}$, $\phi^{\mathbf{N}_{l1,l2}}$)—the l_2-th module of the l_1-th module of an MSS, etc.

3.2 Application of Modular Decomposition in Analysis of Series-Parallel Systems

Series-parallel systems are typical examples of MSSs on which a modular decomposition can be applied recursively until we obtain modules composed of components with just series or parallel topology. For example, the system depicted in Fig. 6 can be decomposed in the following way:

$$\mathbf{N}=\{1,2,3,4,5,6\}\ \text{and}\ \phi(z_1,z_2,z_3,z_4,z_5,z_6)=\psi\left(\phi^{\mathbf{N}_1},z_1,z_6\right) \\ =\min\left(\phi^{\mathbf{N}_1},z_1,z_6\right),$$

where:

$$\mathbf{N}_1=\{2,3,4,5\}\ \text{and}\ \phi^{\mathbf{N}_1}(z_2,z_3,z_4,z_5)=\max\left(\phi^{\mathbf{N}_{1,1}},z_4,z_5\right),$$

where:

$$\mathbf{N}_{1,1}=\{2,3\}\ \text{and}\ \phi^{\mathbf{N}_{1,1}}(z_2,z_3)=\min(z_2z_3).$$

This example illustrates straightness of a recursive application of a modular decomposition on series-parallel systems. However, the question is how it allows us to evaluate reliability of the system, that is, which reliability characteristic, or a function describing the characteristic over time, of a series-parallel system can be calculated based on it. For this purpose, let us firstly consider a series system of n components with structure function $\phi_{ser}(\boldsymbol{z})$ (41) at a fixed time. The probability that the system is in state j, for $j \in \{0, 1, \ldots, m-1\}$, agrees with the probability that exactly one component is in state j and all other components are in a state greater than or equal to j, that is:

$$P_{ser,j}=\Pr\{\phi_{ser}(\mathbf{Z})=j\}=\Pr\left\{\bigvee_{i=1}^{n}\left\{\{Z_i=j\}\bigwedge_{\substack{l=1\\l\neq i}}^{n}\{Z_l\geq j\}\right\}\right\}, \tag{49}$$

where $\wedge$ denotes logical conjunction and $\vee$ agrees with logical disjunction. Computation of this probability can be quite complicated because $\{Z_i=j\}\wedge_{\substack{l=1\\l\neq i}}^{n}\{Z_l\geq j\}$ for $i=1, 2, \ldots, n$ are not mutually exclusive events. To simplify this calculation, we can transform the sum of products from Formula (49) into a sum of disjoint products [29]. This results in the following formula:

$$P_{ser,j}=\Pr\left\{\bigvee_{i=1}^{n}\left\{\{Z_i=j\}\bigwedge_{l=1}^{i-1}\{Z_l>j\}\bigwedge_{l=i+1}^{n}\{Z_l\geq j\}\right\}\right\}. \tag{50}$$

In this case, events $\{Z_i = j\} \wedge_{l=1}^{i-1} \{Z_l > j\} \wedge_{l=i+1}^{n} \{Z_l \geq j\}$ for $i = 1, 2, \ldots, n$ are mutually exclusive and, therefore, we can write:

$$P_{ser,j} = \sum_{i=1}^{n} \Pr\left\{ \{Z_i = j\} \bigwedge_{l=1}^{i-1} \{Z_l > j\} \bigwedge_{l=i+1}^{n} \{Z_l \geq j\} \right\}. \tag{51}$$

Specially, if $j = m - 1$, this formula can be transformed into the following form:

$$P_{ser,m-1} = \Pr\left\{ \bigwedge_{i=1}^{n} \{Z_i = m-1\} \right\}. \tag{52}$$

Using the assumption that the components are independent, Formulae (51), (52) can be rewritten as follows:

$$P_{ser,j} = \begin{cases} \sum_{i=1}^{n} \left(p_{i,j} \left(\prod_{l=1}^{i-1} \sum_{s=j+1}^{m-1} p_{l,s} \right) \left(\prod_{l=i+1}^{n} \sum_{s=j}^{m-1} p_{l,s} \right) \right) & \text{if } j \in \{0, 1, \ldots, m-2\} \\ \prod_{i=1}^{n} p_{i,m-1} & \text{if } j = m-1 \end{cases}. \tag{53}$$

This formula implies that computation of the probability that a series MSS is in state $m - 1$ requires the following number of elementary operations (NEO) (additions, subtractions, and multiplications):

$$\text{NEO}_{(53)}\left(P_{ser,m-1}\right) = n - 1, \tag{54}$$

where the subscript of NEO (i.e., (53)) denotes the investigated formula used for computation of the characteristic placed in the parentheses (i.e., $P_{ser,m-1}$). In case of computing the probability that the system is in state j, for $j = 0, 1, \ldots, m-2$, the next NEO has to be executed:

$$\text{NEO}_{(53)}\left(P_{ser,j}\right) = \sum_{i=1}^{n} (2 + ((i-1)(m-1-j-1) + (i-1-1)) + ((n-i)(m-1-j) + (n-i-1))) + (n-1), \tag{55}$$

where term 2 agrees with 2 multiplications between terms $p_{i,j}$, $\prod_{l=1}^{i-1} \sum_{s=j+1}^{m-1} p_{l,s}$, and $\prod_{l=i+1}^{n} \sum_{s=j}^{m-1} p_{l,s}$ in Formula (53), expression $(i-1)(m-1-j-1)$ with number of additions that are performed in evaluation of expression $\prod_{l=1}^{i-1} \sum_{s=j+1}^{m-1} p_{l,s}$ in Formula (53), expression $(i-1-1)$ with number of multiplications executed during evaluation of expression

$\prod_{l=1}^{i-1}\sum_{s=j+1}^{m-1}p_{l,s}$ in Formula (53), expressions $(n-i)(m-1-j)$ and $(n-i-1)$ with number of additions and multiplications, respectively, performed during evaluation of expression $\prod_{l=i+1}^{n}\sum_{s=j}^{m-1}p_{l,s}$ in Formula (53), and $(n-1)$ with number of additions of expressions of the form of $p_{i,j}\left(\prod_{l=1}^{i-1}\sum_{s=j+1}^{m-1}p_{l,s}\right)\left(\prod_{l=i+1}^{n}\sum_{s=j}^{m-1}p_{l,s}\right)$ in Formula (53).

Formula (55) can be rewritten in the following way:

$$\mathrm{NEO}_{(53)}\left(P_{ser,j}\right)=\sum_{i=1}^{n}(n-1+((i-1)(m-1-j-1))$$
$$+((n-i)(m-1-j)))+(n-1)=n(n-1)\left(m-j-\frac{1}{2}\right)+(n-1), \quad (56)$$

which implies that about $n(n-1)(m-j-0.5)$ elementary operations have to be performed to compute the probability that the system is in state j, for $j=0, 1, \ldots, m-2$.

Using Formula (30), Formula (53) can also be used to compute availability of the series system composed of n components with respect to system state j. Formulae (30), (54), and (56) imply that the following NEO has to be performed to compute the availability for $j=1, 2, \ldots, m-1$:

$$\mathrm{NEO}_{(30),(53)}\left(A_{ser}^{\geq j}\right)=\sum_{h=j}^{m-2}\left(n(n-1)\left(m-h-\frac{1}{2}\right)+n-1\right)+(n-1)+(m-j-1), \quad (57)$$

where the sum expression represents NEO needed to compute the probability that the system is in a state $h\in\{j, j+1, \ldots, m-2\}$, expression $(n-1)$ agrees with NEO needed to calculate the probability that the system is in state $m-1$, and expression $(m-j-1)$ with number of additions in Formula (30).

Formula (57) can be rewritten in the following manner:

$$\mathrm{NEO}_{(30),(53)}\left(A_{ser}^{\geq j}\right)=n(n-1)\frac{(m-j)^2-1}{2}+n(m-j)-1. \quad (58)$$

According to this result, about $n(n-1)((m-j)^2-1)/2$ elementary operations have to be executed to compute availability of the series system using the method based on direct computation of system-state probabilities using Formula (53).

Now let us consider computation of the probability that the series system is in a state greater than or equal to j, that is, direct computation of availability

$A^{\geq j}$ of the system. Clearly, this agrees with the probability that all the components are in a state greater than or equal to j and, therefore, we can write:

$$A_{ser}^{\geq j} = \Pr\{\phi_{ser}(\mathbf{Z}) \geq j\} = \Pr\left\{ \bigwedge_{i=1}^{n} \{Z_i \geq j\} \right\}. \tag{59}$$

Because we assume that the components are independent, we can write:

$$A_{ser}^{\geq j} = \prod_{i=1}^{n} \Pr\{Z_i \geq j\} = \prod_{i=1}^{n} \sum_{s=j}^{m-1} p_{i,s}. \tag{60}$$

As one can see, this computation requires $n(m - j - 1)$ additions and $(n - 1)$ multiplications. So, the total NEO is as follows:

$$\mathrm{NEO}_{(60)}\left(A_{ser}^{\geq j}\right) = n(m-j-1) + (n-1) = n(m-j) - 1. \tag{61}$$

This is significantly less than Formula (58). Furthermore, definition (30) suggests that the probability that an MSS is in state j can be computed using the following formula:

$$P_j = \Pr\{\phi(\mathbf{Z}) = j\} = \begin{cases} 1 - A^{\geq 1} & \text{if } j = 0 \\ A^{\geq j} - A^{\geq (j+1)} & \text{if } j \in \{1, 2, \ldots, m-2\}, \\ A^{\geq m-1} & \text{if } j = m-1 \end{cases}$$

$$j = 0, 1, \ldots, m-1, \tag{62}$$

which implies (based on Formula 61) this method requires the following NEO to compute the probability of the system being in state j:

$$\mathrm{NEO}_{(62),(60)}\left(P_{ser,j}\right) = \begin{cases} n(m-1) & \text{if } j = 0 \\ n(2m - 2j - 1) - 1 & \text{if } j \in \{1, 2, \ldots, m-2\}. \\ n - 1 & \text{if } j = m-1 \end{cases} \tag{63}$$

These numbers are also better than those derived in Formula (56).

Comparing results presented in Formulae (63), (61) with those obtained in Formulae (56), (58), we can conclude that evaluation of state probabilities and availabilities of a series MSS based on direct computation of system availabilities (Formulae 60, 62) is less time-consuming than the evaluation based on direct computation of system-state probabilities (Formulae 30, 53). Therefore, we will prefer this method in the rest of the chapter.

Using the similar procedure as previously discussed, we can prove that the method based on direct computation of system availability is also better in evaluation of a parallel system consisting of n components with structure function $\phi_{par}(\mathbf{z})$ (42). In this case, the availability with respect to system state j, for $j \in \{1, 2, \ldots, m-1\}$, can be computed using the next formula:

$$\begin{aligned} A_{par}^{\geq j} &= 1 - U_{par}^{\geq j} = 1 - \Pr\left\{\phi_{par}(\mathbf{Z}) < j\right\} = 1 - \Pr\left\{\bigwedge_{i=1}^{n} \{Z_i < j\}\right\} \\ &= 1 - \prod_{i=1}^{n} \Pr\{Z_i < j\} = 1 - \prod_{i=1}^{n} \sum_{s=0}^{j-1} p_{i,s}. \end{aligned} \quad (64)$$

As one can see, this formula requires executing $n(j-1)$ additions, $(n-1)$ multiplications, and 1 subtraction. So, totally, the following NEO is needed:

$$\text{NEO}_{(64)}\left(A_{par}^{\geq j}\right) = n(j-1) + (n-1) + 1 = nj. \quad (65)$$

Using Formula (62), Formula (64) can also be used to compute the probability that the parallel system is in state j. Based on Formula (65), the method will require the next NEO:

$$\text{NEO}_{(62),(64)}\left(P_{par,j}\right) = \begin{cases} n+1 & \text{if } j = 0 \\ n(2j+1)+1 & \text{if } j \in \{1, 2, \ldots, m-2\} \\ n(m-1) & \text{if } j = m-1 \end{cases}. \quad (66)$$

The power of variables n, m, and j in Formulae (65), (66) defining NEO to analyze a parallel MSS is same as in the case of Formulae (61), (63) defining NEO to perform reliability analysis of a series MSS. Therefore running time of algorithms computing availability or state probabilities using Formulae (60), (62), and (64) will be similar for series and parallel MSSs composed of the same number of components and modeled using the same number of states.

Obviously, Formulae (60) and (64) can be applied not only to a series or parallel MSS but also to a module of an MSS composed of components arranged in series or parallel. In such a case, the availability of module $(\mathbf{B}, \phi^{\mathbf{B}})$ will be denoted as $A_{(\mathbf{B},\phi^{\mathbf{B}})}{}^{\geq j}$. Furthermore, the module might not be composed only from components but can also contain one or more other modules. In this case, the corresponding random variable Z_i in Formulae (60) and (64), which models behavior of a component, has to be replaced by a random variable modeling behavior of the module. Therefore if we want to compute availability of a system or a module that contains module $(\mathbf{C}, \phi^{\mathbf{C}})$, we use random variable denoted as $Z_{(\mathbf{C},\phi^{\mathbf{C}})}$ to model behavior of this module in Formulae (60) or (64) instead of the corresponding random variable Z_i. These facts, together with a modular decomposition, allow us to develop a fast recursive algorithm for computation of availability with respect to state j of a series-parallel system. This algorithm can be defined in the following two steps:

1. Recursively decompose an MSS into modules until you obtain modules composed of components connected either in a series or parallel way.
2. Based on Formulae (60), (64), and a bottom-up approach (from basic modules to the entire system), compute the availabilities of the modules.

When this algorithm finishes, we obtain availability of the series-parallel system with respect to system state j. If we repeat this algorithm for all possible values of j, that is, for $j=1, 2, \ldots, m-1$, we can find the system-state probabilities using Formula (62).

For illustration, let us return to the system depicted in Fig. 6. According to the modular decomposition performed at the beginning of this section, the most nested module is $(\mathbf{N}_{1,1}, \phi^{\mathbf{N}_{1,1}})=(\{2,3\}, \min(z_2, z_3))$. Because this module is composed of two components arranged in series, we can use Formula (60) in the following manner to compute availability of this module:

$$A^{\geq j}_{\left(\mathbf{N}_{1,1},\phi^{\mathbf{N}_{1,1}}\right)} = \Pr\{Z_2 \geq j\}\Pr\{Z_3 \geq j\}. \tag{67}$$

After we obtain this availability, we can compute the availability with respect to system state j of module $(\mathbf{N}_1, \phi^{\mathbf{N}_1})=(\{2,3,4,5\}, \max(\phi^{\mathbf{N}_1}, z_4, z_5))$. Because its elements (components 4 and 5, and module $(\mathbf{N}_{1,1}, \phi^{\mathbf{N}_{1,1}})$) are arranged parallel, we can use Formula (64) to compute this availability:

$$\begin{aligned} A^{\geq j}_{\left(\mathbf{N}_1,\phi^{\mathbf{N}_1}\right)} &= 1-\Pr\left\{Z_{\left(\mathbf{N}_{1,1},\phi^{\mathbf{N}_{1,1}}\right)} < j\right\}\Pr\{Z_4 < j\}\Pr\{Z_5 < j\} \\ &= 1-\left(1-A^{\geq j}_{\left(\mathbf{N}_{1,1},\phi^{\mathbf{N}_{1,1}}\right)}\right)\Pr\{Z_4 < j\}\Pr\{Z_5 < j\}. \end{aligned} \tag{68}$$

Finally, availability of the entire system depicted in Fig. 6 can be computed using Formula (60) as the system can be viewed as a series arrangement of components 1 and 6 together with module $(\mathbf{N}_1, \phi^{\mathbf{N}_1})$, that is:

$$\begin{aligned} A^{\geq j} &= \Pr\left\{Z_{\left(\mathbf{N}_1,\phi^{\mathbf{N}_1}\right)} \geq j\right\}\Pr\{Z_1 \geq j\}\Pr\{Z_6 \geq j\} \\ &= A^{\geq j}_{\left(\mathbf{N}_1,\phi^{\mathbf{N}_1}\right)}\Pr\{Z_1 \geq j\}\Pr\{Z_6 \geq j\}. \end{aligned} \tag{69}$$

Clearly, if we replace random variables Z_i, for $i=1, 2, \ldots, 6$, modeling behavior of the components and random variables $Z_{\left(\mathbf{N}_{1,1},\phi^{\mathbf{N}_{1,1}}\right)}$ and $Z_{\left(\mathbf{N}_1,\phi^{\mathbf{N}_1}\right)}$ modeling behavior of the modules by time-dependent functions $Z_i(t)$, $Z_{\left(\mathbf{N}_{1,1},\phi^{\mathbf{N}_{1,1}}\right)}(t)$, and $Z_{\left(\mathbf{N}_1,\phi^{\mathbf{N}_1}\right)}(t)$ modeling stochastic behavior of the elements of the system over time, we can use all these formulae also in time-dependent analysis of the system.

4 CASE STUDY: RELIABILITY ANALYSIS OF DISTRIBUTED COMPUTING SYSTEM

In this section, we show how the formulae previously developed can be used in reliability analysis of a distributed computing system depicted in Fig. 7. This system is composed of two powerful computation units that need to communicate with each other during computation. The communication is done using the dual communication network in which two lines are used to connect two network nodes. The network itself is composed of six nodes (each node contains two switches due to dual communication) and lines between them (each line is composed of two sublines working in a parallel way). In this example, we will assume that the lines are perfectly reliable, but the network nodes composed of switches are unreliable, that is, their performance deteriorates over time. Furthermore, we will assume that the switches are nonrepairable. From a reliability point of view, this implies the network can be viewed as a system of six components that correspond to the nodes of the network.

As we previously mentioned, the analyzed network is a dual communication network. Because of that, each node contains two switches (Fig. 8). (We will assume that the switches are independent.) Every switch used in the network can transmit data by 10 Gbit/s if it is working or 0 Gbit/s if it fails. This implies that a node composed of two switches can transmit data by

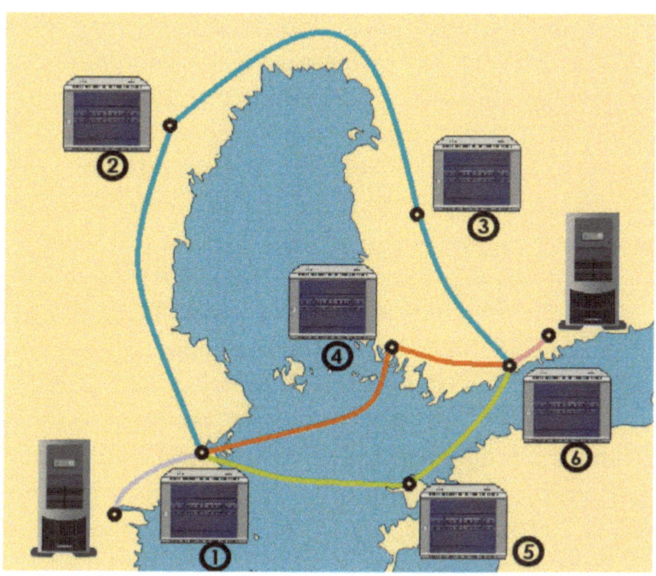

Fig. 7 Distributed computing system and its components (network nodes).

Fig. 8 Internal view on a communication node in the network.

20 Gbit/s (both switches are working), 10 Gbit/s (just one switch is working), or 0 Gbit/s (no switch in the node is working). Based on these assumptions, a network node can be modeled as a three-state component with states 0 (transmission speed 0 Gbit/s), 1 (transmission speed 10 Gbit/s), and 2 (transmission speed 20 Gbit/s).

Communication starts when sending computation unit sends information to the nearest network node. Then the information can be sent through one of three communication lines to the network node nearest to the receiver. The first communication line is a submarine communication cable with one network node, the second is a combination of a submarine and a landline communication cable with one network node, and the last one is a landline communication cable with two network nodes. Lastly, the information is sent to the receiving computation unit. From this, we can conclude that the connection between communication units can be established if they are both connected to the relevant network nodes and at least one of the three communication lines, that is, nodes on these lines, are functioning. It is important to note that all data, from sending computation unit to the receiving one, are sent over the network using a path that has a maximal transmission speed. If more paths satisfy this condition, then one of them is randomly selected. Clearly, a transmission speed at which a specific path operates is determined by a node at the path whose transmission speed is minimal. These characteristics imply that the network can be in one of the next three states: 0 (no data can be transferred, i.e., the transmission speed is 0 Gbit/s), 1 (data can be transferred at half speed, i.e., 10 Gbit/s), or 2 (data can be transferred at full speed, i.e., 20 Gbit/s).

Based on the previous description of the distributed computing system, its topology can be expressed as the reliability block diagram depicted in Fig. 6, whose elements correspond to the nodes of the network. Therefore, the structure function of this system agrees with Formula (43), that is:

$$\phi(\boldsymbol{z}) = \min(z_1, \max(\min(z_2, z_3), z_4, z_5), z_6),$$

where $z_i \in \{0,1,2\}$ for $i=1, 2, \ldots, 6$, and $\phi(\boldsymbol{z}) \in \{0,1,2\}$ for all $\boldsymbol{z} \in \{0,1,2\}^6$. Based on the modular decomposition performed in Section 3.2 and based on Formulae (67)–(69) defining system availability, this function can be transformed into the following system of time-dependent functions defining availability of the distributed computing system over time:

$$\begin{aligned}
A^{\geq j}_{(\mathbf{N}_{1,1}, \phi^{\mathbf{N}_{1,1}})}(t) &= \sum_{s=j}^{2} p_{2,s}(t) \sum_{s=j}^{2} p_{3,s}(t), \\
A^{\geq j}_{(\mathbf{N}_1, \phi^{\mathbf{N}_1})}(t) &= 1 - \left(1 - A^{\geq j}_{(\mathbf{N}_{1,1}, \phi^{\mathbf{N}_{1,1}})}(t)\right) \sum_{s=0}^{j-1} p_{4,s}(t) \sum_{s=0}^{j-1} p_{5,s}(t), \\
A^{\geq j}(t) &= A^{\geq j}_{(\mathbf{N}_1, \phi^{\mathbf{N}_1})}(t) \sum_{s=j}^{2} p_{1,s}(t) \sum_{s=j}^{2} p_{6,s}(t), \\
&\text{for } j \in \{1, 2\}.
\end{aligned} \tag{70}$$

Formula (70) allows us to evaluate how availability of the distributed computing system with respect to a specific system state changes over time. Based on Formulae (62) and (34), it can also be used to find the state probabilities of the system and to evaluate the expected performance of the system over time. However, all these evaluations require knowledge of the time-dependent functions $p_{i,s}(t)$ (for $i=1, 2, \ldots, 6$, $s=0, 1, 2$, and $t \geq 0$), which capture evolution of the state probabilities of the components (network nodes) over time. To find these probabilities, let us develop a time-dependent representation of network node i.

According to the description of the distributed computing system at the beginning of this section, the node is composed of two independent switches. The node can be at time t in one of three states depending on the number of switches working at that time. This implies that the following relationships can be defined:

$$\begin{aligned}
p_{i,2}(t) &= p_{i^{(1)},1}(t) p_{i^{(2)},1}(t), \\
p_{i,1}(t) &= p_{i^{(1)},0}(t) p_{i^{(2)},1}(t) + p_{i^{(1)},1}(t) p_{i^{(2)},0}(t), \\
p_{i,0}(t) &= p_{i^{(1)},0}(t) p_{i^{(2)},0}(t),
\end{aligned} \tag{71}$$

where $p_{i(k),s}(t)$ denotes the probability that switch k, for $k \in \{1,2\}$, in the network node i is in state s, for $s \in \{0,1\}$, at time t. Now we need to find these probabilities for the switches in the network node.

Let each network node (a component of the analyzed system) contain two switches of the same type, and let the switches placed in the nodes be those shown in Table 1. According to data obtained from Siemens web site [30], these switches are characterized by Mean Time To Failures (MTTFs) presented in Table 1. Based on these data, we can compute conditional failure rate $\lambda_{i(k)}$ as follows [8, 9]:

$$\lambda_{i(k)} = \frac{1}{\text{MTTF}_i}. \tag{72}$$

These data are presented in the last column of Table 1. Please note each node is composed of two switches of the same type. So, it has no meaning to distinguish between $\lambda_{i(1)}$ and $\lambda_{i(2)}$. Therefore in what follows, λ_i will denote conditional failure rate of one switch in node i of the network.

According to Formula (18), transition rate $a_{i^*,1\to 0}(t)$ of a binary-state component i^* agrees with its conditional failure rate. This implies that transition rates from state 1 to state 0 are equal to λ_i for each switch in network node i. Because we assume that the switches are nonrepairable, the transition rates from state 0 to state 1 have to be equal to 0 for all switches. Based on these facts, we can obtain transition matrix (19), which models behavior of one switch in node i, that is:

$$\mathbf{A}_{i(k)}(t) = \begin{pmatrix} 0 & 0 \\ \lambda_i & -\lambda_i \end{pmatrix}, \text{for } k \in \{1,2\}, \quad t \geq 0.$$

Table 1 Characteristics of the switches used in the distributed computing system

Component (network node i)	Component data: Switch type	MTTF$_i$ (years)	λ_i (years^{-1})
1	RUGGEDCOM M969	97.1	0.0103
2	RUGGEDCOM M2100	50.8	0.0197
3	RUGGEDCOM RSG2300P	77.9	0.0128
4	RUGGEDCOM RSG920P	14.3	0.0699
5	RUGGEDCOM RS920L	40.1	0.0249
6	RUGGEDCOM RS416	92.8	0.0108

Because the system of differential equations (20) obtained based on this matrix has solution (23), we can write:

$$\left.\begin{array}{l} p_{i^{(k)},0}(t) = 1 - e^{-\lambda_i t} \\ p_{i^{(k)},1}(t) = e^{-\lambda_i t} \end{array}\right., \quad \text{for } k \in \{1, 2\}, \quad t \geq 0. \tag{73}$$

After substituting these formulae into Formula (71), we obtain the following state probabilities for network node i of the distributed computing system:

$$\begin{aligned} p_{i,2}(t) &= e^{-2\lambda_i t}, \\ p_{i,1}(t) &= 2e^{-\lambda_i t}\left(1 - e^{-\lambda_i t}\right), \\ p_{i,0}(t) &= \left(1 - e^{-\lambda_i t}\right)^2, \\ & t \geq 0. \end{aligned} \tag{74}$$

Formula (74) can be substituted into Formula (70). This allows us to obtain functions presenting availabilities of the system over time. Because we assume that no maintenance is performed, according to Formula (40), these functions also agree with reliability functions of the system.

The availability (reliability) functions of the distributed computing system are depicted in Fig. 9. Because the system has three performance levels, its availability can be computed with respect to system state 2 (the *green dashed curve* in Fig. 9) or system state 1 (the *blue solid curve* in Fig. 9). As one can see, the availability with respect to system state 2 decreases relatively fast, and after 2.5 years, its value is less than 0.9 (it is about 0.8953). This

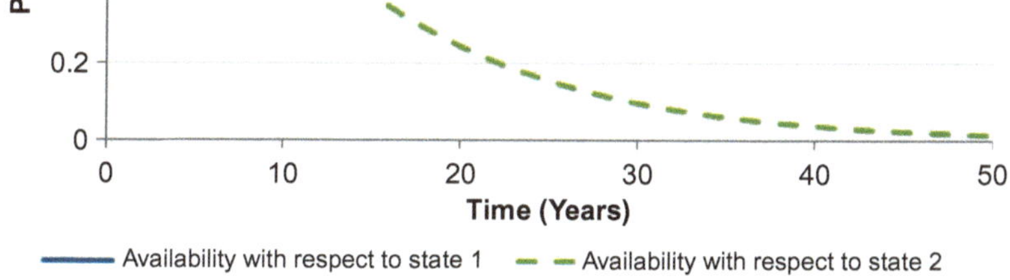

Fig. 9 Changes in the availabilities of the distributed computing system over time.

implies there is a big chance that, after 2.5 years, the system will be not able to transmit data at speed 20 Gbit/s. If we look at the *blue solid line* in Fig. 9, then we can see that the availability of the system with respect to state 1 decreases significantly slower than the availability with respect to system state 2. More precisely, the availability will decrease below value 0.9 after more than 21.5 years, so the system should be able to transmit data at speed 10 Gbit/s significantly longer than at speed 20 Gbit/s.

According to Formula (70), the system availability with respect to state j, for $j = 1, 2$, can be obtained as a product of availabilities with respect to state j of module (subnetwork) $(\mathbf{N}_1, \phi^{\mathbf{N}_1})$ and components (network nodes) 1 and 6 of the entire network. This is illustrated using Figs. 9–12. As one can see,

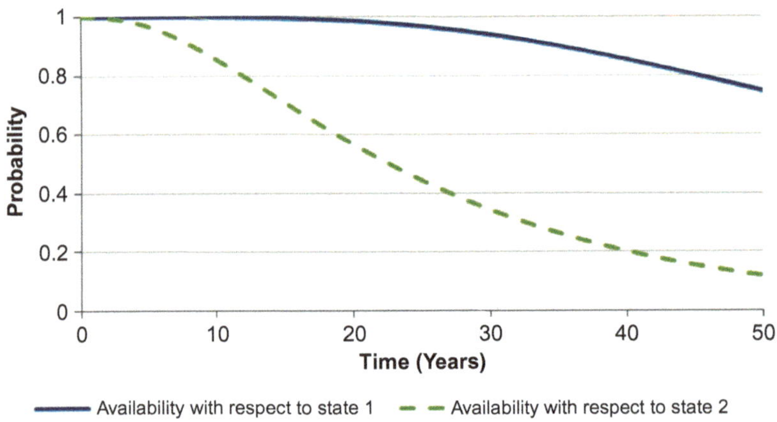

Fig. 10 Changes in the availabilities of module $(\mathbf{N}_1, \phi^{\mathbf{N}1})$ of the system over time.

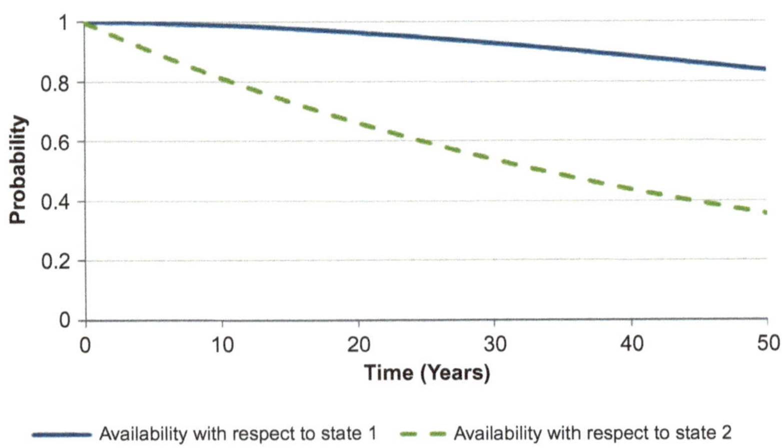

Fig. 11 Changes in the availabilities of component 1 of the system over time.

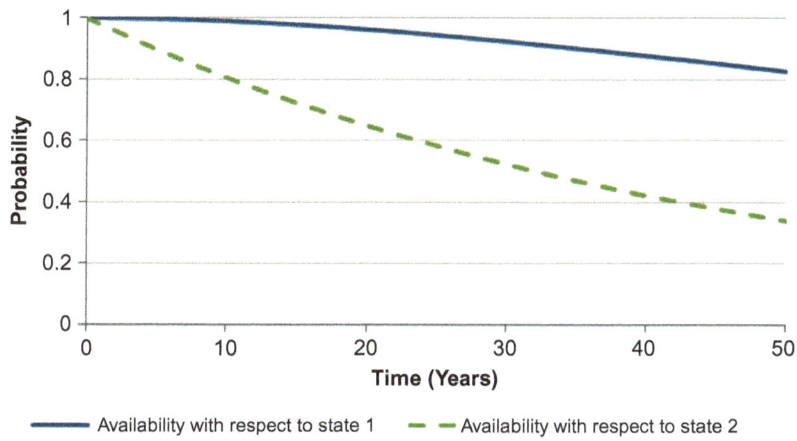

Fig. 12 Changes in the availabilities of component 6 of the system over time.

availabilities of the subnetwork (Fig. 10) are worse than availabilities of network nodes 1 and 6 (Figs. 11 and 12). If we multiply availabilities with respect to state j of these two components together with availability with respect to state j of the subnetwork, we obtain availability with respect to state j of the entire system (Fig. 9). This illustrates how a modular decomposition can be used to find the availability functions of an MSS.

Knowledge of the availabilities functions of the system allows us to find system-state probabilities (Formula 62). Evolution of these probabilities over time is shown in Fig. 13. As one can see, in the first phase of operation of the distributed computing system, the probability that the system is in state 1

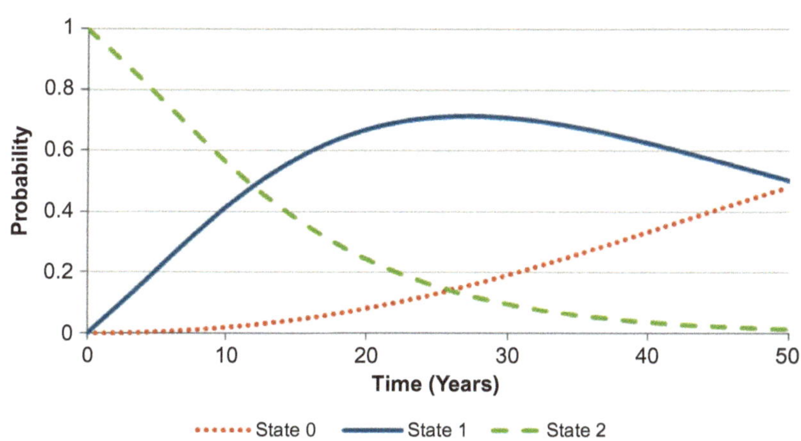

Fig. 13 Changes in the state probabilities of the distributed computing system over time.

(the *blue solid curve*) increases at the expense of the probability that the system is in state 2 (the *green dashed curve*). In this phase, the probability that the system is not able to transmit any data (state 0 represented by the *red dotted curve*) increases very slowly. After 20 years, the probability of this state begins growing faster because the probability that the system will be in state 1 begins decreasing. Based on these functions we can conclude similar results as in the case of the availability functions (Fig. 9), that is, the probability that the system will be able to transmit data at speed 20 Gbit/s decreases after 2.5 years below 0.9, and the probability that the system will not be able to transmit any data exceed value 0.1 after 21.5 years.

Finally, functions representing changes in system availabilities or system-state probabilities over time can be used to find how the expected performance of the system changes over time (Formula 34). In the case of the distributed computing system, the performance agrees with transmission speeds at which data from one computing unit are transmitted over the network to another computing unit. These speeds are 20 (state 2), 10 (state 1), and 0 Gbit/s (state 0). If we combine these speeds with the system availabilities or system-state probabilities functions, we find how the expected transmission speed changes over time. This is presented in Fig. 14. As one can see, in the first phase, the expected speed decreases about 5 Gbit/s in each 10 years. However, when the system lifetime exceeds 20 years, then the expected performance decreases much slower. According to the graph shown in Fig. 14, if the expected transmission speed should not decrease, for example, below 15 Gbit/s, the maintenance of the system has to be performed not later than after 10 years of its nonstop operation.

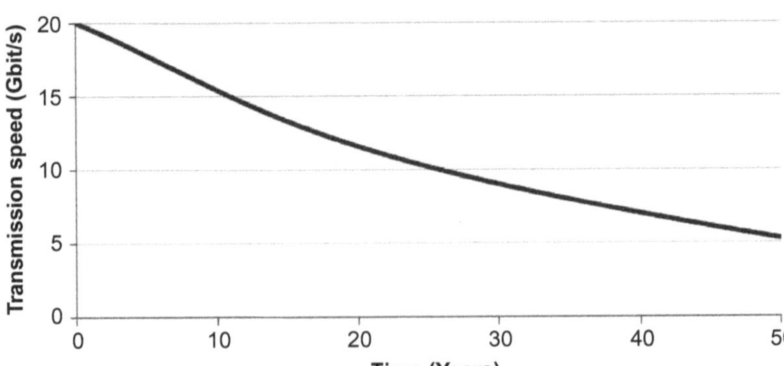

Fig. 14 Decrease in the expected performance (transmission speed) of the distributed computing system over time.

5 CONCLUSION

Evaluation of reliability of MSSs is a challenging task. It includes a lot of steps from which development of a model of the system and computation of global reliability characteristics of the system, such as system-state probabilities or system availabilities, are the essential ones. Knowledge of these and other measures can be used to investigate importance of the components, plan system maintenance, or optimize reliability or availability of the system. This implies that efficient calculations of these and similar measures play a key role in reliability analysis of systems composed of many components. In this chapter, we showed how static and time-dependent versions of these measures can be computed based on the knowledge of system topology and based on some basic information about system components.

MSSs can have different topologies. The topology of a system is captured in the structure function of the system. This function is independent of time. However, as we showed in Section 2, if we assume that the system components are independent, then the structure function can be used to define a stochastic process modeling behavior of the system over time. This process can be obtained by composing the stochastic processes modeling behavior of the system components using the structure function. This allows us to find global reliability characteristics of the entire system as a simple composition of the characteristics of individual system components. This was illustrated in Section 3, where we showed how availabilities and state probabilities of a series-parallel system can be obtained based on the knowledge of the state probabilities of the components and the system structure function without exact knowledge of the stochastic process describing behavior of the entire system.

Another important problem addressed in Section 3 was efficient computation of the system-state probabilities and availabilities at a fixed time. This issue was solved using a modular decomposition. Based on it, we were able to develop the simple fast recursive algorithm for computation of the system availability. Furthermore, in case of series-parallel MSSs, we showed that a modular decomposition is more useful for direct computation of system availabilities than for direct calculation of the system-state probabilities. Knowledge of the system availabilities (in static or time-dependent form) can then be used to find the state probabilities (Formula 62) and the expected performance of the system (Formulae 32, 34) or, in case of nonrepairable systems, system reliability functions (Formula 40). All facts and results presented in Sections 2 and 3 were used in Section 4 to find time-dependent availability and state probabilities functions of the distributed computing system, and to evaluate its expected performance over time.

ACKNOWLEDGMENTS

This work was partly supported by the grants of VEGA 1/0038/16, VEGA 1/0354/17, and APVV SK-FR-2017-003.

REFERENCES

[1] M.R. Choudhury, K. Mohanram, Reliability analysis of logic circuits, IEEE Trans. Comput. Aided Des. Integr. Circuits Syst. 28 (2009) 392–405.
[2] M. Kvassay, E. Zaitseva, V. Levashenko, J. Kostolny, Reliability analysis of multiple-outputs logic circuits based on structure function approach, IEEE Trans. Comput. Aided Des. Integr. Circuits Syst. 36 (2017) 398–411.
[3] L. Haarla, U. Pulkkinen, M. Koskinen, J. Jyrinsalo, A method for analysing the reliability of a transmission grid, Reliab. Eng. Syst. Saf. 93 (2008) 277–287.
[4] P. Praks, V. Kopustinskas, M. Masera, Probabilistic modelling of security of supply in gas networks and evaluation of new infrastructure, Reliab. Eng. Syst. Saf. 144 (2015) 254–264.
[5] E. Zaitseva, in: Reliability analysis methods for healthcare system, 3rd Conference on Human System Interactions (HSI), 2010, IEEE, 2010, pp. 211–216.
[6] E. Zaitseva, V. Levashenko, M. Rusin, in: Reliability analysis of healthcare system, 2011 Federated Conference on Computer Science and Information Systems, FedCSIS 2011, IEEE, 2011, pp. 169–175.
[7] E. Zio, Reliability engineering: old problems and new challenges, Reliab. Eng. Syst. Saf. 94 (2009) 125–141.
[8] M. Rausand, A. Høyland, System Reliability Theory, John Wiley & Sons, Inc., Hoboken, NJ, 2004.
[9] A. Birolini, Reliability Engineering. Theory and Practice, Springer, Berlin, 2017.
[10] A. Lisnianski, G. Levitin, Multi-state System Reliability. Assessment, Optimization and Applications, World Scientific, Singapore, 2003.
[11] A. Lisnianski, I. Frenkel, Y. Ding, Multi-State System Reliability Analysis and Optimization for Engineers and Industrial Managers, Springer-Verlag London Ltd., London, 2010.
[12] B. Natvig, Multistate Systems Reliability Theory with Applications, John Wiley & Sons, Ltd., Chichester, 2011.
[13] M. Kvassay, E. Zaitseva, Topological analysis of multi-state systems based on direct partial logic derivatives, in: A. Lisnianski, I. Frenkel, A. Karagrigoriou (Eds.), Recent Advances in Multi-State Systems Reliability, Springer International Publishing, Cham, 2018, pp. 265–281.
[14] Z.W. Birnbaum, On the importance of different components in a multicomponent system, in: P.R. Krishnaiah (Ed.), Multivariate Analysis, Academic Press, New York, NY, 1969, pp. 581–592.
[15] S. Wu, Joint importance of multistate systems, Comput. Ind. Eng. 49 (2005) 63–75.
[16] W. Kuo, X. Zhu, Importance Measures in Reliability, Risk, and Optimization: Principles and Applications, Wiley, Chichester, 2012.
[17] D.A. Butler, A complete importance ranking for components of binary coherent systems, with extensions to multi-state systems, Nav. Res. Log. Q. 26 (1979) 565–578.

[18] W.S. Griffith, Multistate reliability models, J. Appl. Probab. 17 (1980) 735–744.
[19] G. Levitin, L. Podofillini, E. Zio, Generalised importance measures for multi-state elements based on performance level restrictions, Reliab. Eng. Syst. Saf. 82 (2003) 287–298.
[20] E. Zaitseva, V. Levashenko, J. Kostolny, Importance analysis based on logical differential calculus and binary decision diagram, Reliab. Eng. Syst. Saf. 138 (2015) 135–144.
[21] M. Kvassay, E. Zaitseva, V. Levashenko, Importance analysis of multi-state systems based on tools of logical differential calculus, Reliab. Eng. Syst. Saf. 165 (2017) 302–316.
[22] T. Nakagawa, Stochastic Processes, Springer London, London, 2011.
[23] E. Zaitseva, V. Levashenko, Multiple-valued logic mathematical approaches for multi-state system reliability analysis, J. Appl. Log. 11 (2013) 350–362.
[24] A. Kaufmann, D. Grouchko, R. Cruon, Mathematical Models for the Study of the Reliability of Systems, Academic Press, New York, NY, 1977.
[25] E. El-Neweihi, F. Proschan, Degradable systems: a survey of multistate system theory, Commun. Stat. Theory Methods 13 (1984) 405–432.
[26] A.M. Abouammoh, M.A. Al-Kadi, Component relevancy in multistate reliability models, IEEE Trans. Reliab. 40 (1991) 370–374. 379.
[27] G. Teschl, Ordinary Differential Equations and Dynamical Systems, American Mathematical Society, Providence, RI, 2012.
[28] K. Kolowrocki, Reliability of Large and Complex Systems, Elsevier, London, 2014.
[29] J. Xing, C. Feng, X. Qian, P. Dai, in: A simple algorithm for sum of disjoint products, 2012 Proceedings Annual Reliability and Maintainability Symposium, IEEE, 2012, pp. 1–5.
[30] Siemens, Mean Time Between Failures (MTBF)—List for RUGGEDCOM Products, https://support.industry.siemens.com/cs/document/109479200/mean-time-between-failures-(mtbf)-%E2%80%93-list-for-ruggedcom-products?dti=0&lc=en-WW.

CHAPTER 6

Modeling Correlated Counts in Reliability Engineering

Alessandro Barbiero*
*Department of Economics, Management and Quantitative Methods, Università degli Studi di Milano, Milan, Italy

Abstract

In this chapter, we review the problem of modeling correlated count data. Among the several methods that can be used for this scope, we focus on the copula approach, illustrating its advantages, but also possible limitations and issues arising in the discrete context if compared to the continuous case. After introducing the basic notions about copulas, the construction of a multivariate joint distribution is discussed and pseudo-random simulation and point estimation of copula-based models for count data are then outlined. Results related to minimum and maximum correlation between two assigned discrete marginal distributions are also described and put in connection with the choice of the copula to be used for modeling correlated counts. A numerical example and an application to a real dataset are provided.

Keywords: Copula, Cumulative distribution function, Dependence structure, Discrete variables, Joint distribution, Marginal distributions, Pearson correlation

1 INTRODUCTION

Over the last decades, a large amount of literature on discrete bivariate and multivariate distributions has been accumulated. An extensive account of these distributions was discussed by Kocherlakota and Kocherlakota [1] and Johnson et al. [2]. Focusing on reliability engineering and lifetime analysis, discrete distributions are naturally used for modeling count variables, such as the number of rounds fired by a weapon until the first failure; the number of deaths at a given place over a given time period; the number of cycles prior to the first failure when devices work in cycles; the number of accidents a worker experiences in different time periods; etc. But they can also be usefully employed when the lifetime (of an item, a system, etc.) is measured in days, weeks, months, etc., that is, when the lifetime is measured on a discrete scale. When the researcher has to deal with several phenomena of this type, he needs to take into account and model their statistical association by considering appropriate multivariate distributions.

Advances in System Reliability Engineering
https://doi.org/10.1016/B978-0-12-815906-4.00006-3

For a long time, the multivariate normal distribution has been used even for modeling noncontinuous correlated data. However, this is not a satisfactory choice, especially when one needs to model data with just a few distinct observed values or presenting an excess of zeros, something that is quite common in several engineering fields. Thus the need for using a multivariate discrete model. Sarabia and Gómez-Déniz [3] suggested that, "A multivariate discrete distribution should satisfy some theoretical properties in order to model practical situations. We highlight two of these theoretical properties: a rich enough correlation structure and marginal overdispersion." There are several methods of constructing discrete bivariate (and multivariate) distributions. An exhaustive account is given by Lai [4]. Most of multivariate models for count data take their origin from the multivariate Poisson model. This model allows for positive correlations only and for equidispersed margins (remember that for a Poisson random variable [r.v.], expected value and variance have the same value, coinciding with the parameter λ). Extensions of this model are based on mixtures, which are able to (partially) overcome both problems. Apart from these models, other proposals have been suggested, often limited to the bivariate case because a generalization of higher dimension is not immediate. These models usually have identical marginal distributions or marginal distributions of the same type, and the dependence structure is somewhat limited.

Here we will focus on a method that is particularly popular and flexible when used for modeling multivariate continuous distributions, the copula approach, whose use has been recently adapted to the discrete case. We will highlight the great convenience of this approach as well as differences and limitations with respect to the original context. Copulas offer a great flexibility both in the dependence structure they can induce and in the choice of marginal distributions. Nikoloulopoulos [5] provides a list of desiderata properties of multivariate copula families for modeling multivariate discrete data.

The chapter is structured as follows: In the next section we introduce the concept of copula and explain how it can be exploited for constructing multivariate distributions with arbitrary margins and flexible dependence structures. Section 3 discusses some issues related to Pearson's correlation coefficient, often used as a measure of dependence, which is incorrect, especially when moving far from the multivariate normal distribution, and then in particular when modeling correlated counts. Alternative dependence measures are recalled. Section 4 briefly outlines estimation for copula models, highlighting potential pitfalls one has to take into consideration when handling dependent discrete distributions. Section 5 briefly mentions how copula-based models can include covariates. Section 6 presents two simple numerical examples of modeling bivariate discrete counts, pointing out from a practical perspective what has been theoretically illustrated in the previous sections. Section 7 illustrates an example of fitting several bivariate discrete models to a real dataset. The last section concludes the chapter with a summary and some brief remarks.

2 COPULAS FOR MODELING MULTIVARIATE DISTRIBUTIONS

The theory of copulas, whose origins date back to the first half of the last century, has received great and renewed interest in the last two decades, especially due to the number of applications in the fields of quantitative finance and actuarial mathematics. The copula approach allows us to separate the study of the marginal models from the study of the dependence models, then combine more marginal distributions with a variety of possible dependence structures. About the choice of univariate discrete distributions for modeling count data, we address the reader to the study by Nikoloulopoulos and Karlis [6].

Let us start with the mathematical definition of copula and a theorem that formally establishes its relationship with multivariate distributions.

2.1 Definition of Copula and Sklar's Theorem

A d-dimensional copula is a joint cumulative distribution function (c.d.f.) in $[0, 1]^d$ with standard uniform c.d.f.s U_j, $j = 1, \ldots, d$:

$$C(u_1,\ldots,u_d) := P(U_1 \leq u_1,\ldots,U_d \leq u_d). \tag{1}$$

The importance of copulas in the study of multivariate distribution functions is summarized by the theorem of Sklar [7, 8], which here we briefly reprise.

Let F be a joint c.d.f. with margins $F_1, \ldots, F_d$. Then there exists a copula $C\colon [0,1]^d \to [0,1]$ such that, for all $x_1, \ldots, x_d$ in $\overline{\mathbb{R}} = [-\infty, +\infty]$,

$$F(x_1,\ldots,x_d) = C(F_1(x_1),\ldots,F_d(x_d)). \tag{2}$$

If the margins are continuous, then C is unique, otherwise C is uniquely determined on $\mathrm{Ran}(F_1) \times\cdots\times \mathrm{Ran}(F_d)$, with $\mathrm{Ran}(F_j)$ denoting the range of F_j. Conversely, if C is a copula and $F_1, \ldots, F_d$ are univariate c.d.f.s, then the function F defined in Eq. (2) is a joint distribution function with margins

$F_1, \ldots, F_d$. If the margins $F_1, \ldots, F_d$ are continuous, the unique copula C is given by

$$C(u_1, \ldots, u_d) = F(F_1^{-1}(u_1), \ldots, F_d^{-1}(u_d)), \tag{3}$$

where F_j^{-1} denotes the generalized inverse of the marginal c.d.f. F_j, that is, $F_j^{-1}(t) = \inf\{x \in \mathbb{R} : F_j(x) \geq t\}$. Formulas (2), (3) are fundamental in dealing with copulas. The former shows how joint distributions F are formed by connecting together marginal distributions with copulas C; the latter shows how copulas are extracted from a multivariate c.d.f. with continuous margins.

Sklar's theorem also suggests that, in the case of continuous margins, it is natural to define the notion of the copula of a c.d.f. F with continuous marginal c.d.f.s $F_1, \ldots, F_d$ as the c.d.f. C of $(F_1(X_1), \ldots, F_d(X_d))$.

2.2 Copula Bounds and Fundamental Copulas

We recall that, for any copula C, the following constraint holds for any $(u_1, \ldots, u_d) \in [0, 1]^d$:

$$\max\left(0, \sum_{j=1}^{d} u_j + 1 - d\right) \leq C(u_1, u_2, \ldots, u_d) \leq \min(u_1, u_2, \ldots, u_d); \tag{4}$$

the left and right members of the inequality are called Fréchet lower bound and Fréchet upper bound, respectively [9]. We note that $M(u_1, u_2, \ldots, u_d) := \min(u_1, u_2, \ldots, u_d)$ is itself a copula, named "comonotonicity copula"; $W(u_1, \ldots, u_d) := \max\left(0, \sum_{j=1}^{d} u_j + 1 - d\right)$ is a copula only if $d = 2$, the bidimensional "countermonotonicity copula." In the bivariate case, the comonotonicity copula is the copula associated with two perfectly positively dependent r.v.s X_1 and X_2, satisfying the relationship $X_2 = f(X_1)$ almost surely, with f being a strictly increasing function. The countermonotonicity copula is the copula associated with two perfectly negatively dependent r.v.s X_1 and X_2, satisfying the relationship $X_2 = g(X_1)$ almost surely, with g being a strictly decreasing function. Although Fréchet bounds have been given for a copula, they may be given for any multivariate c.d.f. F with margins $F_1, \ldots, F_d$:

$$\max\left\{\sum_{j=1}^{d} F_j(x_j) + 1 - d, 0\right\} \leq F(x_1 \ldots, x_d) \leq \min\{F_1(x_1), \ldots, F_d(x_d)\};$$

so we have now bounds for F in terms of its own marginal distributions.

The comonotonicity and bivariate countermonotonicity copula are two examples of "fundamental" copulas. Another fundamental copula is the "independence copula," $\Pi(u_1, \dots, u_d) = \prod_{j=1}^{d} u_j$: recalling Eq. (2), we can easily see that, for a continuous joint c.d.f. F, the univariate margins are independent if and only if the copula of F is Π (in fact, the joint c.d.f. factorizes into the product of its marginal c.d.f.s).

2.3 Examples of Implicit and Explicit Copulas

We provide now two examples of parametric copula families, which we will use in the following sections.

Gaussian Copula

If $\boldsymbol{Y} \sim N_d(\boldsymbol{\mu}, \Sigma)$ is a d-dimensional Gaussian r.v., with mean vector $\boldsymbol{\mu}$ and covariance matrix Σ, then its copula is a so-called Gaussian copula. Note that the copula of $\boldsymbol{Y}$ is exactly the same as the copula of $\boldsymbol{X} \sim N_d(\boldsymbol{0}, P)$, where P is the correlation matrix of $\boldsymbol{Y}$. By definition:

$$C_P^{Ga}(u_1, \dots, u_d) = P(\Phi(X_1) \le u_1, \dots, \Phi(X_d) \le u_d) = \Phi_P(\Phi^{-1}(u_1), \dots, \Phi^{-1}(u_d)),$$

where Φ is the standard univariate normal c.d.f., $\Phi(z) = \int_{-\infty}^{z} 1/\sqrt{2\pi} e^{-t^2/2} \mathrm{d}t$, and Φ_P denotes the joint c.d.f. of $\boldsymbol{X}$. The Gaussian copula does not have a simple closed form (it belongs to the so-called class of "implicit copulas") but can be expressed as an integral over the density of $\boldsymbol{X}$; in two dimensions for $\rho \neq \pm 1$ we have that:

$$C_P^{Ga}(u_1, u_2) = \int_{-\infty}^{\Phi^{-1}(u_1)} \int_{-\infty}^{\Phi^{-1}(u_2)} \frac{1}{2\pi\sqrt{1-\rho^2}} e^{-\frac{s_1^2 - 2\rho s_1 s_2 + s_2^2}{2(1-\rho^2)}} \mathrm{d}s_1 \mathrm{d}s_2. \quad (5)$$

Both independence and comonotonicity copulas are special cases of the Gaussian copula. If $P = I_d$, we obtain the independence copula; if $P = J_d$, the $d \times d$ matrix consisting entirely of ones, then we obtain the comonotonicity copula. Also, for $d = 2$ and $\rho = -1$, the Gaussian copula is equal to the countermonotonicity copula. Thus in two dimensions, the Gaussian copula can be thought of as a dependence structure that interpolates between perfect positive ($\rho = 1$) and negative ($\rho = -1$) dependence, where the parameter ρ represents the strength and direction of dependence.

Frank Copula

The one-parameter bivariate Frank copula is defined as

$$C(u_1, u_2) = -\frac{1}{\kappa}\ln\left[1 + \frac{(e^{-\kappa u_1} - 1)(e^{-\kappa u_2} - 1)}{e^{-\kappa} - 1}\right],$$

with $\kappa \neq 0$. The Frank copula, which belongs to the wider class of Archimedean copulas, through the choice of its parameter κ, allows both negative ($\kappa < 0$) and positive ($\kappa > 0$) dependence. For $\kappa \to 0$, we have that the Frank copula reduces to the independence copula; for $\kappa \to \infty$, the Frank copula tends to the upper Fréchet bound; for $\kappa \to -\infty$, the Frank copula tends to the lower Fréchet bound. This means that the Frank copula, like the Gaussian copula, interpolates between perfect positive and perfect negative dependence.

2.4 Derivation of the p.m.f. for a Copula-Based Model with Discrete Margins

While in the continuous case, the derivation of the joint p.d.f. of $(X_1, \ldots, X_d)$ is obtained easily through the use of partial derivatives of the joint c.d.f; for the discrete case, the expression of the joint p.m.f. for the multivariate r.v. with c.d.f. (2) can be derived by using finite differences [10]. Let $\boldsymbol{c} = (c_1, \ldots, c_d)$ be vertices where each c_j is equal to either x_j or $x_j - 1$, x_j nonnegative integer, $j = 1, \ldots, d$. Then the joint p.m.f. of the d-variate discrete r.v. $(X_1, \ldots, X_d)$ is given by

$$p(x_1, \ldots, x_d) = \sum \operatorname{sgn}(\boldsymbol{c}) C(F_1(c_1), \ldots, F_d(c_d)), \tag{6}$$

where the sum is taken over all vertices $\boldsymbol{c}$ and $\operatorname{sgn}(\boldsymbol{c})$ is given by:

$$\operatorname{sgn}(\boldsymbol{c}) = \begin{cases} 1 & \text{if } c_j = x_j \text{ for an even number of } j's \\ -1 & \text{if } c_j = x_j \text{ for an odd number of } j's. \end{cases} \tag{7}$$

For the simplest case ($d = 2$), for a given pair (x_1, x_2) of nonnegative integers, we have $2^2 = 4$ possible vertices $\boldsymbol{c}$ and Eq. (6) becomes

$$\begin{aligned} p(x_1, x_2) &= C(F_1(x_1), F_2(x_2)) + C(F_1(x_1 - 1), F_2(x_2 - 1)) \\ &\quad - C(F_1(x_1), F_2(x_2 - 1)) - C(F_1(x_1 - 1), F_2(x_2)). \end{aligned}$$

From Eq. (6) it is clear that one has to evaluate the copula repeatedly for calculating the joint p.m.f. This means that an analytical closed form for the joint c.d.f. would make the computation of the p.m.f. easier and the copula model for count data more usable. Multivariate elliptical copulas (comprising the Gaussian copula) though providing a flexible structure, allowing for both positive and negative linear correlations, do not have a closed form, then for $d > 2$ the computation of p.m.f. requires repeated multivariate numerical integration; for this reason, their use is not straightforward.

2.5 Simulating Copulas

Based on Sklar's theorem, if $(X_1, \ldots, X_d)$ has copula C and marginal c.d.f.s $F_1, \ldots, F_d$, then one may simulate from $(X_1, \ldots, X_d)$ in the following way:

1. Simulate a random vector $(U_1, \ldots, U_d)$ with c.d.f. C

2. Return the random vector $(X_1, \ldots, X_d) := (F^{-1}(U_1), \ldots, F_d^{-1}(U_d))$

Now, the problem arises on how to simulate from C. Let us first focus on the case $d = 2$. If a bivariate copula $C(u_1, u_2)$ has a sufficiently simple algebraic expression (this occurs for typical bivariate copulas, especially absolutely continuous copulas), then there exists a very efficient, analytical simulation algorithm that can often be applied. It goes under the name of "conditional sampling method" (see, e.g., Mai and Scherer [11, pp. 78–79]). This method is based on the quantity

$$C_{u_2|u_1}(U_2|u_1) := P(U_2 \leq u_2 | U_1 = u_1) = \frac{\partial C(u_1, u_2)}{\partial u_1}, \quad u_2 \in [0, 1],$$

which exists for almost every $u_1 \in (0, 1)$. $C_{u_2|u_1}(U_2|u_1)$ is the c.d.f. of U_2 conditioned on the event that $U_1 = u_1$.

If we want to simulate r.v.s with copula C, we can implement the following steps:

- Simulate independently $U_1 \sim U(0, 1)$ and $V \sim U(0, 1)$
- Compute $U_2 := C^{-1}_{u_2|u_1}(V|u_1)$
- The bivariate r.v. (U_1, U_2) has copula C

If we consider the Frank copula presented in Section 2.3, one can easily derive the conditional c.d.f.

$$C_{U_2|u_1}(u_2|u_1) = \frac{\exp(-\kappa(u_1 + u_2)) - \exp(-\kappa u_1)}{\exp(-\kappa(u_1 + u_2)) - \exp(-\kappa u_1) - \exp(-\kappa u_2) + \exp(-\kappa)},$$

which can be inverted analytically:

$$C^{-1}_{U_2|u_1}(v|u_1) = -\frac{1}{\kappa} \log \left[1 - \frac{1 - \exp(-\kappa)}{1 + (v^{-1} - 1)\exp(-\kappa u_1)}\right].$$

The major shortcoming of the conditional sampling method is that it is very difficult to generalize to dimensions $d \geq 3$.

In practice, the simulation of r.v. for large d is usually only possible along a sufficiently easy stochastic model, which can be exploited to derive a simulation algorithm, for example, Gaussian copula or more generally elliptical copulas, or Archimedean copulas [11, p. 81].

Simulating a Gaussian copula C_P^{Ga} with correlation matrix P can be carried out following these steps:

1. Simulate $\boldsymbol{Z} = (Z_1, \ldots, Z_d) \sim N(\boldsymbol{0}, P)$, a multivariate normal distribution with mean vector $\boldsymbol{0}$ and correlation matrix P
2. Return $\boldsymbol{U} = (\Phi(Z_1), \ldots \Phi(Z_d))$. The r.v. $\boldsymbol{U}$ has copula C_P^{Ga}

The implementation is very straightforward, due to the large availability of software carrying out the simulation of the multivariate normal (step 1 of the previous algorithm). Actually, one of the first attempts to simulate correlated discrete data is due to Cario and Nelson [12], who employed the Gaussian copula (with a summary reference to copula theory) to link discrete variables together with assigned marginal distributions and correlation matrix. The simulation algorithm is named NORTA (NORmal To Anything), which posits that, starting from a multivariate normal distribution, we can construct any other multivariate distribution with arbitrary margins (sharing the Gaussian copula). Indeed, the focus was on how to set the correlation matrix P to ensure the desired correlation matrix $P_{\mathbf{X}}$ on the nonnormal multivariate distribution. This attempt was reprised and developed later by other authors [see, e.g., 13–17].

3 COPULAS AND MEASURES OF DEPENDENCE

Often the researcher is interested in synthesizing the statistical dependence between two r.v.s through a scalar "measure of dependence." Here we focus on two kinds of dependence measures: the usual Pearson's linear correlation and rank correlations.

3.1 Linear Correlation

Pearson's linear correlation coefficient is by far the most popular measure of dependence between r.v.s. For a pair of r.v.s (X_1, X_2) it is defined as

$$\rho_{X_1X_2} = \frac{\mathbb{E}[(X_1 - \mathbb{E}(X_1))(X_2 - \mathbb{E}(X_2))]}{\sqrt{\mathrm{Var}(X_1)\mathrm{Var}(X_2)}}$$

as far as X_1 and X_2 have both finite and non-null variance; it takes values in $[-1, 1]$ and actually measures the strength and direction of the linear relationship between X_1 and X_2. It is the canonical dependence measure in the world of multivariate normal distributions, and more generally for spherical and elliptical distributions; more precisely, the linear correlation (or the correlation matrix, for $d \geq 3$) contains all the information about the dependence structure of the multivariate model. However, empirical research in many applied sciences shows that the distributions of the real world are seldom in this class, and thus Pearson's ρ is not suitable to capture dependence [18]. This is one of its main drawbacks, along with others that denotes as "fallacies" [8], which are often underrated.

A first fallacy is that, given two marginal distributions F_1 and F_2 and a correlation value $\rho \in [-1, +1]$, it is not always possible to construct a joint distribution F with margins F_1 and F_2, whose correlation is equal to the assigned ρ. To better explain it, let us introduce the concept of equality in type for random variables. Two r.v.s V and W (or their distributions) are said to be of the same type if there exist constants $a > 0$ and $b \in \mathbb{R}$ such that $V =^d aW + b$, with "$=^d$" denoting equality in distribution. In other words, distributions of the same type are obtained from one another by location and scale transformations. Then, we can state the following result, concerning "attainable correlations." Let (X_1, X_2) be a random vector with finite-variance marginal c.d.f.s F_1 and F_2 and an unspecified joint c.d.f.; assume also that $\text{var}(X_1) > 0$ and $\text{var}(X_2) > 0$. The following statements hold [8, pp. 204–205].

1. The attainable correlations form a closed interval $[\rho_{\min}, \rho_{\max}]$ with $\rho_{\min} < 0 < \rho_{\max}$.
2. The minimum correlation $\rho = \rho_{\min}$ is attained if and only if X_1 and X_2 are countermonotonic. The maximum correlation $\rho = \rho_{\max}$ is attained if and only if X_1 and X_2 are comonotonic.
3. $\rho_{\min} = -1$ if and only if X_1 and $-X_2$ are of the same type, and $\rho_{\max} = 1$ if and only if X_1 and X_2 are of the same type.

Let us focus for a while on continuous r.v.s. Clearly, the Gaussian family of distributions with mean $\mu \in \mathbb{R}$ and variance $\sigma^2 \in \mathbb{R}^+$ is an example of class of distributions of the same type, because we know that, if $X \sim N(\mu_x, \sigma_x)$, then any linear transformation $Y = aX + b$, with $a \neq 0, b \in \mathbb{R}$, is still normal. Then, thanks to the previous statements, it is possible to join together two any normal distributions into a joint distribution with an assigned correlation $-1 \leq \rho \leq +1$. This result was quite predictable, as we know the bivariate normal distribution satisfies these conditions. Another less trivial example: let X_1 and X_2 be exponential r.v.s with parameters λ_1 and λ_2, respectively. Are the minimum and maximum attainable correlation values equal to − 1 and + 1? We have that $\rho_{\max} = +1$ because X_2 is equal in distribution to $\lambda_1 X_1/\lambda_2$. In fact, if the d.f. of X_1 is $f_1(x) = \lambda_1 \exp(-\lambda_1 x)$, the d.f. of the transformation $\lambda_1 X_1/\lambda_2$ is $f_2(x) = \lambda_2 \exp(-\lambda_2 x)$. In other terms, the exponential r.v.s "scale." However, $-1 < \rho_{\min} < 0$, because it cannot be $X_1 =^d -aX_2 + b$ for any $a > 0$ and b (both X_1 and X_2 are defined on $\mathbb{R}^+$). It can be shown that $\rho_{\min} = 1 - \pi^2/6 \approx= -0.645$, for any choice of λ_1, λ_2.

For nonnegative count r.v.s, it is then clear that the maximum correlation is + 1 if and only if X_1 and X_2 are identically distributed; whereas the minimum correlation can never be − 1. Determining analytically $\rho_{\min}$ (and $\rho_{\max}$, in case of nonidentical margins) for a pair of discrete r.v.s is a challenging task; for the geometric distribution, one can refer to Huber and Maric [19], where it is shown that, for two identical geometric margins with parameter θ, $\rho_{\min}$ is equal to $\theta - 1$ if $\theta \geq 1/2$, whereas, if $\theta < 1/2$, a numerical procedure is sketched to recover $\rho_{\min}$.

If dealing with two discrete r.v.s with finite support, the values $\rho_{\min}$ and $\rho_{\max}$ can be computed by building the cograduation and countergraduation tables (see Ferrari and Barbiero [16] for an example of calculation). An empirical and straightforward method that can be employed to numerically derive the extremal correlations $\rho_{\min}$ and $\rho_{\max}$ between two (discrete, continuous, or mixed-type) r.v.s is the following [20]:

1. Generate two random samples from the two univariate distributions independently, using a large number of observations (e.g., $n = 100,000$).
2. Sort the two samples in the same direction, and compute the sample correlation, which corresponds to the upper bound $\rho_{\max}$.
3. Sort the two samples in opposite directions (i.e., in ascending order for one of the variables, and in descending order for the other). Then, compute the sample correlation, which corresponds to the lower bound $\rho_{\min}$.

The rationale of this algorithm clearly relies on the result about attainable correlations (statement 2). For illustrative purpose, we used this algorithm, by setting $n = 10^7$, to compute the range $[\rho_{\min}, \rho_{\max}]$ for six possible combinations of discrete margins. We considered three common families of count distributions, namely, geometric, Poisson, and negative binomial (a generalization of the geometric). The results are reported in Table 1. Note that the extremal correlations can be quite far from the corresponding limits − 1 and + 1 of the correlation coefficient. Actually, as can be easily observed considering the combination of two negative binomials, as the discrete distributions tend to resemble a continuous one (i.e., as most of the probability mass tends to spread over a huge number of integers and not to concentrate

Table 1 Correlation range $[\rho_{min}, \rho_{max}]$ for six possible combinations of discrete margins

(X_1, X_2)	Geo(1/4)	Pois(4)	NegBin(10, 2/3)
Geo(1/2)	[−0.549, 0.978]	[−0.772, 0.891]	[−0.743, 0.919]
Pois(2)		[−0.912, 0.965]	[−0.907, 0.968]
NegBin(5, 1/2)			[−0.909, 0.989]

on a few values), then the extremal correlations tend to their corresponding limits $\pm$ 1.

Another fallacy of the linear correlation can be outlined as follows. Given two margins F_1 and F_2 and a feasible linear correlation ρ (where by feasible we mean comprised between the ρ_{min} and ρ_{max} previously discussed), the joint distribution F having margins F_1 and F_2 and correlation ρ is not unique. In other terms, the marginal distributions and pairwise correlations of an r.v. do not univocally determine its joint distribution. McNeil et al. [8] presented an example, where two normal distributions can be linked together forming two different joint distributions with the same correlation, by simply selecting two different copulas (a Gaussian copula and a linear combination of Fréchet-bound copulas). We will provide an analogous example related to the discrete case in Section 7.

3.2 Rank Correlations

The first fallacy of Pearson's correlation can be overcome, when handling continuous distributions only, by employing two other dependence measures, namely, Spearman's rho [21] and Kendall's tau τ [22], also known as "rank correlations."

Spearman's rho between two r.v.s X_1 and X_2 with marginal c.d.f.s F_1 and F_2, respectively, is defined as

$$\rho_S(X_1, X_2) = \rho(F_1(X_1), F_2(X_2));$$

whereas the definition of Kendall's tau is

$$\rho_\tau(X_1, X_2) = \mathbb{E}(\text{sgn}(X_1 - X_2)(\tilde{X}_1 - \tilde{X}_2)),$$

where sgn is the usual sign function, taking value + 1 or − 1 according to whether its argument is positive or negative, and $(\tilde{X}_1, \tilde{X}_2)$ is an independent copy of (X_1, X_2). In higher dimensions, the Spearman's rho matrix for the general d-variate random vector $\boldsymbol{X} = (X_1, \ldots, X_d)^T$ is given by

$\rho_S(\boldsymbol{X}) = \rho(F_1(X_1), \ldots, F_d(X_d))$; the Kendall's tau matrix of $\boldsymbol{X}$ may be written as $\rho_\tau(\boldsymbol{X}) = \text{cov}(\text{sgn}(\boldsymbol{X} - \tilde{\boldsymbol{X}}))$, where $\tilde{\boldsymbol{X}}$ is an independent copy of $\boldsymbol{X}$.

For a bivariate sample $(\boldsymbol{x}_1, \boldsymbol{x}_2)$, with $\boldsymbol{x}_j = (x_{ij})^T$, $i = 1, \ldots, n$, $j = 1, 2$, the sample version of Spearman's rho is $\hat{\rho}_S(\boldsymbol{x}_1, \boldsymbol{x}_2) = \rho(\text{rank}(\boldsymbol{x}_1), \text{rank}(\boldsymbol{x}_2))$, where the function "rank" assigns a number from 1 to n corresponding to the position of x_{1j} (or x_{2j}) in an ascending order. The sample version of Kendall's tau can be defined as

$$\hat{\rho}_\tau(\boldsymbol{x}_1, \boldsymbol{x}_2) = \frac{\text{Number of concordant pairs} - \text{Number of discordant pairs}}{\binom{n}{2}},$$

where two points in $\mathbb{R}^2$, denoted by (x_{1t}, x_{2t}) and (x_{1u}, x_{2u}), are said to be concordant if $(x_{1t} - x_{1u})(x_{2t} - x_{2u}) > 0$ and to be discordant if $(x_{1t} - x_{1u})(x_{2t} - x_{2u}) < 0$.

Differently from Pearson's ρ, these two measures (i) are both able to capture monotone dependence between two random variables; (ii) in the continuous case, their value depend only on the copula C and not on the margins (i.e., they are "margin-free"); and (iii) they take the value 1 when the margins are comonotonic and the value -1 when they are countermonotonic. These three properties do not keep holding when we move to the discrete case [23]. Indeed, with discrete variables, the definition itself of these two measures leads to ambiguity, due to the stepwise nature of c.d.f. and then the presence in the sample data of the so-called "ties" (observations having the same value). Depending on the choice of margins, the two rank correlations may or may not span the entire interval $[-1, 1]$ (see, e.g., Denuit et al. [24, Chapter 5]). Several rescaled versions of Kendall's tau and Spearman's rho were introduced in an attempt to correct it, but none of them is margin-free, however, and some of them do not reach the bounds ± 1 [23].

4 INFERENCE FOR COPULA MODELS

Let us consider a multivariate copula-based discrete model whose joint c.d.f. has the following representation:

$$F(x_1, \ldots, x_d; \theta_1, \ldots, \theta_d, \theta) = C(F_1(x_1; \theta_1), \ldots, F_d(x_d; \theta_d); \theta), \qquad (8)$$

where F_j is the marginal c.d.f. corresponding to the jth margin, characterized by marginal parameter θ_j, $j = 1, \ldots, d$, and θ is the copula parameter (θ and the θ_j can be scalar or vectors). Henceforth, we suppose that the functions F_j and C are known, except for the values of their parameters; this means that we move within a parametric framework; for nonparametric and semiparametric cases, we address the reader to Joe [25, pp. 247–251]. The p.m.f. p of $(X_1, \ldots, X_d)$ can be derived recalling Eq. (6). Now, if a d-variate random sample of size n, $[x_{ij}]$, $i = 1, \ldots, n, j = 1, \ldots, d$, is available, the log-likelihood functions for the univariate margins are

$$\ell_j(\theta_j) = \sum_{i=1}^{n} \log p_j(x_{ij};\theta_j), \quad j = 1, \ldots, d, \tag{9}$$

where p_j is the marginal p.m.f. of X_j; and the total log-likelihood function can be written as

$$\ell(\theta, \theta_1, \ldots, \theta_d) = \sum_{i=1}^{n} \log p(x_{i1}, \ldots, x_{id}; \theta_1, \ldots, \theta_d, \theta). \tag{10}$$

Parameter estimates can be simultaneously recovered maximizing the log-likelihood function of Eq. (10), thus implementing a full maximum likelihood estimation (MLE). Such a maximization can be usually solved only numerically. Alternatively, efficient estimation of the model parameters is succeeded by the inference function of margins (IFM) method, which consists of a two-step approach. At the first step, the univariate log-likelihoods of Eq. (9) are maximized independently of the copula parameter; at the second step, the joint log-likelihood of Eq. (10) is maximized over θ with the values of univariate parameters θ_j fixed as estimated at the first step. The parameter estimation is thus decomposed into two smaller problems: fitting the marginal distributions (as if they were independent) and then fitting the existing dependence structure. From a computational cost perspective, estimation by IFM method becomes more advantageous than full MLE as the dimension d increases. Asymptotic efficiency of the IFM estimator has been studied by Joe [25] for a number of multivariate models and is overall shown to be highly efficient compared to standard maximum likelihood, except for extreme cases near the Fréchet bounds.

Another method uses empirical estimates of either Spearman's or Kendall's rank correlation to infer an estimate for the copula parameter. One needs a theoretical relationship between one of the rank correlations and the copula parameter and substitute empirical values of the rank correlation into this relationship to get estimates of the copula parameter. If, for example, for a certain (bivariate) parametric copula family we have that $\rho_\tau = T(\theta)$, for some function T, and T is invertible, so that $\theta = T^{-1}(\rho_\tau)$, then an estimate for θ is $T^{-1}(\hat{\rho}_\tau)$, with $\hat{\rho}_\tau$ being the sample version of ρ_τ. This method is usually referred to as method of moments.

5 ISSUES WITH COPULAS FOR DISCRETE DATA

The copula concept is slightly less natural for multivariate discrete distributions. As seen when discussing Sklar's theorem, this is because there is more than one copula that can be used to join the margins to form the joint c.d.f., as the following example shows [8].

Copulas of Bivariate Bernoulli Let (X_1, X_2) have a bivariate Bernoulli distribution satisfying

$$P(X_1=0,X_2=0)=1/8, \quad P(X_1=1,X_2=1)=3/8,$$
$$P(X_1=0,X_2=1)=2/8, \quad P(X_1=1,X_2=0)=2/8.$$

Clearly, $P(X_1 = 0) = P(X_2 = 0) = 3/8$ and the marginal distributions F_1 and F_2 of X_1 and X_2 are the same. From Sklar's theorem—see Eq. (2)—we know that $P(X_1 \leq x_1, X_2 \leq x_2) = C(P(X_1 \leq x_1), P(X_2 \leq x_2))$ for all x_1, x_2 and some copula C. Because $\mathrm{Ran}(F_1)=\mathrm{Ran}(F_2)=\{0,3/8,1\}$, clearly the only constraint on C is that $C(3/8, 3/8) = 1/8$. Any copula fulfilling this constraint is a copula of (X_1, X_2), and there are infinitely many such copulas.

This fact does not cause any issues in the modeling/simulation step, because Formula (2) always returns a valid joint c.d.f. even if some of the F_j are discrete; some problems arise at the estimation stage, when one has to make inference about the parameter of the copula that is assumed to link the margins. The fact that from two or more correlated discrete r.v.s the copula that can be extracted is not unique (indeed, there are infinite copulas that can be extracted) naturally poses an unidentifiability problem when one needs to estimate the dependence structure. This issue leads to some important consequences; from a practical point of view, the most important is that inference for the dependence parameter θ under a parametric copula model should not resort to rank-based approach but rather to MLE. Resuming and citing Genest and Nešlehová [23], "copula models provide a viable approach to the construction of multivariate distributions with given margins, even in the discrete case," and "When dealing with count data, however, modeling and interpreting dependence through copulas is subject to caution. Furthermore, inference (and particularly rank-based inference) for copula parameters from discrete data is fraught with difficulties." For a detailed discussion of this topic, which is still debated among the statistical community, we address the reader to the above-mentioned thorough work [23] and to the recent paper by Faugeras [26], where the author seems much more reluctant than Genest and Nešlehová [23] to extend copulas from the continuous to the discrete case.

6 REGRESSION MODELS

Copula-based models are susceptible to the introduction of explanatory variables. Model parameters can be regressed toward different sets of covariates, increasing the goodness of fit of the model to the data and at the same time its complexity [27, 28]. Covariates are commonly used for the marginal parameters; however, because the interest of the researcher often lies on studying the dependence structure rather than the marginal properties and copula measures the association between marginal distributions, the use of covariates in its parameters, allowing for direct modeling of association, has been recently increasing. Following the approach of Nikoloulopoulos and Karlis [10], if we consider a bivariate copula-based parametric model for count r.v.s Y_1 and Y_2, their joint c.d.f. F, by slightly adapting Eq. (8), can be written as:

$$F(y_1, y_2; \theta_1, \theta_2, \theta) = C(F_1(y_1; \theta_1), F_2(y_2; \theta_2), \theta).$$

Suppose the data are (y_{ij}, x_{ij}), $i = 1, \ldots, n$, $j = 1, 2$, with x_{ij} a vector of covariates for the ith observation associated with the jth random component. One can easily introduce covariates x_{ij} on the copula-based parametric model by assuming that the jth margin is $y_{ij} \sim F_j(;\theta_{ij})$, with $\theta_{ij} = (\mu_{ij} = g(\beta_j^T x_{ij}), \gamma_j)$ where μ_{ij} denotes the mean parameterized by a suitable link function g to accommodate the covariates, β_j the vector of the regression coefficients, and γ_j the vector of marginal parameters not depending on covariates. Furthermore, one can introduce a regression part for the copula parameter θ, which can be specified through an appropriate covariate function on θ, $s(\theta_i) = b^T x_i$, with parameter vector b. Possible covariate functions for the copula parameter of several copula families are reported in Nikoloulopoulos and Karlis [10, Table 1]. For parameter estimation, one can use the standard MLE method, maximizing the log-likelihood function, which can be written as

$$L(\beta_1, \gamma_1, \beta_2, \gamma_2, b) = \sum_{i=1}^{n} \log p(y_{i1}, y_{i2}; \beta_1, \gamma_1, \beta_2, \gamma_2, b),$$

where p is the joint p.m.f. (6).

7 NUMERICAL EXAMPLE

In this example, focusing on the simple bivariate case, we show how to construct different bivariate models with the same margins X_1 and X_2 but different dependence structure, that is, how we can connect the same margins through different (families of) copulas. Let us consider two univariate r.v.s X_1, which we assume to be binomial, and X_2, which we assume to follow the Poisson law. The p.m.f. of X_1 is

$$p_1(x_1; n, p) = \binom{n}{x_1} \theta^{x_1} (1-\theta)^{n-x_1}, \quad x_1 = 0, 1, \ldots, n; \ n \in \mathbb{N}, 0 < \theta < 1;$$

the p.m.f. of X_2 is

$$p_2(x_2; \lambda) = \frac{\lambda^{x_2} e^{-\lambda}}{x_2!}, \quad x_2 = 0, 1, 2, \ldots, \quad \lambda > 0.$$

We set the marginal parameters as follows: $n = 4$, $\theta = 0.5$, and $\lambda = 2$. As copulas linking the two margins, we consider the Gaussian and the Frank copula. We describe how to recover the p.m.f. and compute Pearson's correlation, and compare the results under both dependence structures, varying the dependence parameter (ρ for the former copula, κ for the latter).

If X_1 and X_2 are connected through the Gaussian copula with parameter ρ, Eq. (5), then the probability values of the joint p.m.f. can be computed numerically, as double integrals over rectangles of the joint normal p.d.f. For example, computing the probability $p(0, 0)$ according to Eq. (6) results in computing the following:

$$p(0,0) = F(0,0) = C(F_1(0), F_2(0); \rho) = \int_{-\infty}^{\Phi^{-1}(F_1(0))} \int_{-\infty}^{\Phi^{-1}(F_2(0))} \phi_2(s, t; \rho) \mathrm{d}s \mathrm{d}t \tag{11}$$

For our example, $F_1(0) = 0.0625$ and then $\Phi^{-1}(0.0625) = -1.534121$; $F_2(0) = 0.1353353$ and then $\Phi^{-1}(0.1353) = -1.10152$. If we set the dependence parameter ρ equal to 0.9, the integral on the right-hand side of Eq. (11) is equal to 0.05594503. The other probability values can be calculated the same way. Computation of the double integral can be performed through the function `pmvnorm` comprised in the `R` package `mvtnorm`. Table 2 displays (part of) the joint p.m.f. when $\rho = 0.9$ (we truncated the values of the Poisson margin at $x_2 = 4$).

For the bivariate discrete r.v. (X_1, X_2), it is possible to compute Pearson correlation $\rho_{X_1 X_2}$, which we expect to be different from the correlation of the Gaussian copula ρ. To do that, we need to compute the mixed moment $\mathbb{E}(X_1 X_2)$, preferably employing a joint p.m.f. computed over a larger support grid than that of Table 2. To tackle the infinite support of the Poisson variate and numerically compute the value of the mixed moment $\mathbb{E}(X_1 X_2)$

Table 2 Joint p.m.f. of the r.v. (X_1, X_2) with binomial and Poisson margins connected by Gaussian copula with $\rho = 0.9$

(x_1, x_2)	0	1	2	3	≥ 4	Total
0	0.05595	0.00649	0.00007	0.00000	0.00000	0.0625
1	0.07427	0.14545	0.02916	0.00111	0.00001	0.25
2	0.00512	0.11478	0.18662	0.06237	0.00611	0.375
3	0.00001	0.00394	0.05453	0.11146	0.08006	0.25
4	0.00000	0.00000	0.00029	0.00551	0.05670	0.0625
Total	0.13534	0.27067	0.27067	0.18045	0.14288	1

with a small error margin, we can follow the expedient suggested by Barbiero and Ferrari [29, 30] and (temporarily) truncate it to a very high quantile; in this case taking a threshold equal to $x_{2\max} = 13$ is reasonable, because $F_2(x_{2\max}) \approx 1$. Then,

$$\mathbb{E}(X_1X_2) = \sum_{i=0}^{4}\sum_{j=0}^{\infty} i \cdot j \cdot p(i,j) \approx \sum_{i=0}^{4}\sum_{j=0}^{\infty} i \cdot j \cdot p^{*}(i,j),$$

where

$$p^{*}(i,j) = \begin{cases} p(i,j) & \text{if } j < x_{2\max} \\ \sum\limits_{j=x_{2\max}}^{\infty} p(i,j) & \text{if } j \geq x_{2\max} \end{cases}$$

is the joint p.m.f. of the truncated version of the bivariate discrete r.v.

In Fig. 1, we graphically displayed the relationship between the correlation parameter ρ^{Ga} of the bivariate Gaussian copula, which takes all the values in $[-1, +1]$, and the correlation coefficient $\rho_{X_1X_2}$ between the two discrete margins $X_1 \sim$Binom$(n = 4, \theta = 0.5)$ and $X_2 \sim$Pois$(\lambda = 2)$. The function $\rho_{X_1X_2} = G(\rho^{Ga})$ is a strictly increasing function passing through the origin. From Fig. 1, it is also evident how the minimum and maximum correlation between the two discrete margins connected by a Gaussian copula are not -1 and $+1$, but -0.92171 and $+0.92171$. This empirically confirms the result given in Section 3, because the binomial and Poisson are clearly not distributions of the same type. It is also evident that, for $\rho^{Ga} = 0$, also $\rho_{x_1x_2} = 0$ (for $\rho = 0$, the Gaussian copula reduces to the independence copula) and that the relationship between $\rho_{x_1x_2}$ and ρ^{Ga} is almost linear over the entire interval $[-1, +1]$. More importantly, we have that $|\rho_{X_1X_2}| \leq |\rho|$, which was empirically noticed by Cario and Nelson [12]

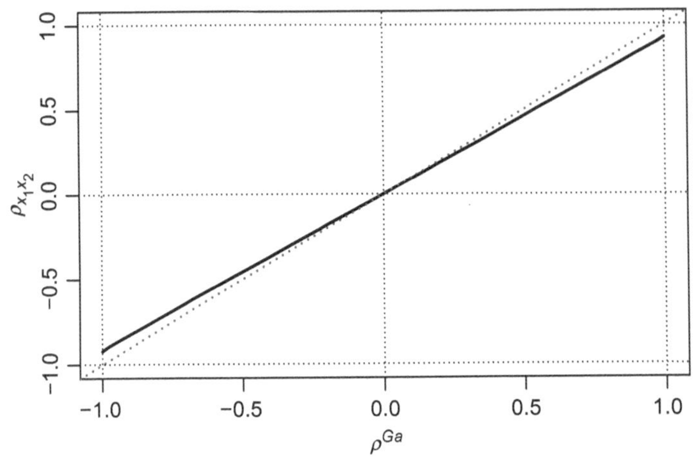

Fig. 1 Relationship (represented by the *solid line*) between the correlation parameter ρ^{Ga} of the Gaussian copula and the correlation of the binomial (with $n = 4$, $\theta = 0.5$) and the Poisson ($\lambda = 2$) margins connected by the same copula. The *red dashed line* indicates the I–III quadrant bisector.

and is a more general result due to findings of Klaassen and Wellner [31], and reprised by Kotz and Mari [32, p. 155]. For example, if we set $\rho^{Ga} = 0.9$, we have that $\rho_{X_1X_2} = 0.8259$.

We can develop an analogous exercise by using the Frank copula, parameterized by $\kappa \in \mathbb{R}$, which spans the entire dependence spectrum, like the Gaussian copula. As we did for this latter case, we can reconstruct the joint p.m.f. of the r.v. (X_1, X_2) recalling again Eq. (6). For example, letting $\kappa = 14.21$, to compute $p(0, 0)$ we calculate

$$\begin{aligned} p(0,0) &= F(0,0) = C(F_1(0), F_2(0)) \\ &= -\frac{1}{14.21}\log\left[1 + \frac{(e^{-14.21\cdot 0.00625} - 1)(e^{-14.21\cdot 0.13534} - 1)}{e^{-14.21} - 1}\right] = 0.04914. \end{aligned}$$

Table 3 displays the joint p.m.f. for $\kappa = 14.21$. As done before, by building the p.m.f. for several values of κ, we can reconstruct the relationship between the correlation coefficient $\rho_{X_1X_2}$ and the dependence parameter κ, which is displayed in Fig. 2. Note that $\rho_{X_1X_2}$ is an increasing function of κ, passing through the origin (when the Frank copula coincides with the independence copula, the two margins are independent and then uncorrelated); when κ goes to $+\infty$ $(-\infty)$, which corresponds to the comonotonicity (countermonotonicity) case, then ρ tends asymptotically but quite

Table 3 Joint p.m.f. of the r.v. (X_1, X_2) with binomial and Poisson margins connected by Frank copula with $\kappa = 14.21$

(x_1, x_2)	0	1	2	3	≥ 4	Total
0	0.04914	0.01305	0.00031	0.00001	0.00000	0.0625
1	0.08151	0.15245	0.01566	0.00036	0.00003	0.125
2	0.00466	0.10394	0.21277	0.04837	0.00526	0.375
3	0.00002	0.00121	0.04094	0.12062	0.08720	0.125
4	0.00000	0.00002	0.00099	0.01110	0.05039	0.0625
Total	0.13534	0.27067	0.2707	0.18045	0.14288	1

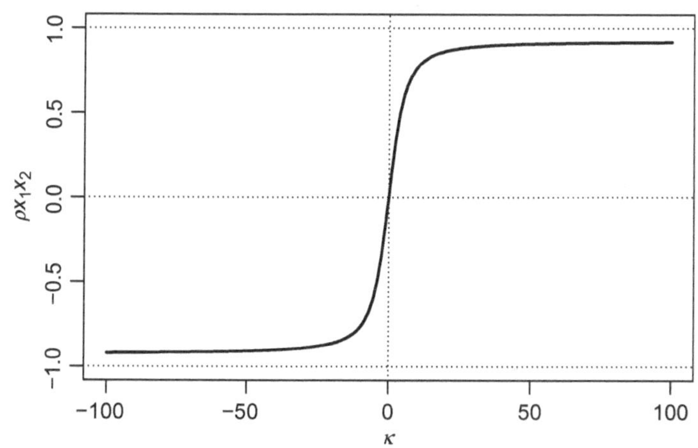

Fig. 2 Relationship (represented by the *solid line*) between the correlation parameter κ of the Frank copula and the correlation between the binomial (with $(n = 4, \theta = 0.5)$) and the Poisson ($\lambda = 2$) margins connected by the same copula.

slowly to the limit $\rho_{\max}$ ($\rho_{\min}$). The value $\kappa = 14.21$ we employed before in the example of computation of $p(x_1, x_2)$ yields a correlation equal to 0.8259, the same induced by the Gaussian copula with $\rho^{Ga} = 0.9$.

The comparison between Tables 2 and 3 shows that, though sharing the same value of linear correlation, the two joint distributions are quite different (one can just consider the (0, 0) probabilities). Through this example, we showed that it is possible to construct two bivariate discrete distributions with the same choice of margins and the same value of (linear) correlation but with a different joint distribution. This is a practical counterexample that indicates how Pearson ρ is in general unable to characterize the dependence structure of a bivariate distribution (its second fallacy discussed in Section 3.1).

8 AN APPLICATION TO REAL DATA

In this section, we fit several bivariate copula-based discrete models to a dataset taken from the literature. The data, considered by Mitchell and Paulson [33] (see Table 4), consist of the number of aborts by 109 aircrafts in two (first $= x_1$, second $= x_2$) consecutive 6 months of a 1-year period. Summary statistics for the dataset are $\overline{x}_1 = 0.624$, $\overline{x}_2 = 0.725$, $\text{var}(x_1) = 1.024$, $\text{var}(x_2) = 1.062$. The sample correlation coefficient between x_1 and x_2 is $\hat{\rho}_{x_1 x_2} = -0.1609$, which denotes a negative dependence. To fit these data, we adopt the copula approach described in this chapter: we separate the modeling of the margins from the modeling of the dependence structure. As to the marginal distributions, it was shown that the geometric distribution could be a plausible model for both x_1 and x_2; then this can be a first choice. Alternatively, we can fall back on one-parameter discrete Lindley distribution [34], whose p.m.f. is $p(x;\theta) = \sum_{i=0}^{1}(-1)^i(1+\theta(x+1)/(1+\theta)e^{-\theta(x+i)})$, for $x = 0, 1, 2, \ldots$, with $\theta > 0$. We can add a further degree of complexity by considering the discrete Weibull distribution [35], which can be regarded as a generalization of the geometric distribution. The expression of the p.m.f. of the discrete Weibull distribution with parameters q and β is $p(x;q,\beta) = q^{x^\beta} - q^{(x+1)^\beta}$, for $x = 0, 1, 2, \ldots$, with $0 < q < 1$ and $\beta > 0$. If $\beta = 1$, the discrete Weibull distribution reduces to a geometric distribution with parameter $\theta = 1 - q$. As for the modeling of the dependence structure, here we consider the Gaussian and the Frank copula, which are able to handle negative dependence. We considered the bivariate models displayed in Table 5 (just a selection of all possible combinations), for which we computed the MLEs for all the parameters, the maximum value of the log-likelihood function ℓ, and the value of the Akaike Information Criterion (AIC) index, $2r - 2\ell$, where r is the number of model parameters. In terms of AIC, the best model among the

Table 4 Bivariate distribution of the data taken from Mitchell and Paulson [33]: Number of flight aborts by 109 aircrafts in the first and second consecutive 6 months of a 1-year period

$x_1 \setminus x_2$	0	1	2	3	4	Total
0	34	20	4	6	4	68
1	17	7	0	0	0	24
2	6	4	1	0	0	11
3	0	4	0	0	0	4
4	0	0	0	0	0	0
5	2	0	0	0	0	2
Total	59	35	5	6	4	109

Table 5 Synthetic results about the bivariate models fitted to the data taken from [33]

x_1	p_1	θ_1	q_1	β_1	x_2	p_2	θ_2	q_2	β_2	Copula	ρ	τ	ℓ	AIC
Geo	0.6132	–	–	–	Geo	0.5806	–	–	–	Ga	−0.2261	–	−244.4287	494.8574
Geo	0.6157	–	–	–	Geo	0.5769	–	–	–	F	–	−1.1158	−244.7584	495.5168
DL	–	1.3456	–	–	DL	–	−1.2453	–	–	Ga	−0.2140	–	−244.5413	495.0826
DL	–	1.3533	–	–	DL	–	1.2359	–	–	F	–	−1.0744	−244.9184	495.8368
DW	–	–	0.3763	0.9660	DW	–	–	0.4534	1.1370	Ga	−0.2353	–	−243.7156	497.4312
DW	–	–	0.3746	0.9695	DW	–	–	0.4551	1.1290	F	–	−1.1536	−244.1309	498.2618
Geo	0.6156	–	–	–	DW	–	–	0.4528	1.1365	Ga	−0.2302	–	−243.7573	495.5145
Geo	0.6182	–	–	–	DW	–	–	0.4548	1.1286	F	–	−1.1343	−244.1652	496.3303

Notes: For each model (consisting of a parametric distribution for x_1, a parametric distribution for x_2, and a parametric copula for their dependence structure), the values of the parameter estimates, the log-likelihood function and AIC are computed. *Geo*, geometric; *DW*, discrete Weibull; *DL*, discrete Lindley; *Ga*, Gaussian; *F*, Frank.

Table 6 Theoretical joint frequencies for Mitchell and Paulson data [33] under the copula-based model with geometric margins and Gaussian copula

(x_1, x_2)	0	1	2	3	≥ 4	Total
0	35.12	17.60	7.95	3.50	2.67	66.84
1	16.51	5.84	2.17	0.82	0.51	25.85
2	6.93	2.01	0.68	0.24	0.14	10.00
3	2.83	0.70	0.22	0.07	0.04	3.87
4	1.14	0.25	0.07	0.02	0.01	1.50
≥5	0.75	0.14	0.04	0.01	0.01	0.94
Total	63.28	26.54	11.13	4.67	3.37	109

Notes: Cell borders highlight the cells' groupings.

eight considered here is the one linking two geometric margins through a Gaussian copula (first line of Table 5). The corresponding theoretical joint frequencies are displayed in Table 6. The second best model is the one linking two discrete Lindley distributions through the Gaussian copula (third row). The results indicate that complicating the model by fitting the two-parameter discrete Weibull distribution to the margins is not relatively convenient; as expected, the value of the log-likelihood function increases, but also the AIC increases; note that all the estimated β values, for both x_1 and x_2, are very close to 1, the value for which the discrete Weibull degenerates into a geometric distribution. We reaffirm that the models of Table 5 represent just a small part of the infinite copula-based models that can be devised by exploiting the construction principle expressed by Eq. (2).

The estimates of Table 5 were derived according to the full MLE method; alternatively, one can implement the IFM method (see Section 4), which is computationally more convenient. For the "best" model, it works as follows: one first computes the MLEs of the parameters θ_1 and θ_2, of the two geometrically distributed margins, as if they were independent, which are equal to $\hat{\theta}_{1,IFM} = 1/(1+\bar{x_1}) = 0.6158$ and $\hat{\theta}_{2,IFM} = 1/(1+\bar{x_2}) = 0.5798$, and then one can maximize the log-likelihood with respect to the dependence parameter only, plugging in the two previous estimates: $\hat{\rho}_{IFM} = -0.2251$. The three estimates derived through the IFM method are very close to the corresponding full MLEs.

We can compute the customary chi-squared statistic as an absolute measure of goodness of fit for the "best" model. To do it, we have first to group cells and sum up the values of the theoretical contingency table of Table 6, to ensure a minimum frequency of 5 for each grouping. A possible grouping is displayed there. Then we calculate $\chi^2 = \sum_{g=1}^{G}(e_g - o_g)^2/e_g = 5.408$, where e_g and o_g are the expected and observed frequency of the gth grouping, respectively, $g = 1,\ldots, G = 8$. The corresponding P-value of the chi-squared test statistic, under the null hypothesis that the observed bivariate sample actually comes from the selected bivariate model, is .248, which attests the model fits the data adequately.

9 CONCLUSIONS

In this chapter we reviewed the statistical modeling of correlated count data via copulas. After presenting the basic notions of copulas, we discussed how copula-based models can be constructed for modeling multivariate correlated count data, how to recover the joint probability mass function, how to simulate pseudorandom samples, and how to make inference, cautioning about possible issues arising from the discrete nature of variables. A numerical example is presented that practically shows how to derive the joint probability function and the linear correlation for two bivariate models sharing the same margins but having different dependence structures. Finally, a dataset taken from the literature has been analyzed and fitted using several copula-based distributions, whose goodness of fit has been compared through a customary index.

Rather than discussing all literature on the modeling of correlated count data, which would have been a very arduous task, as it is quite scattered in many and recent works, we preferred to highlight and linger on several connotative points, illustrating them through some numerical examples.

Although a very intuitive and flexible tool for building up multivariate discrete distributions, copulas are, however, problematic when the dimension increases; the joint probability function may be cumbersome to compute and pseudorandom simulation can be not straightforward. Moreover, the discrete nature of the data adds some questions related to estimation. Statistical research is facing these issues to spread and facilitate the use of copulas for building appropriate stochastic models for real data.

REFERENCES

[1] S. Kocherlakota, K. Kocherlakota, Bivariate Discrete Distributions, Wiley Online Library, 1992.
[2] N.L. Johnson, S. Kotz, N. Balakrishnan, Discrete Multivariate Distributions, Wiley Online Library, 1997.

[3] J.M. Sarabia, E. Gómez-Déniz, Multivariate Poisson-beta distributions with applications, Commun. Stat. Theory Methods 40 (6) (2011) 1093–1108.

[4] C.-D. Lai, Constructions of discrete bivariate distributions, in: N. Balakrishnan, E. Castillo, J.M.A. Sarabia (Eds.), Advances in Distribution Theory, Order Statistics, and Inference, Springer, New York, NY, 2006.

[5] A.K. Nikoloulopoulos, Copula-based models for multivariate discrete response data, Copulae in Mathematical and Quantitative Finance, Springer, New York, NY, 2013, pp. 231–249.

[6] A.K. Nikoloulopoulos, D. Karlis, On modeling count data: a comparison of some well-known discrete distributions, J. Stat. Comput. Simul. 78 (3) (2008) 437–457.

[7] A. Sklar, Fonctions de répartition à *n* dimensions et leurs marges, Publ. Inst. Stat. Univ. Paris 8 (1959) 229–231.

[8] A.J. McNeil, R. Frey, P. Embrechts, Quantitative Risk Management: Concepts, Techniques and Tools, Princeton University Press, Princeton, 2005.

[9] M. Fréchet, Sur les tableaux de corrélation dont les marges sont données, Ann. Univ. Lyon. Sect. A Sci. Math. Astron. 3 (14) (1951) 53–77.

[10] A.K. Nikoloulopoulos, D. Karlis, Modeling multivariate count data using copulas, Commun. Stat. Simul. Comput. 39 (1) (2010) 172–187.

[11] J. Mai, M. Scherer, Financial Engineering With Copulas Explained, Palgrave Macmillan, New York, NY, 2014.

[12] M.C. Cario, B.L. Nelson, Modeling and Generating Random Vectors With Arbitrary Marginal Distributions and Correlation Matrix, Department of Industrial Engineering and Management Sciences, Northwestern University, Evanston, IL, 1997. Tech. Rep.

[13] L.-F. Lee, On the range of correlation coefficients of bivariate ordered discrete random variables, Economet. Theor. 17 (1) (2001) 247–256.

[14] H.V. Ophem, A general method to estimate correlated discrete random variables, Economet. Theory 15 (2) (1999) 228–237.

[15] L. Madsen, D. Dalthorp, Simulating correlated count data, Environ. Ecol. Stat. 14 (2) (2007) 129–148.

[16] P.A. Ferrari, A. Barbiero, Simulating ordinal data, Multivar. Behav. Res. 47 (4) (2012) 566–589.

[17] L. Madsen, D. Birkes, Simulating dependent discrete data, J. Stat. Comput. Simul. 83 (4) (2013) 677–691.

[18] C. Genest, J.G. Nešlehová, Modeling dependence beyond correlation, in: J.L. Lawless (Ed.), Statistics in Action: A Canadian Outlook, CRC Press, London, 2014, pp. 59–78.

[19] M. Huber, N. Maric, Minimum correlation for any bivariate Geometric distribution, Alea 11 (1) (2014) 459–470.

[20] H. Demirtas, D. Hedeker, A practical way for computing approximate lower and upper correlation bounds, Am. Stat. 65 (2) (2011) 104–109.

[21] C. Spearman, The proof and measurement of association between two things, Am. J. Psychol. 15 (1) (1904) 72–101.

[22] M.G. Kendall, A new measure of rank correlation, Biometrika 30 (1/2) (1938) 81–93.

[23] C. Genest, J. Nešlehová, A primer on copulas for count data, ASTIN Bull. 37 (2) (2007) 475–515.

[24] M. Denuit, J. Dhaene, M. Goovaerts, R. Kaas, Actuarial Theory for Dependent Risks: Measures, Orders and Models, John Wiley & Sons, New York, NY, 2006.

[25] H. Joe, Dependence Modelling With Copulas, CRC Press, Boca Raton, FL, 2015.

[26] O.P. Faugeras, Inference for copula modeling of discrete data: a cautionary tale and some facts, Depend. Model. 5 (1) (2017) 121–132.

[27] R. Winkelmann, Econometric Analysis of Count Data, fifth ed., Springer-Verlag, Berlin, 2008.

[28] A.C. Cameron, P.K. Trivedi, Regression Analysis of Count Data, 53, Cambridge University Press, Cambridge, 2013. vol.
[29] A. Barbiero, P.A. Ferrari, Simulation of correlated Poisson variables, Appl. Stoch. Model. Bus. Ind. 31 (5) (2015) 669–680.
[30] A. Barbiero, P.A. Ferrari, An R package for the simulation of correlated discrete variables, Commun. Stat. Simul. Comput 46 (7) (2017) 5123–5140.
[31] C.A.J. Klaassen, J.A. Wellner, Efficient estimation in the bivariate normal copula model: normal margins are least favourable, Bernoulli 3 (1) (1997) 55–77.
[32] S. Kotz, D.D. Mari, Correlation and Dependence, Imperial College Press, London, 2001.
[33] C.R. Mitchell, A.S. Paulson, A new bivariate negative binomial distribution, Naval Res. Logist. Q. 28 (1981) 359–374.
[34] E. Gómez-Déniz, E. Calderín-Ojeda, The discrete Lindley distribution: properties and applications, J. Stat. Comput. Simul. 81 (11) (2011) 1405–1416.
[35] T. Nakagawa, S. Osaki, The discrete Weibull distribution, IEEE Trans. Reliab. 24 (5) (1975) 300–301.

CHAPTER 7

Statistical Inference of an Imperfect Repair Model with Uniform Distributed Repair Degrees

Soufiane Gasmi*
*Optimization, Modeling and Decision Support (OMAD), University of Tunis—Tunis National Higher School of Engineering, Tunis, Tunisia

Abstract

The main purpose of this work is to introduce the inference of Weibull intensity parameters, which are used in the general repair model. This intensity is often used to describe failure models that analyze the reliability of different types of repairable systems. We consider an imperfect repair model as one that is not perfect ("as good as new") as in renewal process and not minimal ("as bad as old") as in a nonhomogeneous Poisson process but lies between these border cases. When a failure occurs, our repairable system will be restored with a uniform distributed degree of repair including as special cases minimal, perfect, and imperfect repair models. To determine the estimation of the model parameters, the maximum likelihood estimator is considered. For the scale and shape parameter estimators of the Weibull intensity, simultaneous confidence regions based on the likelihood ratio statistics are developed.

Keywords: Reliability, Weibull intensity, Imperfect repair, Maximum likelihood estimation, Virtual age

1 INTRODUCTION

In this chapter we discussed the parameter estimation of a general repair model introduced by Last and Szekli [1] using a Weibull intensity. The most commonly used models for the failure process of a repairable system are known as minimal repair or "as bad as old" and perfect repair or "as good as new." It is well known in practice that the reality is between these two boundary cases. The repair may not yield a functioning system that is as good as new, and the minimal repair assumption seems to be too pessimistic in repair strategies. Therefore, it is seen that the imperfect repair is of great

Advances in System Reliability Engineering
https://doi.org/10.1016/B978-0-12-815906-4.00007-5

significance in practice. The repair effect in this research is expressed by a reduction of the system virtual age introduced by Kijima et al. [2] and Kijima [3]. Here we can note that most of the models concerning the modeling of repairable systems identify the minimal repair and the imperfect repair actions. Naturally, this popular assumption is a very unreal one. The model that we will introduce is more flexible than many other models. It includes the special cases of classical perfect repair, minimal repair, imperfect repair, and general repairable systems of Kijima [3]. When the system fails, a repair is allowed, which is performed with a uniform distributed degree of repair. This assumption gives more flexibility to model repairable systems. Statistical studies of parameter estimation in systems with various degrees of repair using virtual age models are given by Bathe and Franz [4], Stadje and Zuckerman [5], Gasmi et al. [6], Baxter et al. [7], Doyen and Gaudoin [8], and Gasmi [9–11]. Recently, some results on repairable systems were given by Jiawen et al. [12], Ghnimi et al. [13], Tanwar et al. [14], Dijoux et al. [15], Rashad and Nagwa [16], and Li and Ding [17].

The failure intensity used in this research is a Weibull intensity, which is especially used as a failure model analyzing the reliability of different types of systems and can characterize the probabilistic behavior of a large number of real phenomena. Estimations of the unknown parameters included in the failure intensity and used in the general repair model is an interesting problem in reliability analysis. Our main objective is the development of confidence estimations for the parameters of the general repair model in the case of a uniform distributed degree of repair. We will particularly use the Kijima's type II imperfect repair model to describe the age reduction after repair actions.

2 DESCRIPTION OF THE GENERAL MODEL

Consider a model introduced by Last and Szekli [1]. Repair activity for such a model served to adjust the virtual age of the system. The general model includes many special cases of classical "as bad as old" model, "as good as new" model, imperfect repair models, Kijima's type I and type II models [3], the very quoted model of Brown and Proschan [18], and the model of Stadje and Zuckerman [5]. We will further assume that our system considers only unplanned repair actions.

The machine (item) starts working with an initial prescribed failure rate $\lambda_1(t) = \lambda(t)$. Let t_1 denote the random time of the first failure. At this time, t_1, the item will be repaired with the random degree z_1. The degree of repair is between 0 and 1, where the case of 1 corresponds to the "as bad as old" repair and the case of 0 to the "as good as new" repair. Following the repair action, the virtual age of the item at time t_1 is decreased to $\nu_1 = z_1 t_1$. We can remark that by such repair action the age of the item will be reduced. We consider that the distribution of the time until the next failure has the failure intensity $\lambda_2(t)$: $= \lambda(t - t_1 + \nu_1)$. Assuming now that for $(k \geq 1)$, t_k is the time of the kth failure and, that z_k is the corresponding degree of repair at that time. After repair action the failure intensity during the $(k + 1)$th sojourn is given by $\lambda_{k+1}(t) = \lambda(t - t_k + \nu_k)$, $t_k \leq t < t_{k+1}$, $k \geq 0$, where ν_k is from Kijima's type II imperfect repair model $\nu_k(t) = z_k(\nu_{k-1} + (t_k - t_{k-1}))$, $k \geq 1$, $t_0 = 0$, and $\nu_0 = 0$, that is, the repair resets the intensity of failure proportional to the virtual age. The process $\nu(t) = t - t_k + \nu_k$, $t_k \leq t < t_{k+1}$, $k \geq 1$ is defined as the virtual age process [9, 11]. Fig. 1 illustrates a realization of the virtual age process. We note that at times t_1 and t_4 imperfect repairs appear, at time t_2 and t_3 we have, respectively, "as good as new" repair action and "as bad as old" repair action. Fig. 2 shows an illustration of the corresponding failure intensity.

We now want to show how many basic models can be considered as special cases of the general repair model.

(i) The "as bad as old" model: By considering the repair degrees equal to 1, then the virtual age ν_k is equal to t_n for all $k \geq 1$. The consecutive failure rates are given by $\lambda_{k+1}(t) = \lambda(t)$, $t_k \leq t < t_{k+1}$. In this case, the obtained failure repair process is a nonhomogeneous Poisson process with the intensity function $\lambda(t)$.

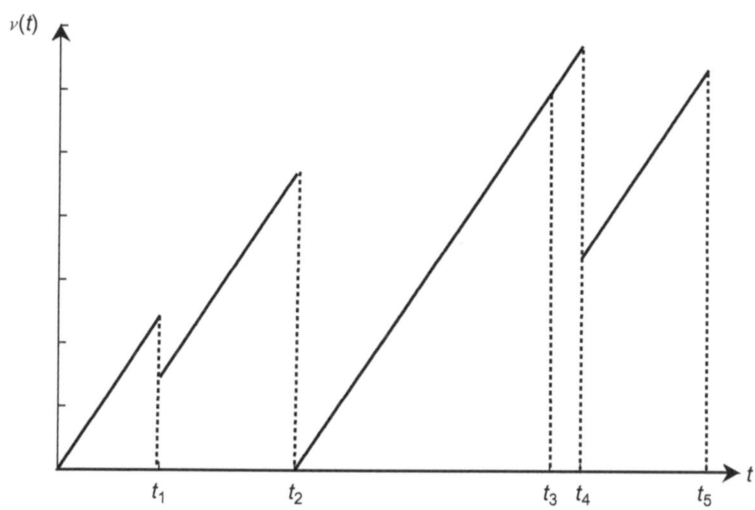

Fig. 1 Illustration of the virtual age process.

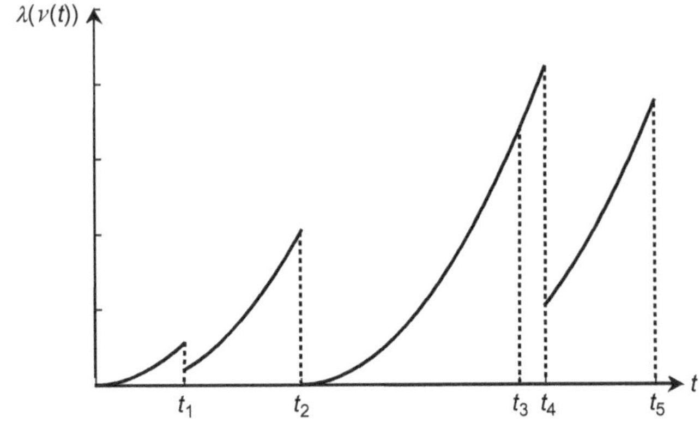

Fig. 2 Illustration of the failure intensity.

(ii) The "as good as new" model: If we consider that the consecutive repair degrees are deterministically equal to 0, then $\nu_k = 0$ for all $k \geq 1$, and the consecutive failure rates are given by $\lambda_{k+1}(t) = \lambda(t - t_k)$, $t_k \leq t < t_{k+1}$. The obtained failure repair process is a renewal process.

(iii) The Brown and Proschan model [18]: For a sequence (z_k), $k = 1, \ldots, n$ of independent and identically distributed (i.i.d.) random variables independent of $(t_1, \ldots, t_k)$ that verifies $P(z_k = 0) = p$ and $P(z_k = 1) = 1 - p$, $p \in (0, 1)$. The failure rate process is given by $\lambda_{k+1}(t) = \lambda(t - t_k + z_k D_k)$, where D_k is the time from t_k to the last perfect repair. We note that, as known after repair, the item is renewed with probability p and restarted in the same state as before failure with probability $(1 - p)$.

(iv) The Kijima's type I model [3]: By taking a sequence of i.i.d. random variables $(a_k)_{k \geq 1}$ uniformly distributed on $[0, 1]$. If we assume that (a_k) is independent of (t_k) for $k = 1, \ldots, n$, then $\nu(t_k) = \nu(t_{k-1}) + a_k x_k$, with $x_k = t_k - t_{k-1}$, $t_0 = 0$, and $a_0 = 0$. The consecutive failure rate is determined by

$$\lambda_{k+1}(t) = \lambda(t - t_k + a_k x_k + \cdots + a_1 x_1), \quad t_k \leq t < t_{k+1}.$$

If we insert in the general model

$$z_k := 1 - \frac{(1 - a_k)x_k}{a_1 x_1 + \cdots + a_{k-1} x_{k-1} + x_k} \quad k \geq 1,$$

we obtain the Kijima's type I model.

(v) The Kijima's type II model [3]: This model is also based on the virtual age notion. We use the same condition as in model I of Kijima. But here we assume that $\nu(t_k) = a_k(\nu(t_{k-1}) + x_k)$, ($t_0 = 0$ and $a_0 = 0$), and therefore $\lambda_{k+1}(t) = \lambda(t - t_k + a_k x_k + a_k a_{k-1} x_{k-1} + \cdots + a_k a_{k-1} \ldots a_1 x_1)$, $t_k \leq t < t_{k+1}$.

If we insert in the general model $z_k := a_k$, $k \geq 1$ we obtain the Kijima's type II model.

We can note that, in both Kijima's models, the consecutive failure rates depend on the failure counting process and the whole history of the repair degree processes.

(vi) The Stadje and Zuckerman model [5]: Unlike the Kijima's model II, the z_k's are, in this case, dependent on the underlying virtual age process $(\nu(t))$.

By assuming z_k: =; $\gamma(\nu(t_k-))/\nu(t_k-)$, where $\nu(t_k-)$ denotes the virtual age of the working item just before a failure and γ is a deterministic function that fulfils $\gamma(x) \leq x$, then we obtain the Stadje and Zuckerman model.

3 ESTIMATION OF THE PARAMETERS

Throughout this chapter we assume that repair times are very small, therefore they will be ignored. Moreover, our approach assumes that the repairs affect the failure intensity at any instant following a virtual age process from Kijima's type II. We will further assume that, after failure, one of the three following cases is possible: "as good as new," "as bad as old," or an imperfect repair with uniform distributed repair degree. We suppose that the failure intensity of the system is from Weibull type $\lambda(t,\alpha,\beta) = \frac{\beta}{\alpha^\beta} t^{\beta-1}$, $\alpha > 0$, $\beta > 0$. Our objective now is to estimate the two unknown parameters α and β. We can define the following three models:

Model 1: If all repair degrees $z_k \in]0, 1[$, $k = 1, \ldots, N(t)$, we then have an imperfect repair (*IR*) model.

Model 2: If all repair degrees $z_k = 0$, $k = 1, \ldots, N(t)$, only "as good as new" repairs appear in the model, and the obtained process is a renewal process (*RP*).

Model 3: If all repair degrees $z_k = 1$, $k = 1, \ldots, N(t)$, only minimal repairs appear in the model, and we obtain a nonhomogeneous Poisson process (*NHPP*).

The log-likelihood (LL) function for the failure-repair process is of the form presented by Liptser and Shiryayev [19]:

$$\ln L(t,\alpha,\beta) = \sum_{k=1}^{N(t)} \ln \lambda(\nu_{k-1} + t_k - t_{k-1}) - \int_0^t \lambda(\nu(s))ds. \qquad (1)$$

After substitution of the Weibull failure intensity, we then get the following log-likelihood function:

$$\ln L(t,\alpha,\beta) = (\beta-1)\sum_{k=1}^{N(t)} \ln(\nu_{k-1}+t_k-t_{k-1}) + N(t)(\ln\beta - \beta\ln\alpha) - \frac{1}{\alpha^{\beta}}K_1(t,\alpha,\beta), \qquad (2)$$

where $K_1(t,\alpha,\beta) = \sum_{k=1}^{N(t)}\{(\nu_{k-1}+t_k-t_{k-1})^{\beta} - \nu_k^{\beta}\} + (t - t_{N(t)} + \nu_{N(t)})^{\beta}$ and $N(t)$ denotes the number of failures until t. Using the standard maximum likelihood approach to maximize Eq. (2) we can see that it is possible to explicitly determine the scale parameter, and we then obtain the usual results for the Power-Law process:

$$\hat{\alpha} = (K_1(t,\alpha,\beta)/N(t))^{1/\hat{\beta}}. \qquad (3)$$

We remark that by setting Eq. (3) into Eq. (2), some analytical simplifications of the structure are possible. The remaining log-likelihood function is then:

$$\ln L(t,\beta) = (\beta-1)\sum_{k=1}^{N(t)} \ln(\nu_{k-1}+t_k-t_{k-1}) + N(t)(\ln\beta - 1 - \ln K_1(t,\alpha,\beta)). \qquad (4)$$

The estimation of the shape parameter β can be found by numerically solving the following equation:

$$\frac{1}{\hat{\beta}} + \frac{1}{N(t)}\sum_{k=1}^{N(t)} \ln(\nu_{k-1}+t_k-t_{k-1}) - \frac{K_2(t,\alpha,\beta)}{K_1(t,\alpha,\beta)} = 0, \qquad (5)$$

where

$$K_2(t,\alpha,\beta) = \sum_{k=1}^{N(t)}\{(\nu_{k-1}+t_k-t_{k-1})^{\beta}\ln(\nu_{k-1}+t_k-t_{k-1}) - \nu_k^{\beta}\ln\nu_k\} + (t-t_{N(t)}+\nu_{N(t)})^{\beta}\ln(t-t_{N(t)}+\nu_{N(t)}).$$

4 SIMULTANEOUS CONFIDENCE REGIONS

The most efficient way to calculate simultaneous confidence regions for two or more parameters is based on the likelihood ratio. If the parameters of the lifetime distribution are estimated with the maximum likelihood method, then it is possible to calculate, for instance, the reliability of the product.

It is well known (see Barndorff-Nielsen and Blaesild [20]) that in general the log-likelihood ratio

$$LQ = 2\{\ln L(t,\hat{\alpha},\hat{\beta}) - \ln L(t,\alpha,\beta)\} \tag{6}$$

is asymptotically χ^2-distributed with two degrees of freedom, where $\hat{\alpha},\hat{\beta}$ are the maximum likelihood estimations of the parameters of interest α and β. This fact can be used to determine simultaneous confidence regions. The simultaneous confidence region is defined by the inequality $LQ \le \chi^2_{1-\mu,2}$, where $\chi^2_{1-\mu,2} = -2\ln\mu$, which is the $(1-\mu)$-quantile of the χ^2-distribution with two degrees of freedom.

Now, we observe n independent failure repair processes. Let $r \in \{1, 2, \ldots, n\}$; we introduce the following notations:

- $N^r(t)$: The number of failures until t for the rth failure repair process,
- $t_{r,1},\ldots,t_{r,N^r(t)}$: Failure times of the rth failure repair process,
- $\nu_{r,1},\ldots,\nu_{r,N^r(t)}$: The virtual ages of failures until t for the rth failure repair process, and
- $L_r(t, \alpha, \beta)$: The likelihood function of the rth failure repair process.

The log-likelihood function of the rth failure repair process is defined as before in Eq. (2). We get the following LL function:

$$\begin{aligned}\ln L_r(t,\alpha,\beta) &= (\beta-1)\sum_{k=1}^{N^r(t)} \ln(\nu_{r,k-1} + t_{r,k} - t_{r,k-1}) + N^r(t)(\ln\beta - \beta\ln\alpha) \\ &\quad - \frac{1}{\alpha^\beta}K_{r,1}(t,\alpha,\beta),\end{aligned}$$

$$\text{where } K_{r,1}(t,\alpha,\beta) = \textstyle\sum_{k=1}^{N^r(t)}\{(\nu_{r,k-1} + t_{r,k} - t_{r,k-1})^\beta - \nu^\beta_{r,k}\} + (t - t_{N^r(t)} + \nu_{r,N^r(t)})^\beta. \tag{7}$$

By using the log-likelihood ratio, we then obtain the simultaneous confidence region for n independent failure repair processes as follows:

$$\begin{aligned}&\frac{1}{\alpha^\beta}\sum_{r=1}^{n} K_{r,1}(t,\alpha,\beta) + (\hat{\beta} - \beta)\sum_{r=1}^{n}\sum_{k=1}^{N^r(t)} \ln(\nu_{r,k-1} + t_{r,k} - t_{r,k-1}) \\ &+ (\ln\hat{\beta} - \ln\beta - 1 - \hat{\beta}\ln\hat{\alpha} + \beta\ln\alpha)\sum_{r=1}^{n} N^r(t) = -n\ln\mu.\end{aligned} \tag{8}$$

5 SIMULATION STUDY

Based on a large simulation study, we present in this section some numerical results. This study is considered to apply the previous theoretical results to

Algorithm 1 An Algorithm for Generating Data of the IR Process and Following the Kijima's Type II Model

1 Initialize $k = 1$, $t_0 = 0$, $\nu_0 = 0$, and $d_k = 0$.
2 Generate a uniform distribution random variable u on the interval (0, 1).
3 Evaluate $t_1 = -\alpha \ln^{1/\beta}(1 - u)$.
4 Generate a uniform distribution random variable u on the interval (0, 1).
5 Solve the equation $(d_k + \nu_{k-1})^\beta - (\nu_{k-1})^\beta + \alpha^\beta \ln(1 - u) = 0$.
6 Evaluate $t_k = t_{k-1} + d_k$.
7 If $t_k < t$, then $k \leftarrow k + 1$ and go to *step 4*.
8 Store the vector of failure times $T = (t_0, t_1, \ldots, t_{N(t)})$.

simulated lifetime data and has been made by writing some computer programs with Matlab.

The simulation study to generate data of the IR process and following the Kijima's type II repair model with uniform distributed degrees of repair has been done according to Algorithm 1.

In the following example, let $\alpha = 1.2$ and $\beta = 3$. Table 1 illustrates two samples: the first is the uniform distributed degree of repair and the second is the corresponding failure times. We obtain $N(t) = 10$ failures until the time $t = 10$.

Now, parameter estimations of α and β are illustrated. A sample 1 was observed until $t = 100$. Let $\alpha = 1.2$, $\beta = 3$, and $r = 10$, where r is the number of simulations. The estimations of α, β, and LL from data of sample 1 are given in Table 2.

Table 3 illustrates the squared errors MSE of $\hat{\alpha}$ and $\hat{\beta}$.

Based on the results presented in Table 3, we could remark that, if the number of simulations s increases, then the mean squared errors of $\hat{\alpha}$ and $\hat{\beta}$ decrease. Simulations are carried out with different sets of parameters. From one set of parameters, the bias and variance of the estimators are estimated by their empirical version on 200 replicates. The estimation results are then given in Table 4.

Table 1 Failure times and repair degrees for $\alpha = 1.2$ and $\beta = 3$

z_k	0.4159	0.4900	0.3486	0.6980	0.1740	0.2247	0.8906	0.1236
t_k	1.3515	1.8118	2.4047	3.3637	3.4218	4.1991	4.6912	5.6913
z_k	0.1357	0.4553						
t_k	7.9375	8.0719						

Table 2 Estimations of α, β, and LL from data of sample 1

$\hat{\alpha}$	$\hat{\beta}$	LL
1.1907	2.9978	− 25.0579
1.2014	2.9880	− 24.9123
1.2186	3.0934	− 25.0803
1.2216	3.1723	− 26.7767
1.1970	3.1603	− 25.3764
1.2158	2.9855	− 26.9009
1.1792	2.9560	− 24.5424
1.2004	2.9841	− 26.0829
1.1855	3.0615	− 25.9890
1.2104	2.9219	− 26.1210

Table 3 MSE of $\hat{\alpha}$ and $\hat{\beta}$

s	50	100	500	1000
$MSE(\hat{\alpha})$	0.0022	0.0018	0.0017	0.0015
$MSE(\hat{\beta})$	0.0993	0.0542	0.0514	0.0504

Table 4 Estimation results considering an average of 200 simulations

	$\hat{\alpha}$	$\hat{\beta}$
Estimation	1.1903	2.9896
Empirical mean	1.2001	3.0370
Empirical variance	0.0016	0.0541

Fig. 3 illustrates the LL function with respect to α and β in case of the model (first line in Table 2). For fixed β equal 3, we obtain in Fig. 4 the graph of the LL function with respect to α. By setting α equal 1.2, Fig. 5 illustrates the graph of the LL function with respect to β.

By using the likelihood ratio statistic, we obtain in Fig. 6 simultaneous confidence regions of the parameter estimations $\hat{\alpha} = 1.2030$ and $\hat{\beta} = 3.0375$ by giving $\mu = 0.05$ for $n = 50$ (curve in line) and $n = 100$ (curve in dash).

Following the previous analysis and the results of Fig. 6, we could remark that the simultaneous confidence region based on the likelihood ratio for $n = 100$ is smaller than that for $n = 50$. In the case of $n = 100$, the parameter α varies between 1.124 and 1.270, and the parameter β varies between 2.6 and 3.4.

In Fig. 7 simultaneous confidence regions of the parameter estimations $\hat{\alpha} = 1.4955$ and $\hat{\beta} = 3.4975$ by giving $\mu = 0.05$ for $n = 50$ (curve in line) and $n = 100$ (curve in dash) are illustrated.

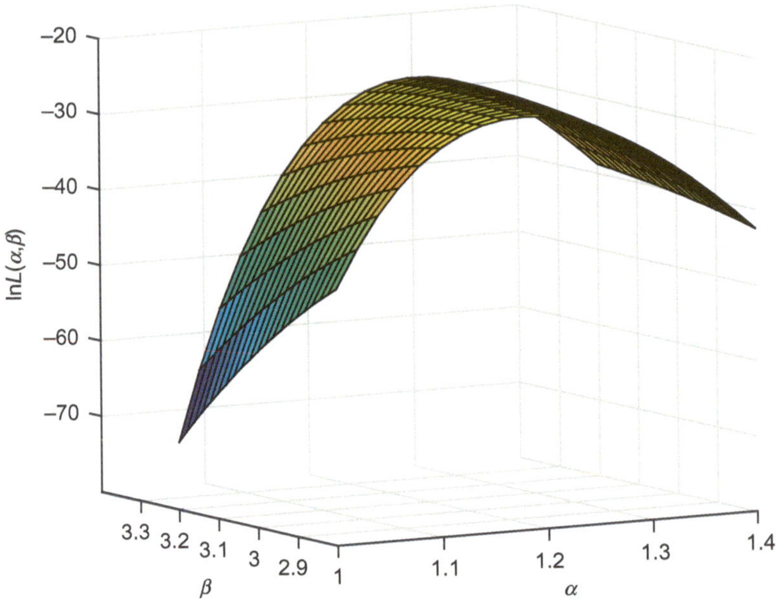

Fig. 3 Graph of the LL function for one sample with respect to α and β.

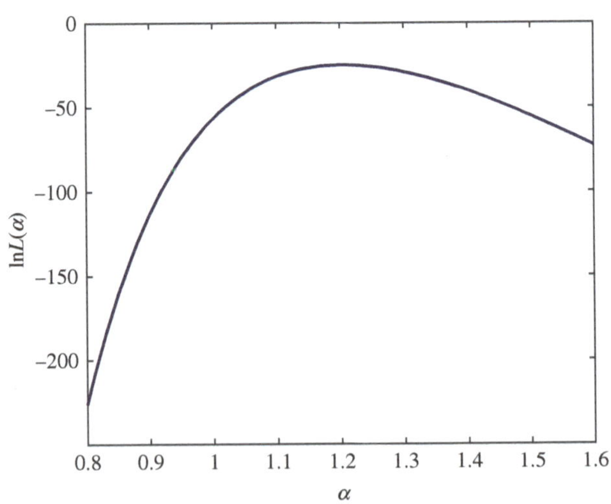

Fig. 4 Graph of the LL function for one sample with respect α.

Here too, following the previous analysis and the results of Fig. 7, we could remark that the simultaneous confidence region based on the likelihood ratio for $n = 100$ is smaller than that for $n = 50$. In the case of $n = 100$, the parameter α varies between 1.399 and 1.581, and the parameter β varies between 2.920 and 3.999.

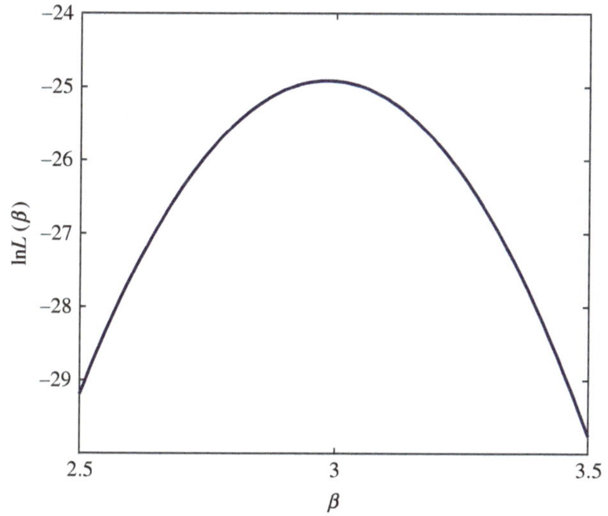

Fig. 5 Graph of the LL function for one sample with respect β.

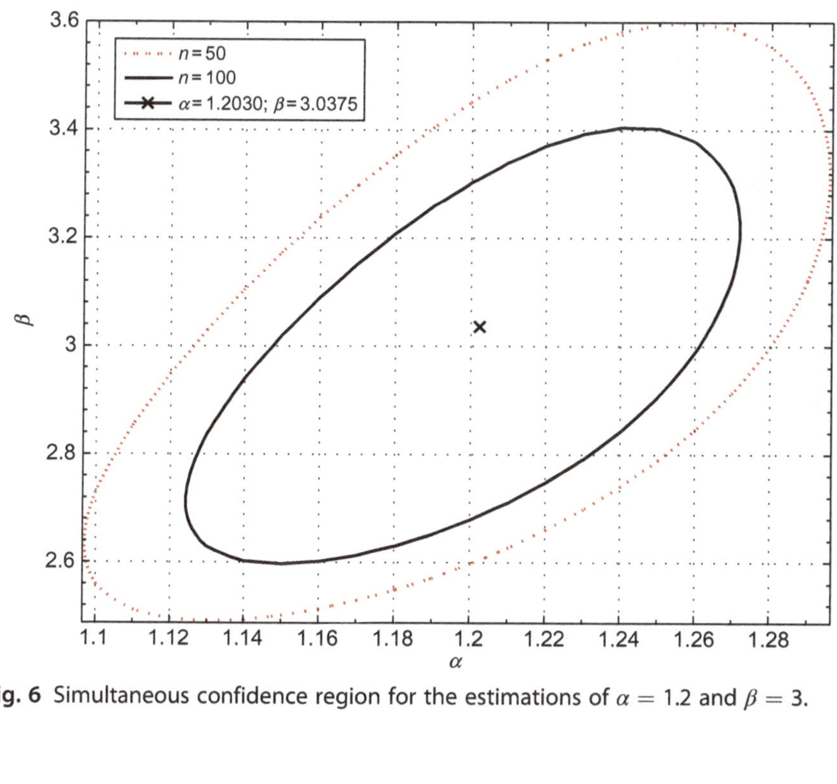

Fig. 6 Simultaneous confidence region for the estimations of $\alpha = 1.2$ and $\beta = 3$.

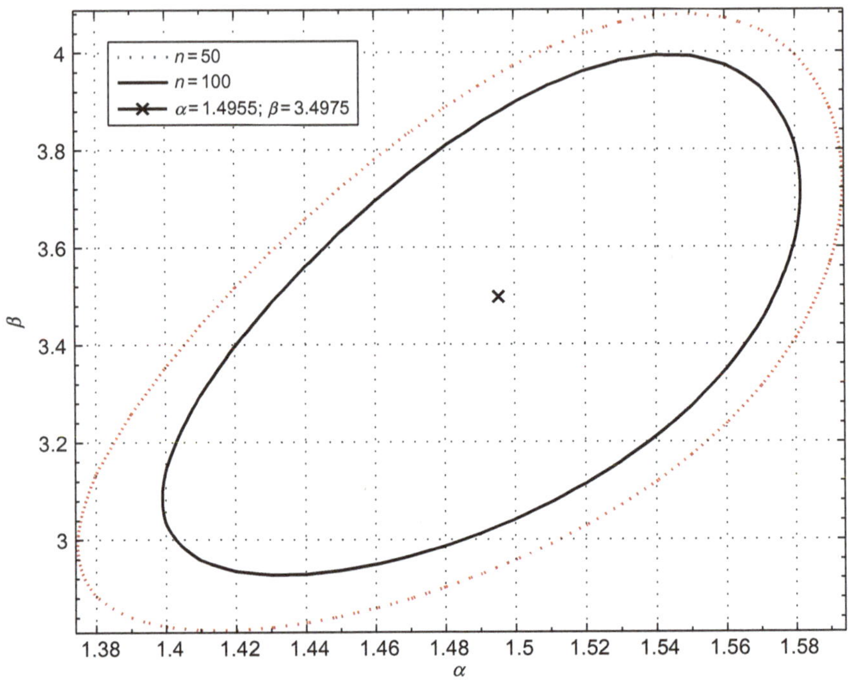

Fig. 7 Simultaneous confidence region for the estimations of $\alpha = 1.5$ and $\beta = 3.5$.

6 ILLUSTRATIVE EXAMPLE

Consider now well-known data on airplane air-conditioning failures on a fleet of Boeing aircraft (Plane 7914) giving in Hollander and Wolfe [21].

We note that, for the data from Plane 7914, the failure number is $N(t) = 24$. The rest time after the last failure is assumed to be equal to zero. Our objective is to compare the model with uniform distributed degrees of repair, denoted by Model 1 and models using a fixed degree of repair. Here it should be noted that the degrees of repair are simulated and followed a uniform distribution. The Models 2 and 3 are defined as before in Section 3. If the repair degree z_k is equal to 0.5 for all $k \in \{1, \ldots, N(t)\}$, we then obtain Model 4.

For comparison purpose, the mean square of the difference between the empirical *cdf* and the fitted *cdf*, say *MSD*, have been used. Note that *MSD* is computed by the relation:

$$MSD = \frac{1}{N(t)} \sum_{k=1}^{N(t)} \left[\hat{F}_k - F_{E,k}\right]^2,$$

where $\hat{F}_k$ and $F_{E,k}$ are, respectively, the empirical and the estimated *cdf* computed at the cumulative failure times t_k.

Table 5 Interfailure times from Plane 7914 and simulated degrees of repair (Model 1)

i	1	2	3	4	5	6	7	8	9	10	11	12
x_i	50	44	102	72	22	39	3	15	197	188	79	88
z_k	0.83	0.63	0.54	0.65	0.73	0.09	0.87	0.01	0.29	0.18	0.93	0.07
i	13	14	15	16	17	18	19	20	21	22	23	24
x_i	46	5	5	36	22	139	210	97	30	23	13	14
z_k	0.58	0.64	0.65	0.86	0.05	0.81	0.53	0.69	0.21	0.54	0.70	0.96

Table 6 Estimations of α, β, *LL*, and *MSD* from data of Plane 7914

	$\hat{\alpha}$	$\hat{\beta}$	***LL***	***MSD***
Model: 1	56.8875	0.9224	− 123.8080	0.0010
Model: 2	65.4087	1.0338	− 123.9937	0.0014
Model: 3	82.9261	1.0880	− 123.9455	0.0057
Model: 4	55.8538	0.9139	− 123.7917	0.0011

Table 5 gives the interfailure times for Plane 7914 and the simulated uniform distributed degrees of repair used in Model 1.

The ML estimates of the parameters α, β, the *MSD*, and the *LL* values were numerically evaluated for Models 1, 2, 3, and 4. The results are given in Table 6. The empirical, the estimated cdf, and the 95% lower and upper confidence bounds for the cdf of the data from Plane 7914 are represented for all introduced models. The results are shown in Fig. 8.

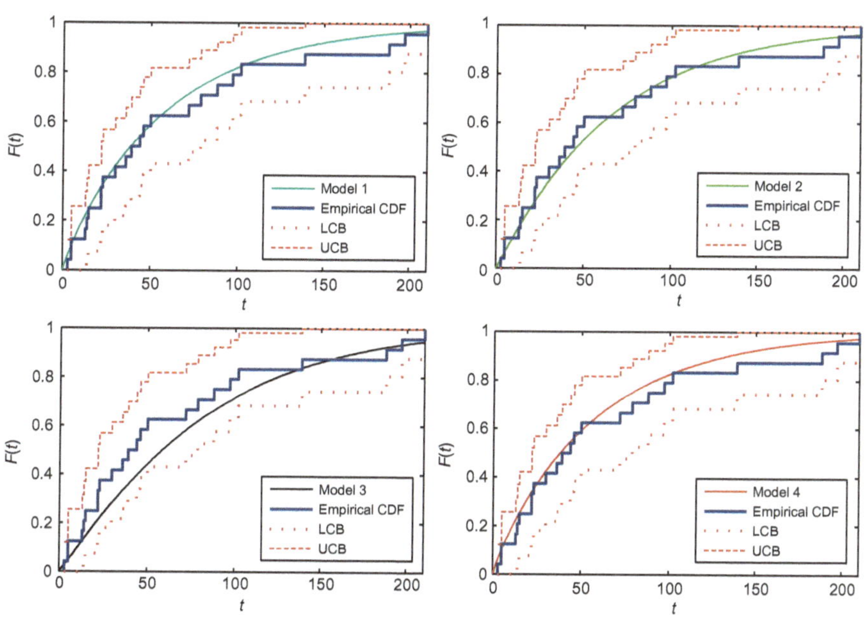

Fig. 8 *Cdf* and empirical *cdf* of data from Plane 7914 for Models 1–4.

Based on the results presented on Table 6 and Fig. 8, we can conclude that Model 4 fits the data better than Models 2 and 3, but Model 1 with simulated set of uniform distributed degrees of repair fits the data better than Model 4. We can see that the *MSD* for the Model 1 ($MSD = 0.0010$) is smaller than the *MSD* for all other models. Of course, this result depends on the simulated set of uniform distributed repair degrees.

7 CONCLUSIONS

In this chapter we developed statistical inference of a general repair model with various repair degrees based on the virtual age principle of Kijima's type II by using a Weibull intensity. The maximum likelihood estimators and the confidence regions based on the likelihood ratio statistics are obtained. The obtained theoretical results have been applied on a set of simulated data. Furthermore, for illustrative purposes, well-known and much discussed data on airplane air conditioning failures on a fleet of Boeing aircraft (Plane 7914) have been analyzed. The results obtained indicate that the Model 1 with uniform distributed degrees of repair fits the data better than the Model 2 (renewal process), Model 3 (nonhomogeneous Poisson process), and Model 4 with fixed degree of repair equal to 0.5.

REFERENCES

[1] G. Last, R. Szekli, Stochastic comparison of repairable systems by coupling, J. Appl. Probab. 35 (1998) 348–370.

[2] M. Kijima, H. Morimura, Y. Suzuki, Periodical replacement problem without assuming minimal repair, Eur. J. Oper. Res. 37 (1988) 194–203.

[3] M. Kijima, Some results for repairable systems with general repair, J. Appl. Probab. 26 (1989) 89–102.

[4] F. Bathe, J. Franz, Modelling of repairable systems with various degrees of repair, Metrika 43 (1996) 149–164.

[5] W. Stadje, D. Zuckerman, Optimal maintenance strategies for repairable systems with general degrees of repair, J. Appl. Probab. 28 (1991) 384–396.

[6] S. Gasmi, C.E. Love, W. Kahle, A general repair, proportional-hazards, framework to model complex repairable systems, IEEE Trans. Reliab. 52 (1) (2003) 26–32.

[7] L. Baxter, M. Kijima, M. Tortorella, A point process model for the reliability of a maintained system subject to general repair, Commun. Stat. Stoch. Models 12 (1) (1996) 37–65.

[8] L. Doyen, O. Gaudoin, Classes of imperfect repair models based on reduction of failure intensity or virtual age, Reliab. Eng. Syst. safety 84 (2004) 45–56.

[9] S. Gasmi, Parameter estimation in an alternating repair model, J. Stat. Plan. Inference 141 (11) (2011) 3605–3616.

[10] S. Gasmi, Statistical Inference of an Imperfect Maintenance Model, Adv. Appl. Stat. 30 (2012) 93–120.

[11] S. Gasmi, Estimating parameters of a log-linear intensity for a repairable system, Appl. Math. Model. 37 (2013) 4325–4336.

[12] H. Jiawen, J. Zuhua, L. Haitao, Preventive maintenance of a single machine system working under piecewise constant operating condition, Reliab. Eng. Syst. Safety 168 (2017) 105–115.

[13] S. Ghnimi, S. Gasmi, A. Nasr, Reliability parameters estimation for parallel systems under imperfect repair, Metrika 80 (3) (2017) 273–288.

[14] M. Tanwar, R. Rai, N. Bolia, Imperfect repair modeling using Kijima type generalized renewal process, Reliab. Eng. Syst. Safety 124 (2014) 24–31.

[15] Y. Dijoux, M. Fouladirad, D.T. Nguyen, Statistical inference for imperfect maintenance models with missing data, Reliab. Eng. Syst. Safety 154 (2016) 84–96.

[16] M.S. Rashad, M.M. Nagwa, On estimation of modified Weibull parameters in presence of accelerated life test, J. Stat. Theory Pract. 12 (3) (2018) 542–560.

[17] X. Li, W. Ding, A stochastic model on one-unit repairable systems with multiple degenerative states, J. Syst. Sci. Complex. (2017), https://doi.org/10.1007/s11424-017-6306-2.

[18] M. Brown, F. Proschan, Imperfect repair, J. Appl. Probab. 20 (1993) 851–859.

[19] R.S. Liptser, A.N. Shiryayev, Statistics of Random Processes, vol. 2, Springer, New York, NY, 2000.

[20] O.E. Barndhoff-Nielsen, P. Blaesild, A note of the calculation of Bartlett adjustments, J. R. Stat. Soc. Ser. B 48 (1968) 353–358.

[21] M. Hollander, D.A. Wolfe, Nonparametric Statistical Methods, John Wiley and Sons, New York, NY, 1973.

CHAPTER 8

Method of Fuzzy Perfectness in Human Reliability Analysis: Selection of Performance Conditions

Alexander Rotshtein*, Ludmila Pustylnik†, Brian A. Polin‡
*Department of Industrial Engineering, Jerusalem College of Technology, Jerusalem, Israel
†Department of Mechanical Engineering, Afeka Tel-Aviv Academic College of Engineering, Tel Aviv, Israel
‡Jerusalem College of Technology, Jerusalem, Israel

Abstract

Within the framework of the cognitive reliability and error analysis method (CREAM), we propose a systematic procedure for the comparison of various performance conditions, that is, the factors affecting human reliability without the actual calculation of the probability of human error. The proposed method is based on the idea of fuzzy perfection of performance conditions and the theory of multicriteria selection of alternatives under fuzziness. The alternatives are vectors of human performance conditions that must be compared. The criteria are levels of perfection of performance conditions calculated using specially introduced membership functions. To test the proposed method, we use probabilities of human errors for the five different scenarios calculated by fuzzy CREAM.

Keywords: Fuzzy perfectness, Human reliability analysis, Performance conditions, Selection of alternatives

1 INTRODUCTION

Human errors are the cause of many accidents and disasters in transport, industry, and other human-machine systems. The probability of human operator error depends on many factors determining the operating conditions, such as working place arrangements, level of training, intensity of work, degree of fatigue, etc. The creation of optimal operating conditions, to minimize the probability of human error, is the most important part of human-machine system design. To formulate this problem in terms of classical mathematical programming, we must define:

Advances in System Reliability Engineering
https://doi.org/10.1016/B978-0-12-815906-4.00008-7

(a) human performance conditions as control variables
(b) probability of human error as the objective function, depending on performance conditions
(c) restrictions to all human performance conditions

The main difficulties of such an approach are as follows:

1. Many human performance conditions have a qualitative character and may be estimated only by experts.
2. The restrictions to values of human performance conditions do not have crisp borders and are fuzzy.

These difficulties led to the widespread usage of the trial and error method in system analysis [1]; an option is proposed and estimated, then modifications are introduced, the estimation is repeated, etc. This is shown in Fig. 1A, where M is a model of prediction of error probability (q) depending on the vector ($\mathbf{X}$) of human performance conditions, and I is an improvement of performance conditions.

The prediction model (M) can be obtained from training sample "input ($\mathbf{X}$)-output (q)" using empirical modeling methods: regression analysis, neural networks, fuzzy rules, etc. It is not easy to build such a training sample correctly because of the subjectivity involved in estimations of human performance conditions; different experts on different days can differently estimate the same conditions. That is why the precise prediction of error probability (q) is a very complicated problem. High complexity is *incompatible* with high accuracy [2].

Recently, a fuzzy cognitive reliability and error analysis method (fuzzy CREAM [3, 4]) has become popular. This method allows for the calculation of human error probability using fuzzy logic evidence from if-then rules and a defuzzification procedure. The disadvantages of this method are as follows:

1. The process of modeling is very time-consuming because of large number of fuzzy if-then rules.
2. There is a risk of using of contradicting rules.
3. According to the incompatibility principle [2], we cannot guarantee the precise prediction of human error probability.

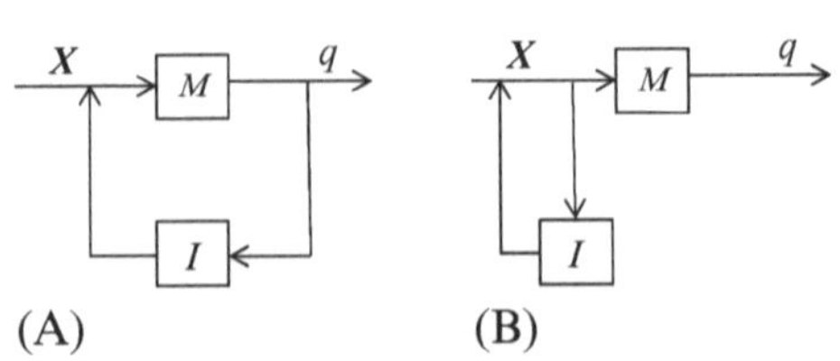

Fig. 1 Strategies of reliability-based design.

In this chapter, we propose a method for the selection of the performance conditions affecting human reliability without the actual prediction of human error probability, as shown on Fig. 1B. The method is based on the specially introduced criteria of fuzzy perfection and the theory of decision-making under uncertainty [5]. To test the proposed method, we use the human error probabilities in five different scenarios previously obtained by fuzzy CREAM [3, 4].

The structure of the paper is as follows.

In Section 2, we consider the existing approaches for the evaluation of human reliability and the main principles of fuzzy CREAM, which provide error probabilities for the five scenarios of human activity. The material contained in this this section is not our novel innovation, but its inclusion is necessary to understand the advantages of the method proposed in the next section.

Section 3 describes the *method of fuzzy perfection*, which provides the set of human performance conditions affecting human reliability, without the time-consuming modeling of error probability. The proposed method is tested using the results described in Section 2.

2 KNOWN APPROACHES

To clarify the relationship "performance conditions (input) → probability of human error (output)", various observation-based modeling methods may be used: regression analysis, neural networks, fuzzy if-then rules, etc. The early publications on human-machine reliability [6] use experimental design and regression models: polynomial [7] and functional nonlinear [8, 9].

Since the latter part of the last century, fuzzy logic of Zadeh [10] has been widely used in reliability theory [11–14]. Using Zadeh's fuzzy logic, Rotshtein [12] proposed the evaluation of human-operator reliability by the $L(\mathbf{X})$-form of fuzzy numbers: $[P, \overline{P}, L(\mathbf{X})]$, where P and $\overline{P}$ are the lowest and highest levels of probability of human correct action; $L(\mathbf{X})$ is a fuzzy term (low, average, high, etc.) that depends on the vector $(\mathbf{X})$ of affected factors and describes the level (or class) of reliability in the interval $[P, \overline{P}]$. The fuzzy term $L(\mathbf{X})$ is calculating by evidence from fuzzy if-then rules, such as: IF the level of operator's training is *high* and the intensity of work is *lower than average*, and the degree of fatigue is *low*, THEN the probability of correct action is *low*.

The same ideas have been used in fuzzy CREAM [3,4], which is now the most popular method in the human reliability analysis of systems with high level of danger (transportation, nuclear power plants, etc.).

The fuzzy CREAM method is based on the original CREAM [15] method using the following principles:

2.1 Classes of Reliability

For the qualitative evaluation of human reliability, the CREAM method uses four classes, presented in Table 1. Each class corresponds to a control mode, characterized by the human ability to anticipate the future situation, and an interval of the probability of human error. These classes of reliability are considered as fuzzy sets formalized by membership functions (Fig. 2) on the universe [0, 1].

2.2 Human Performance Conditions

Nine common performance conditions, presented in Table 2, are considered as the factors affecting human reliability. Fuzzy terms (X_{ij}), which are formalized by membership functions, are used for the qualitative evaluation of human performance conditions (x_i). The examples of these membership functions for the two different universes are shown in Fig. 3. Each level of the quality of performance condition has one of the three types of effects influencing to human reliability: reducing (−), insignificant (0), or improving (+).

2.3 Generator of If-Then Rules

CREAM diagram 15 (Fig. 4) is used to determine the class (c_i) of human reliability for a given vector of qualitative evaluations of performance conditions. This diagram is a presentation of the formula $c_i = \varphi(N_-, N_+)$, where N_- and N_+ are the sums of the performance conditions reducing (−) and improving (+) the human reliability.

As an example, let us consider some situation S defined by the following combination of the terms:

$$S = \{(x_1 = X_{12}) \text{ and } (x_2 = X_{22}) \text{ and } (x_3 = X_{32}) \text{ and } (x_4 = X_{42}) \text{ and } (x_5 = X_{51}) \\ \text{ and } (x_6 = X_{63}) \text{ and } (x_7 = X_{72}) \text{ and } (x_8 = X_{83}) \text{ and } (x_9 = X_{93})\}.$$

In this situation, we have two negative influences (X_{12} and X_{51}) and two positive influences (X_{63} and X_{83}), that is, $N_- = 2$, $N_+ = 2$. Using the CREAM diagram, we can see that situation S corresponds to the class of reliability c_3 (opportunistic), that is, IF S, THEN $q \varepsilon c_3$.The total number of rules IF <situation>, THEN <class of reliability>, generated by the CREAM diagram is $N = 4^3 \cdot 3^6 = 46656$.

Table 1 Classes of reliability

Class	Control mode	Ability of anticipation	Error probability	Intervals of error probability
c_1	Scramble	Absent	Very high	$1 \times 10^{-1} < q < 1 \times 10^{0}$
c_2	Opportunistic	Low	High	$1 \times 10^{-2} < q < 0.5 \times 10^{0}$
c_3	Tactical	Medium	Medium	$1 \times 10^{-3} < q < 1 \times 10^{-1}$
c_4	Strategic	High	Low	$0.5 \times 10^{-5} < q < 1 \times 10^{-2}$

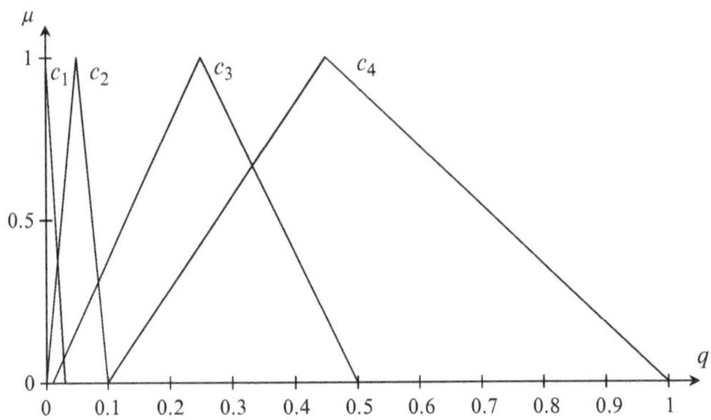

Fig. 2 Membership functions of the classes of error probability.

2.4 Fuzzy Logic Evidence

The crisp value of human error probability can be obtained from the fuzzy if-then rules generated by CREAM diagram and Mamdani's fuzzy evidence [16]. Let k_j be the number of rules in class c_j, which are numbered as follows: $j_1, j_2, \ldots, jk_j$, where $j=1, \ldots, 4$. The system of fuzzy rules approximating the dependence of the human error probability (q) with influencing factors x_i can be written as:

$$\text{IF}\left(x_1=a_1^{j1}\right)\text{ and }\left(x_2=a_2^{j1}\right)\ldots\text{and }\left(x_9=a_9^{j1}\right)\text{ or }\left(x_1=a_1^{j2}\right)\text{ and }\left(x_2=a_2^{j2}\right)$$

$$\ldots\text{and }\left(x_9=a_9^{j2}\right)\ldots\text{or }\left(x_1=a_1^{jk_j}\right)\text{ and }\left(x_2=a_2^{jk_j}\right)\ldots\text{and }\left(x_9=a_9^{jk_j}\right),$$

$$\text{THEN } q=c_j,$$

where a_i^p is a fuzzy term from Table 2, which estimates the variable x_i in the conjunction with number

$$p=k_j.$$

Using the substitutions and = min, or = max, and defuzzification procedure by center of gravity method [16], we obtain the crisp value of error probability (q^*) for a given vector ($x_1, x_2, \ldots, x_9$) of qualitative estimations of human performance conditions:

$$q^*=\int_0^1 \mu_c(q)\cdot q\cdot dq\Big/\int_0^1 \mu_c(q)\cdot dq$$

Table 2 Common performance conditions

Performance condition	Factor	Universal set	Terms (symbols)	Effect
Adequacy of organization	x_1	[0, 100]	Deficient (X_{11})	–
			Inefficient (X_{12})	–
			Efficient (X_{13})	0
			Very efficient (X_{14})	+
Working conditions	x_2	[0, 100]	Incompatible (X_{21})	–
			Compatible (X_{22})	0
			Advantageous (X_{23})	+
Adequacy of man-machine interface (MMI)	x_3	[0, 100]	Inappropriate (X_{31})	–
			Tolerable (X_{32})	0
			Adequate (X_{33})	0
			Supportive (X_{34})	+
Availability of procedures/plans	x_4	[0, 100]	Inappropriate (X_{41})	–
			Acceptable (X_{42})	0
			Appropriate (X_{43})	+
Number of simultaneous goals	x_5	[0, 100]	More than actual capacity (X_{51})	–
			Matching current capacity (X_{52})	0
			Fewer than actual capacity (X_{53})	0
Available time	x_6	[0, 100]	Continuously inadequate (X_{61})	–
			Temporarily inadequate (X_{62})	0
			Adequate (X_{63})	+
Time of the day	x_7	[0, 24]	Night (X_{71})	–
			Day (X_{72})	0
			Night (X_{73})	–
Adequacy of training and experience	x_8	[0, 100]	Inadequate (X_{81})	–
			Adequate with limited experience (X_{82})	0
			Adequate (X_{83})	+
Crew collaboration quality	x_9	[0, 100]	Deficient (X_{91})	–
			Inefficient (X_{92})	0
			Efficient (X_{93})	0
			Very efficient (X_{94})	+

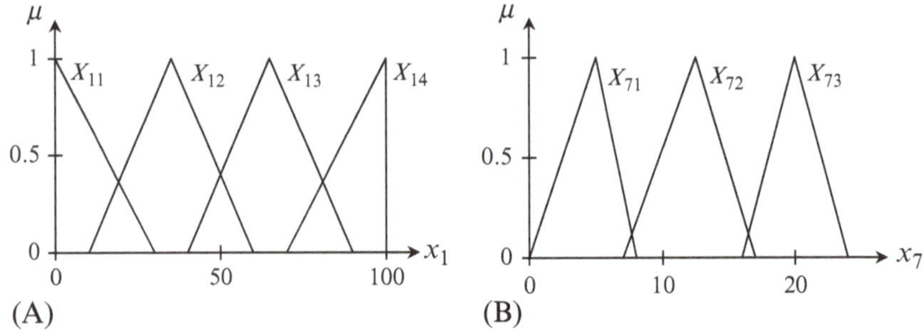

Fig. 3 Membership functions of fuzzy terms of variables x_1 and x_7.

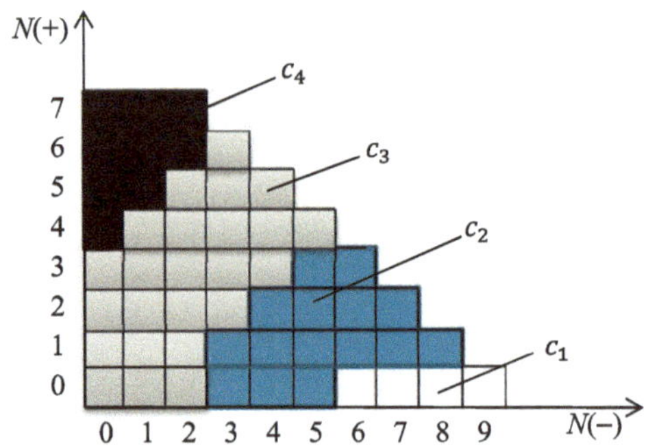

Fig. 4 CREAM diagram.

$$\mu_c(q) = \max \begin{cases} \min\left[w_1, \mu^{C_1}(q)\right] \\ \dots \\ \min\left[w_4, \mu^{C_4}(q)\right], \end{cases}$$

$$w_j = \max \begin{cases} \min\left[\mu^{j1}(x_1), \dots, \mu^{j1}(x_9)\right] \\ \dots \\ \min\left[\mu^{jk_j}(x_1), \dots, \mu^{jk_j}(x_9)\right], \end{cases}$$

where $\mu^{c_j}(q)$ is membership function of error probability q to class of reliability c_j, $\mu^P(x_i)$ is membership function of performance condition x_i to term $p=k_j$, $i=1, \dots, 9$, $j=1, \dots, 4$ (Figs. 2 and 3).

The results of fuzzy CREAM [3] application to five scenarios $(s_1, \dots, s_5)$ with different expert values of human performance conditions $(x_1, x_2, \dots, x_9)$

Table 3 Results of fuzzy CREAM application

	x_1	x_2	x_3	x_4	x_5	x_6	x_7	x_8	x_9	q
s_1	22	30	50	40	60	70	4	50	70	1.00×10^{-2}
s_2	90	90	90	90	90	90	12	90	90	9.81×10^{-4}
s_3	5	17	42	38	78	45	22	56	78	6.33×10^{-2}
s_4	10	10	10	10	10	10	2	10	10	2.02×10^{-1}
s_5	10	12	14	12	15	16	20	18	20	1.91×10^{-1}

are presented in Table 3. These results show that, for human error probabilities q_i, we have the following inequality: $q_2 < q_1 < q_3 < q_5 < q_4$,that is, the best scenario is s_2 and the worst is s_4.

The main disadvantage of the fuzzy CREAM methodology [3,4] is high complexity caused by large number of fuzzy if-then rules. These rules can contain incompatible combinations of fuzzy terms evaluating the performance conditions. For example, it is not logical to have:

(a) the adequacy of organization is *very efficient* ($x_1 = X_{14}$) and the working conditions are *incompatible* ($x_2 = X_{21}$)

(b) the number of simultaneous goals is *more than actual capacity* ($x_5 = X_{51}$) and the available time is *adequate* ($x_6 = X_{63}$), etc.

3 METHOD OF FUZZY PERFECTION

3.1 Perfection as Criterion

For the evaluation of each performance condition, we propose to use a fuzzy term, which we call perfection.

For a formal description of fuzzy perfection of *i*th performance condition, we use the membership function $\pi(x_i) \in [0, 1]$ characterizing the level of closeness of variable x_i to a certain ideal: 0 is the smallest perfection, 1 is the greatest one. "*Excellence*" or "*correctness*" might be synonyms with fuzzy perfection [17].

Fig. 5 depicts examples of different cases of fuzzy borders between perfect and nonperfect values of an arbitrary variable x, estimated by an expert. Panel (A) of Fig. 5 describes the case when, as the value of x increases, there is a transition from absolute nonperfection ($x < a_1$) to absolute perfection ($x > a_2$). Panel (B) describes a fuzzy transition from absolute perfection ($x < b_1$) to absolute nonperfection ($x > b_2$). Panel (C) describes a fuzzy

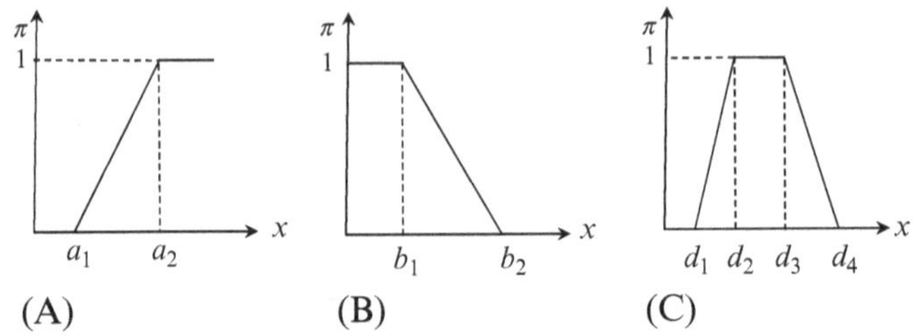

Fig. 5 Options of the borders between perfection and nonperfection.

transition from absolute nonperfection ($x < d_1$) to absolute perfection ($d_2 < x < d_3$) and again to absolute nonperfection ($x > d_4$).

To describe the borders between perfect (1) and nonperfect (0) values of a variable x, we can use the piecewise—quadratic membership functions [18] shown in Fig. 6.

Panel (A) on Fig. 6 describe a fuzzy transition "nonperfection (0)-perfection (1)" by the formula:

$$\pi_1(x) = \begin{cases} 0, & \text{for } x \in [0, a_1] \\ 2\left(\frac{x-a}{1-a}\right)^2, & \text{for } x \in \left[a, \frac{a+1}{2}\right] \\ 1 - \left(\frac{x-1}{1-a}\right)^2, & \text{for } x \in \left[\frac{a+1}{2}, 1\right] \end{cases},$$

where the parameter $a \in [0, 1]$ indicates the subjective selection of the a in such a way that for $x > a$ the degree of perfection is positive, that is, $\pi_1(a) > 0$.

Panel (B) on Fig. 6 describes a fuzzy transition "perfection (1)-nonperfection (0)" by the formula:

$$\pi_2(x) = \pi_1(1-x).$$

Panel (C) on Fig. 6 describe a fuzzy transition "nonperfection (0)-perfection (1)-nonperfection (0)" by the formula:

$$\pi_3(x) = \begin{cases} \frac{1}{2(p-\beta-d_1)}(x-d_1)^2, & \text{for } x \in [a_1, p-\beta] \\ -\frac{1}{2\beta^2}(x-p)^2 + 1, & \text{for } x \in [p-\beta, p+\beta] \\ \frac{1}{2(p+\beta-d_2)}(x-d_2)^2, & \text{for } x \in [p+\beta, d_2] \end{cases},$$

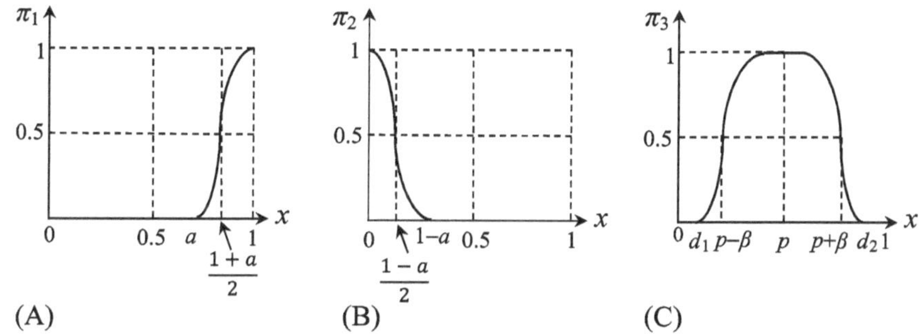

Fig. 6 Piecewise-quadratic membership functions.

where d_1, d_2, and β are subjective parameters such that

$$\pi_3(x \leq d_1) = 0, \pi_3(x \geq d_2) = 0,$$

$$p = \frac{1}{2}(d_1 + d_2),$$

$$\pi_3(p - \beta) = \pi_3(p + \beta) = \frac{1}{2}.$$

For the simplification of calculation of fuzzy perfection of common performance conditions using the CREAM methodology, we propose the following membership functions shown in Fig. 7:

$$\pi(x_i) = \frac{x_i}{100}, \quad \text{for } x_i \in [0, 100], \quad i = 1, \ldots, 9 \ (i \neq 7) \tag{1}$$

$$\pi(x_7) = \begin{cases} \dfrac{x_7}{100}, & \text{for } x_7 \in [0, 8] \\ 1, & \text{for } x_7 \in [8, 16] \\ \dfrac{24 - x_7}{8}, & \text{for } x_7 \in [16, 24] \end{cases}. \tag{2}$$

These models are based on the following assumptions:

(a) The membership degree of a variable x_i to the term perfection increases linearly with increasing the value of variable x_i for all $i = 1, \ldots, 9$ $(i \neq 7)$.

(b) The time of day (x_7) most suitable for reliable human activity is from 8 to 16 h.

3.2 The Rule of Scenarios Comparison

Let $S = \{s_1, s_2, \ldots, s_m\}$ be a set of scenarios to be compared from the human reliability point of view. The performance conditions for the scenario $s_j \in S$ can be described by the vector $(x_{1j}, x_{2j}, \ldots, x_{9j})$, where x_{ij} is the expert value of ith performance condition in the scenario $s_j \in S$, $i = 1, \ldots, 9$; $j = 1, \ldots, m$.

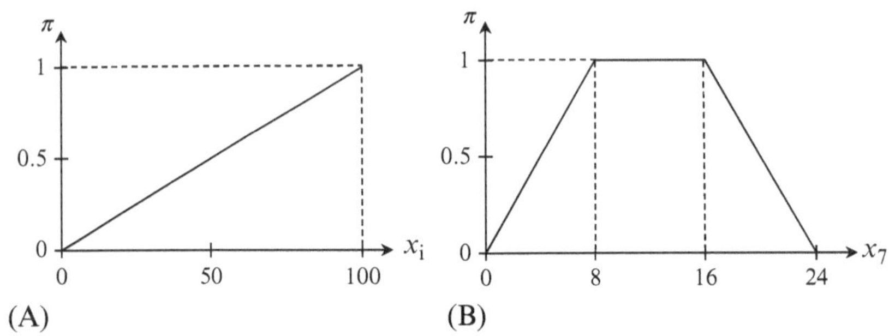

Fig. 7 Membership functions of fuzzy perfection of performance conditions.

To compare the scenarios from the set S, we use the set of criteria $C=\{C_1, C_2, \ldots, C_9\}$, where C_i is the perfection of ith performance condition from Table 2.

Let us present the criterion C_i as a fuzzy set on the universal set of scenarios $S=\{s_1, s_2, \ldots, s_m\}$:

$$C_i=\left\{\frac{\pi(x_{i1})}{s_1}, \frac{\pi(x_{i2})}{s_2}, \ldots, \frac{\pi(x_{im})}{s_m}\right\}, \tag{3}$$

where $\pi(x_{ij})$ is a level of perfection of ith performance condition in the scenario s_j. To calculate $\pi(x_{ij})$, we use the membership functions of fuzzy perfection presented in Fig. 6.

According to the principle of Bellman-Zadeh [5], the best scenario belongs to the intersection (∩) of fuzzy sets of criteria:

$$s_{opt} \in D=C_1 \cap C_2 \cap \cdots \cap C_9, \tag{4}$$

that is, the best scenario is simultaneously the best by all of criteria C_i, $i=1, \ldots, 9$. In the theory of decision-making under fuzziness [5], there is a replacement: $\cap \rightarrow \min$. Therefore the fuzzy set D looks as follows:

$$D=\left\{\frac{\min\left[\pi(x_{11}), \ldots, \pi(x_{91})\right]}{s_1}, \frac{\min\left[\pi(x_{12}), \ldots, \pi(x_{92})\right]}{s_2}, \ldots, \frac{\min\left[\pi(x_{1m}), \ldots, \pi(x_{9m})\right]}{s_m}\right\}.$$

As a best decision, we have to select the scenario s_{opt} with maximal level of perfection, that is,

$$\pi\left(s_{\text{opt}}\right)=\max_{j=1,\ldots,m} \min\left[\pi\left(x_{1j}\right), \ldots, \pi\left(x_{9j}\right)\right]$$

This allows the formulation of a simple rule of comparison of scenarios in reliability-based design of human performance conditions. Suppose we have to compare two scenarios $s_j \in S$ and $s_k \in S$ with known vectors $(x_{1j}, \ldots, x_{9j})$ and $(x_{1k}, \ldots, x_{9k})$ of expert evaluations of performance conditions, and unknown probabilities q_j, q_k of human errors, respectively. The rule of comparison is:

$$\text{IF } \min\left[\pi\left(x_{1j}\right), \ldots, \pi\left(x_{9j}\right)\right] > \min\left[\pi(x_{1k}), \ldots, \pi(x_{9k})\right], \text{THEN } q_j < q_k.$$

Note. To increase the sensitivity of this rule, we can use the algebraic multiplication $(\cdot)$ instead of standard intersection (min), that is:

$$\text{IF } \left[\pi\left(x_{1j}\right) \cdot \pi\left(x_{2j}\right) \ldots \cdot \pi\left(x_{9j}\right)\right] > \left[\pi(x_{1k}) \cdot \pi(x_{2k}) \ldots \cdot \pi(x_{9k})\right], \text{THEN } q_j < q_k.$$

It can be interpreted as a comparison of serial structures in classic reliability theory.

3.3 Example 1

Let us consider the comparison of the five scenarios shown in Table 3. By substitution of the data into the membership functions (1), (2), we get the perfection of performance conditions as a fuzzy set defined on the universal set of scenarios $\{s_1, \ldots, s_5\}$:

$$C_1 = \{0.22/s_1, 0.9/s_2, 0.15/s_3, 0.1/s_4, 0.1/s_5\}$$
$$C_2 = \{0.3/s_1, 0.9/s_2, 0.17/s_3, 0.1/s_4, 0.12/s_5\}$$
$$C_3 = \{0.4/s_1, 0.9/s_2, 0.38/s_3, 0.1/s_4, 0.12/s_5\}$$
$$C_4 = \{0.5/s_1, 0.9/s_2, 0.42/s_3, 0.1/s_4, 0.14/s_5\}$$
$$C_5 = \{0.6/s_1, 0.9/s_2, 0.78/s_3, 0.1/s_4, 0.15/s_5\}$$
$$C_6 = \{0.7/s_1, 0.9/s_2, 0.45/s_3, 0.1/s_4, 0.16/s_5\}$$
$$C_7 = \{0.5/s_1, 1.0/s_2, 0.25/s_3, 0.25/s_4, 0.5/s_5\}$$
$$C_8 = \{0.5/s_1, 0.9/s_2, 0.56/s_3, 0.1/s_4, 0.18/s_5\}$$
$$C_9 = \{0.7/s_1, 0.9/s_2, 0.78/s_3, 0.1/s_4, 0.2/s_5\}$$

The intersection $D = C_1 \cap C_2 \cap \cdots \cap C_9$ is the fuzzy set

$$D = \{0.22/s_1, 0.9/s_2, 0.15/s_3, 0.1/s_4, 0.1/s_5\},$$

which shows that the most perfect scenario is s_2, the second place is s_1, the third place is s_3, the fourth and fifth places are s_4 and s_5, i.e., for the probabilities of human errors we have the inequality: $q_2 < q_1 < q_3 < q_5 = q_4$. This result is little bit different from the fuzzy CREAM result [3] because $q_5 = q_4$, instead of $q_5 < q_4$.

Note. Application of algebraic multiplication ($\cap = *$) instead of standard intersection ($\cap = \min$) generates the inequality: $q_2 < q_1 < q_3 < q_5 < q_4$, which coincides with the result received by fuzzy CREAM [3].

3.4 The Importance of Performance Conditions

In Section 3.2, we considered the case when all of the performance conditions have the same importance. According to Yager [19], it is necessary to use concentration or dilation of the membership functions to take into account the importance of fuzzy criteria. If $\alpha_i \geq 0$ is a weight of the criterion C_i, then according to Yager [19], Formula (4) should be converted to the form:

$$D = (C_1)^{\alpha_1} \cap (C_2)^{\alpha_2} \cap \cdots \cap (C_9)^{\alpha_9}, \tag{5}$$

where

$$(C_i)^{\alpha_i} = \left\{ \frac{[\pi(x_{i1})]^{\alpha_i}}{s_1}, \frac{[\pi(x_{i2})]^{\alpha_i}}{s_2}, \ldots, \frac{[\pi(x_{im})]^{\alpha_i}}{s_m} \right\}. \tag{6}$$

For the evaluations of weights α_i, we use the worst-case method proposed in Ref. [20]:

Let R_1, R_2, ..., R_9 be the ranks of the criteria C_1, C_2, ..., C_9. Let us suppose that the higher the weight of criterion (α_j), the higher the rank (R_j). It can be formalized by the relation:

$$\frac{\alpha_1}{R_1} = \frac{\alpha_2}{R_2} = \ldots = \frac{\alpha_l}{R_l} = \ldots \frac{\alpha_9}{R_9} \tag{7}$$

Let us also suppose that C_l is the least important criterion with weight α_l and rank R_l. Let us express the weight of all criteria relative to the least important criterion using relation (7):

$$\alpha_1 = \alpha_l \frac{R_1}{R_l}, \alpha_2 = \alpha_l \frac{R_1}{R_l}, \ldots, \alpha_9 = \alpha_l \frac{R_1}{R_l}. \tag{8}$$

According to Yager [19], the sum of the weights is equal to the number of the criteria (in our case, Eq. 9),
that is, $\alpha_1 + \alpha_2 + \ldots + \alpha_9 = 9$. Therefore from Eq. (8) we have:

$$\alpha_l = \frac{9}{\frac{R_1}{R_l} + \frac{R_2}{R_l} + \ldots + \frac{R_9}{R_l}}. \tag{9}$$

Relations (9), (8) allow one to calculate the unknown weight of human performance conditions based on the expert ranks comparison of all the criteria (R_i) with the least important one (R_l) using the 9-point Saaty scale [21]:

$$\frac{R_i}{R_l} = \begin{cases} 1, & \text{criteria } C_i \text{ and } C_l \textit{ equally} \text{ important} \\ 3, & C_i \text{ slightly } \textit{moreimportant} \text{ then } C_l \\ 5, & C_i \textit{ stronglymoreimportant} \text{ then } C_l \\ 7, & C_i \text{ very } \textit{stronglymoreimportant} \text{ then } C_l \\ 9, & \text{if } C_i \text{ absolutely } \textit{moreimportant} \text{ then } C_l \\ 2,4,6,8 - \textit{intermediate} \text{ values}, & \end{cases}$$

where $i = 1, 2, \ldots, 9$.

3.5 Example 2

Let us continue example 1 described in Section 3.3. We suppose that the following expert information is known:

- the least important performance condition is the time of day (C_7).
- $\frac{R_1}{R_7} = 2, \frac{R_2}{R_7} = 1, \frac{R_3}{R_7} = 1, \frac{R_4}{R_7} = 1, \frac{R_5}{R_7} = 7, \frac{R_6}{R_7} = 1, \frac{R_7}{R_7} = 1, \frac{R_8}{R_7} = 1, \frac{R_9}{R_7} = 1.$

Using Formulas (9), (8), we obtain:

$$\alpha_7 = \frac{9}{\sum_{i=1}^{9} {}^{R_i}/_{R_7}} = \frac{9}{2+1+1+1+7+1+1+3+1} = 0.5,$$

$$\alpha_1 = \alpha_7 \cdot \frac{R_1}{R_7} = 0.5 \cdot 2 = 1,$$

$$\alpha_2 = \alpha_7 \cdot \frac{R_2}{R_7} = 0.5 \cdot 1 = 0.5,$$

$$\alpha_3 = \alpha_4 = \alpha_6 = \alpha_7 = \alpha_9 = \alpha_2 = 0.5,$$

$$\alpha_5 = \alpha_7 \cdot \frac{R_5}{R_7} = 0.5 \cdot 7 = 3.5,$$

$$\alpha_8 = \alpha_7 \cdot \frac{R_8}{R_7} = 0.5 \cdot 3 = 1.5.$$

Taking into account these weights, we can write the fuzzy sets (6) as follows:

$$(C_1)^1 = \left\{ {}^{0.22}/_{s_1}, {}^{0.9}/_{s_2}, {}^{0.15}/_{s_3}, {}^{0.1}/_{s_4}, {}^{0.1}/_{s_5} \right\}$$

$$(C_2)^{0.5} = \left\{ {}^{0.55}/_{s_1}, {}^{0.95}/_{s_2}, {}^{0.41}/_{s_3}, {}^{0.32}/_{s_4}, {}^{0.35}/_{s_5} \right\}$$

$$(C_3)^{0.5} = \left\{ {}^{0.63}/_{s1}, {}^{0.95}/_{s2}, {}^{0.62}/_{s3}, {}^{0.32}/_{s4}, {}^{0.35}/_{s5} \right\}$$

$$(C_4)^{0.5} = \left\{ {}^{0.71}/_{s1}, {}^{0.95}/_{s2}, {}^{0.65}/_{s3}, {}^{0.32}/_{s4}, {}^{0.37}/_{s5} \right\}$$

$$(C_5)^{3.5} = \left\{ {}^{0.17}/_{s1}, {}^{0.69}/_{s2}, {}^{0.42}/_{s3}, {}^{3.2\cdot 10^{-4}}/_{s_4}, {}^{2.3\cdot 10^{-3}}/_{s_5} \right\}$$

$$(C_6)^{0.5} = \left\{ {}^{0.84}/_{s1}, {}^{0.95}/_{s2}, {}^{0.67}/_{s3}, {}^{0.32}/_{s4}, {}^{0.4}/_{s5} \right\}$$

$$(C_7)^{0.5} = \left\{ {}^{0.71}/_{s1}, {}^{1}/_{s2}, {}^{0.5}/_{s3}, {}^{0.5}/_{s4}, {}^{0.71}/_{s5} \right\}$$

$$(C_8)^{1.5} = \left\{ {}^{0.35}/_{s1}, {}^{0.85}/_{s2}, {}^{0.42}/_{s3}, {}^{0.03}/_{s4}, {}^{0.076}/_{s5} \right\}$$

$$(C_9)^{0.5} = \left\{ {}^{0.84}/_{s1}, {}^{0.95}/_{s2}, {}^{0.88}/_{s3}, {}^{0.32}/_{s4}, {}^{0.45}/_{s5} \right\}.$$

After the intersection of these fuzzy sets, we obtain the fuzzy decision set:

$$D = \left\{ {}^{0.17}/_{s1}, {}^{0.69}/_{s2}, {}^{0.15}/_{s3}, {}^{3.2\cdot 10^{-4}}/_{s_4}, {}^{2.3\cdot 10^{-3}}/_{s_5} \right\},$$

which gives us the scenarios' ratings according to the level of their perfection: s_1 is the best scenario, s_2 is the second best, s_3 is third, s_5 is fourth,and s_4 in fifth place. For the probabilities of human errors, we have the inequality:

$$q_2 < q_1 < q_3 < q_5 < q_4.$$

We find that the different importance of performance conditions has not affected the ranking of scenarios s_1, s_2, and s_3 but has changed the ranking of s_4 and s_5.

A comparison between our proposed method and the known fuzzy CREAM method, from the point of view of human performance conditions selection, is presented in Table 4. Besides the substantial simplification of the mathematical model, the proposed method provides one very important advantage that does not exist in the known fuzzy CREAM method. This advantage is mentioned in line 4 of Table 4.

Table 4 Comparison of two methods

Criteria	Fuzzy CREAM	Fuzzy perfection
Number of performance conditions	9	9
Total number of membership functions	30	9
Total number of fuzzy rules	46,656	1
Using of additional performance conditions	Difficult	Easy

It consists on the possibility of taking into account an additional performance condition or factor affecting human reliability. This does not require a time-consuming changing of the system of fuzzy IF-THEN rules. We need only to define the membership function of fuzzy perfection for the *additional* performance conditions and use the same procedure of scenarios comparison.

4 CONCLUSIONS

The objective of reliability-based design of human performance conditions is to select such conditions, which minimize the probability of human error under existing restrictions. The most popular method of calculation of the probability of human error for the given vector of performance conditions is fuzzy CREAM. The main disadvantages of this method are a huge amount of fuzzy if-then rules and the possibility of using contradicting rules.

In this chapter, we proposed a simple procedure of performance conditions selection without complicated modeling of human error probability. The procedure is based on the specially introduced membership functions of fuzzy perfection of performance conditions affecting human reliability and the theory of decision-making in a fuzzy environment. For the testing of the proposed methodology, we used five scenarios with different performance conditions scores and known probabilities of human errors obtained by fuzzy CREAM. The similarity of results, generated by a much simpler method, favors the method of fuzzy perfection for the selection of performance conditions in human reliability analyses.

REFERENCES

[1] V. Glushkov, Introduction to Computer—Based Information and Management Systems, Technika, Kiev, 1974 320 p. (in Russian).
[2] L. Zadeh, The concept of a linguistic variable and its application to approximate reasoning, part 1 and 2, Inform. Sci. 8 (1975) 199–249 301–357.
[3] M. Konstandinidou, Z. Nivolianitou, C. Kiranoudis, N. Markatos, A fuzzy modeling application of CREAM methodology for human reliability analysis, Reliab. Eng. Syst. Saf. 91–6 (2006) 706–716.
[4] M. Marseguerra, Z. Enrico, M. Librizzi, Human reliability analysis by fuzzy "CREAM" Risk Anal. 27 (1) (2007) 137–154.
[5] R.E. Bellman, L.A. Zadeh, Decision making in a fuzzy environment, Manag. Sci. 17 (1970) 141–164.
[6] A.I. Gubinskij, Reliability and Quality Functioning of Ergatic Systems, Nauka, Leningrad, 1982 (in Russian).
[7] A.I. Gubinskij, V.G. Evgrafov, Ergonomic Design of Marine Control Systems, Sudostroenie, Leningrad, 1977 (in Russian).

[8] A.P. Rotshtein, P.D. Kuznezov, Design of Error Free Man-Machine Technologies, Technika, Kiev, 1992 (in Russian).
[9] A.I. Gubinskij, V.G. Evgrafov, M. Mashinostroenie (Eds.), Information Management Man-Machine Systems: Research, Development, Testing: Handbook, 1993 528 c. (in Russian).
[10] L. Zadeh, Outline of a new approach to the analysis of complex systems and decision processes, IEEE Trans. Syst. Man Cybern. SMC-3 (1973) 28–44.
[11] K.-Y. Cay, Introduction to Fuzzy Reliability, Kluwer Academic Publishers, Dordrecht, 1996.
[12] A. Rotshtein, Fuzzy reliability analysis of man-machine systems, in: T. Onisawa, J. Kacprzyk (Eds.), Reliability and Safety Analyses under Fuzziness. Studies in Fuzziness, vol. 4, Physica-Verlag, A. Springer-Verlag Company, 1995, pp. 43–60.
[13] L.V. Utkin, I.B. Shubinsky, Lubavich Publ., Saint Petersburg, 2000 (in Russian).
[14] T. Onisawa, An application of fuzzy concepts to modelling of reliability analysis, Fuzzy Sets Syst. (1990) 267–286.
[15] E. Hollnagel, Cognitive Reliability and Error Analysis Method (CREAM), Elsevier Science Ltd., Amsterdam, 1998.
[16] E. Mamdani, Application of fuzzy algorithms for control of simple dynamic plant, Proc. IEEE 121 (12) (1974) 1585–1588.
[17] A. Rotshtein, Algebra of algorithms and fuzzy logic in system reliability analyses, J. Comput. Syst. Sci. Int. 49 (2) (2010) 254–264.
[18] G. Bojadziev, M. Bojadziev, Fuzzy Sets, Fuzzy Logic, Applications, World Scientific, Singapore, 1995.
[19] R.R. Yager, Fuzzy decision making including unequal objectives, Fuzzy Sets Syst. 1 (1978) 87–95.
[20] A. Rotshtein, E. Shnaider, M. Schneider, A. Kandel, Fuzzy multicriteria selection of alternatives: the worst-case method, Int. J. Intell. Syst. 25 (9) (2010) 948–957.
[21] T.L. Saaty, The Analytic Hierarchy Process, McGraw-Hill International, New York, NY, 1980.

CHAPTER 9

System Reliability Assessment Through Bayesian Network Modeling

Rong Pan*, Dongjin Lee*, Petek Yontay†, Luis Mejia Sanchez‡
*School of Computing, Informatics, and Decision Systems Engineering, Arizona State University, Tempe, AZ, United States of America
†Intel, Inc., Chandler, AZ, United States of America
‡Cummins Inc., Columbus, IN, United States of America

Abstract

Using a Bayesian network (BN) model to evaluate system reliability is presented in this chapter. The BN model is a graphical model that generalizes the deterministic system reliability structure typically represented by a reliability block diagram or a fault tree, because BN allows inclusion of the uncertainty in system structure to be in the model. The information fusion for establishing BN models is discussed. We use an example to illustrate how to utilize this modeling technique to make design decisions for a complex system.

Keywords: System reliability structure, Bayesian network, Bayesian inference, Sensitivity analysis

1 SYSTEM RELIABILITY

Although the rapid progress of engineering design and information technology has helped the development of advanced engineering systems, it also introduces additional difficulties in analyzing system reliability. Because of the increased complexity in system reliability structure and in component/subsystem interfaces, many unexpected failure modes could occur, and their behaviors are interdependent. To avoid potential losses due to bad design decisions made based on an inaccurate reliability assessment, it is critical to develop a new cutting-edge reliability assessment method that considers the intricacy of system structure and the component/system function interactions inside a system.

The main challenge in analyzing a complex system is the failure uncertainty introduced by the incomplete knowledge of the system. This is particularly true at a system's design and development stage. The uncertainty

https://doi.org/10.1016/B978-0-12-815906-4.00009-9

prevents us from decomposing system reliability into subsystem/component reliability in a deterministic manner, such as series or parallel systems. Consequentially, some common reliability analysis tools such as fault tree (FT) and reliability block diagram (RBD) become inadequate. To overcome this challenge, we introduce Bayesian network (BN) modeling for system reliability assessment in this chapter. Specifically, we explain the connections and differences between BN and other graphical system analysis tools. We focus on the information acquisition for establishing BN models at the system design stage and how to utilize this modeling technique to make design decisions for a complex system.

The rest of this chapter is organized as follows: First, the BN model is introduced and compared with a series or parallel system. Second, we discuss how to quantify some parameters in the BN model during a system's design stage. Third, we provide an example to illustrate the utility of a BN model for system design improvement. Lastly, some recent developments on how to extend discrete BN models to continuous or mixed BN models are presented.

2 BAYESIAN NETWORK

BN is also called belief network, and it is used to represent the knowledge about an uncertain domain [1]. Typically, a BN model has a set of nodes connected by directional arcs/edges. These nodes are random variables, and the directional arcs/edges specify the conditional probability dependencies between nodes. The whole graph is called a directed acyclic graph (DAG), as a directional cycle among nodes is not permitted by a BN model. The conditional probability relationships defined in a BN graph are then used for defining the joint probability function of all random variables in the graph.

Traditional graphical models for evaluating system reliability include RBD and FT, and each of them has its own specific use. RBD is more straightforward for representing the reliability-wise relationship between one component to another one; whereas FT is good at reasoning a system failure and diagnosing it to lower-level component problems. However, the use of BN models in system reliability assessment has been shown to have significant advantages over these traditional models, because the arcs between BN nodes represent probabilistic relationships [2]. A typical RBD or FT can only model deterministic relationships between system

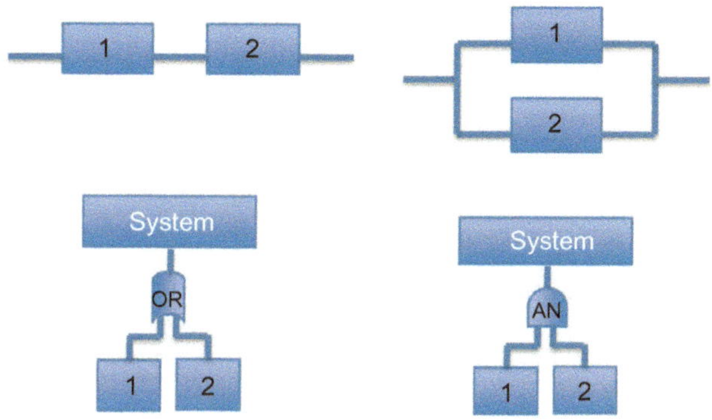

Fig. 1 Two-component series system (left) and two-component parallel system (right).

Table 1 The system-component relationships in a series system (left) and in a parallel system (right)

Component 1	Component 2	System	Component 1	Component 2	System
1	1	1	1	1	1
1	0	1	1	0	0
0	1	1	0	1	0
0	0	0	0	0	0

Here, "0" represents a working state and "1" represent a failure state.

and subsystems/components. For example, the two RBDs or their corresponding FTs shown in Fig. 1 are a two-component series and parallel system, respectively. The system's state is entirely determined by the two components' states as presented in Table 1. These system-component relationships are fixed, without any uncertainty.

However, for a complex system, these system-component relationships could become unstable due to any unknown factors that may influence the system behavior or the interactions between components. In general, it is the lack of knowledge at the system level causing the deterministic representation of system-component relationship inadequate. Probabilistic fault trees have been proposed to tackle this problem, but they are not the most direct and effective way to incorporate any probabilistic relationships between system and components. The conditional probabilities defined in a BN model allow easy capturing this type of system structure uncertainty.

Mathematically, BN is a compact graphical representation of a multivariate joint distribution function. Its nodes and arcs encode a set of conditional independence statements. Therefore for the joint probability density function, it can be found to be the product of all conditional probability density functions of each node given all its predecessor nodes (or parent nodes); that is

$$f(X_1, X_2, \ldots, X_n) = \prod_i f(X_i|\, Pre(X_i)) \tag{1}$$

where $Pre(X_i)$ represents the predecessor variables of variable X_i, and $f(X_i|\,Pre(X_i))$ is the conditional probability density function of variable X_i given its predecessor variables.

Furthermore, to complete a BN model, the quantitative part, which provides the distribution specifications, needs to be specified. This includes the specification of conditional probability tables (CPTs) of the nodes in the model.

Fig. 2 illustrates a simple BN DAG, along with the joint probability distribution of the whole system. Nodes 1 and 2 represent the states of two components, whereas Node 3 represents the state of the system. Suppose only binary states (working or failure) are considered, the conditional relationships between the system node and component nodes are given in Table 2.

The parameters $p_{11}, p_{10}, p_{01}, p_{11}$ in Table 2 have values between 0 and 1. One may notice that, when their values are 1, 0, 0, 0, respectively, the structure of the whole system can be identified as a parallel structure; whereas when they are 1, 1, 1, 0, it becomes a series structure. This is a discrete BN model, as each node is taking a specific number of discrete states. Therefore the discrete BN model is a probabilistic structure generalization of RBD or FT.

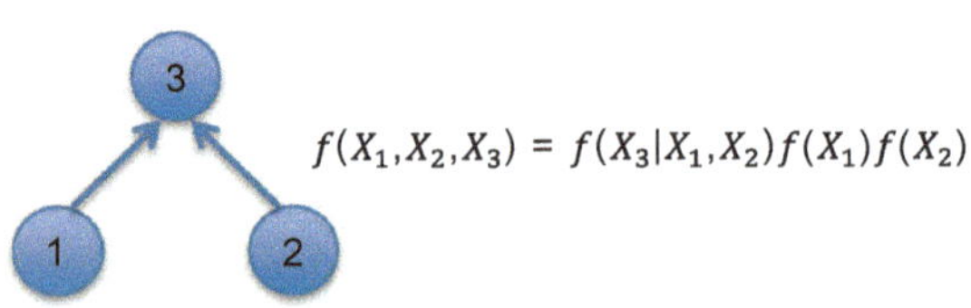

Fig. 2 A simple BN model.

Table 2 The CPT for the simple BN model

State of X_1	State of X_2	$P(X_3=1\|X_1, X_2)$	$P(X_3=0\|X_1,X_2)$
1	1	p_{11}	$1-p_{11}$
1	0	p_{10}	$1-p_{10}$
0	1	p_{01}	$1-p_{01}$
0	0	p_{00}	$1-p_{00}$

In summary, a discrete BN model consists of a DAG that represents the system structure and multiple CPTs that quantify the probabilistic relationships among the nodes in DAG. Overall, this model specifies the multivariate distribution that governs the system's behavior.

The BN-based inference may follow either a bottom-up or top-down manner. For example, given the failure probabilities for the two bottom nodes, X_1 and X_2, we can derive the failure probability distribution of the top node, X_3, which is given by

$$f_3 = f_1 f_2 p_{00} + (1 - f_1) f_2 p_{01} + f_1 (1 - f_2) p_{10} + (1 - f_1)(1 - f_2) p_{11} \tag{2}$$

where f_1, f_2, and f_3 are the failure probabilities of X_1, X_2, and X_3, respectively.

In general, by utilizing the multivariate distribution represented by the BN model, Eq. (2) can be derived from

$$f(X_3) = \sum_{X_1} \sum_{X_2} f(X_1, X_2, X_3) = \sum_{X_1} \sum_{X_2} f(X_3 | X_1, X_2) f(X_1) f(X_2) \tag{3}$$

where $f(X_1), f(X_2), f(X_3)$ are the corresponding marginal distributions.

Conversely, to infer a bottom node such as X_1, we may sum over the other two nodes

$$f(X_1) = \sum_{X_2} \sum_{X_3} f(X_1, X_2, X_3) = \sum_{X_2} \sum_{X_3} f(X_1 | X_2, X_3) f(X_2) f(X_3) \tag{4}$$

However, the conditional distribution of X_1 given X_2 and X_3 is not straightforward. It can derived based on the Bayes' theorem:

$$f(A | B) = \frac{f(B | A) f(A)}{f(B)} \tag{5}$$

Applying Eq. (5) to a three-node BN model, we obtain

$$f(X_1 | X_2, X_3) = \frac{f(X_2, X_3 | X_1) f(X_1)}{f(X_2, X_3)} = \frac{f(X_3 | X_2, X_1) f(X_2 | X_1) f(X_1)}{f(X_2, X_3)}$$

$$= \frac{f(X_3 | X_2, X_1) f(X_2) f(X_1)}{\sum_{X_1} f(X_3 | X_2, X_1) f(X_2) f(X_1)} \tag{6}$$

The last equation utilizes the independency of X_1 and X_2, which is shown in the DAG.

In system reliability analysis, the bottom-up approach is viewed as an evidential inference; that is, we may use the chance of each failure cause (bottom node) to infer the chance of a failure mode (top node). The top-down approach can be viewed as a causal inference; that is, we may trace a system failure event to its possible causes by utilizing the conditional dependencies of nodes embedded in the BN model. However, it has been proven that the inferential problems in BN are generally NP-hard problems [3]. Nonetheless, several approaches and algorithms exist to exploit the network structure for such inferential problems.

In a system design framework, probabilistic inferences can be made by queries about a particular design variable or a set of variables that designers want to answer. A simple query could be "what is the marginal distribution of a particular variable" when the designer is assessing the reliability of a design concept. A little more complicated one can be a conditional query that helps the designer make a decision on the condition of the results from other variables. Sensitivity analysis is often implemented to investigate whether or not the design specifications meet the proposed system requirements.

Both exact inference methods and approximation methods have been used to solve these queries [4]. The exact inference is derived from joint probability functions and conditional probability functions represented by BN. They include enumeration, belief propagation, variable elimination, and clustering/joint tree algorithms. However, not every problem can be answered by an exact inference method; approximate inference such as stochastic simulation/sampling methods, Markov chain Monte Carlo methods, genetic algorithms, neural networks, simulated annealing, and mean field theory are often used on big, complex network and non-Gaussian or discrete-continuous mixed distributions.

3 BAYESIAN INFERENCE

The BN model explained in Section 2 can be used to describe a system's reliability structure when uncertainties exist on the causal relationship of failures from the lower-level components to subsystems, and to higher-level system reliability. However, the parameters in CPTs have to be quantified before any top-down causal inferences or bottom-up evidential inferences can be made. In this section, we discuss how to quantify these parameters by using data observed from the system and its components, as well as how to infuse expert opinions into the parameter quantification.

The Bayesian inference technique needs to be introduced first. Note that the causal or evidential inference methods described in Section 2 are not the same as the Bayesian inference to be introduced in this section. The BN causal inference method utilizes the Bayes' theorem to derive the probabilistic state of each node based on the observations from other nodes, whereas the Bayesian inference to be explained in this section refers to the posterior estimation of the parameters in CPTs.

To quantify CPT parameters, we may rely on the subjective assessments from domain experts or derive parameter values by using system data. Bayesian inference methods can combine these two approaches to obtain balanced parameter estimation. Consider the BN model in Fig. 2, which is a two-component system. Given that there are two states for each component and the CPT as shown in Table 2, there are four parameters, p_{11}, p_{10}, p_{01}, p_{00}, that need to be specified. Although it is uncommon in practice, let us assume a naïve scenario—a complete historical dataset of the states of all nodes (including all components and the system) are available. This is possible only when we have log data from multiple identical systems, and these systems are continuously monitored by sensors.

Therefore each entry in the database is an observed instance of system state and component state; it is straightforward to obtain the conditional system failure probability given any combination of its components' states. For example,

$$p_{11} = \Pr(X_3 = 1 | X_1 = 0, X_2 = 1) = \frac{\sum_k I_k(X_3 = 1, X_1 = 0, X_2 = 1)}{\sum_k I_k(X_1 = 0, X_2 = 1)} \quad (7)$$

where the denominator is the total number of instances of the specific combination of X_1 and X_2, and the numerator is the number of instances of system failure with these component state combinations.

In reality, there could be many component variables related to a system failure, thus the number of combinations of component states will explode. This is troublesome because the log file could be highly fragmented, and some component variable combinations may never appear. To overcome this problem, we can infuse expert opinions into Eq. (7). We may ask domain experts how likely a system failure could happen for a given component state combination. This is equivalent to assigning a prior distribution to the conditional system failure probability number.

Let A in Eq. (5) be the probabilistic distribution of a system's state and B be the observations from the log file, then the Bayes' theorem says that the posterior distribution of A is proportional to the product of the likelihood of observing relevant events and the prior distribution of A. This is the foundation of Bayesian inference.

Note that the system's state is binary in this example, so a Bernoulli distribution is suitable for modeling each system observation or each entry in the log file. It can be shown that the conjugated prior distribution to the parameter in Bernoulli distribution is a Beta prior distribution such as $Beta(a, b)$. Then, the posterior estimation of p_{ij} is given by

$$p_{ij} = \frac{\sum_k I_k(X_3 = 1, X_1 = i, X_2 = j) + a}{\sum_k I_k(X_1 = i, X_2 = j) + (a + b)} \tag{8}$$

However, how to derive the parameter values of the beta prior distribution from expert opinions? We can ask domain experts the following questions:

Q1: In your opinion, how frequently could this particular combination of component states happen?

Q2: In your opinion, what is the likelihood of system failure given this particular combination of component states?

The answer to the first question is related to the sample size for a particular event and the answer to the second question is about the conditional probability. For a beta distribution, the sum of its parameters, $a + b$, can be interpreted as the equivalent sample size, and the ratio, $a/(a + b)$, is the conditional probability of system failure. Therefore the prior distribution can be obtained from the domain expert's answer to these questions. By combining expert assessments and historical data through Eq. (8), a robust estimation of conditional probability table can be obtained.

So far we have described a naïve approach to obtaining the parameters in CPT. By this approach, we assume that a complete record of system and component states from multiple copies of systems is available. However, the maintenance log in an organization often contains failure events only. For example, given a checklist such as Table 3, one can see that system failure is once due to both components' failures, and once due to component 2's failure only.

Table 3 System failure records

X_1	X_2	X_3
X	X	X
	X	X
…	…	…

In such case, one can calculate the probability of component state given the system state; that is, $\Pr(X_1, X_2 | X_3)$, but not the probability of system state given the component state combination. To obtain the conditional probability of system state given the component state combination, we need to know the probability of occurrence of component state combination and the marginal distribution of system failure. Sanchez and Pan [5] proposed an enhanced parenting process for deriving the probability of a failure mode given a failure cause. Through this process, expert opinions on the system or component state's marginal distribution are solicited. Although they assumed that the effects of failure causes were independent to each other, this approach can be easily generalized to remove the independency assumption.

Furthermore, notice that in the first approach we chose to use the conjugation property of beta and Bernoulli distributions, thus the posterior distribution of CPT parameter can be analytically obtained. However, the posterior distribution resulting from a complex Bayesian model often cannot be written in a closed form. This difficulty has hindered the adoption of Bayesian reliability assessment in the past. However, since the 1990s, advances in Bayesian computing through Markov chain Monte Carlo (MCMC) have facilitated inference based on samples from the targeted posterior distribution [6]. In the reliability literature, several authors have embedded the MCMC procedure into system reliability analysis (see Refs. [7–14]).

4 A SYSTEM DESIGN EXAMPLE

In this section, we describe a hazard reduction device, which is installed in the automobile exhaustion system, and its development process in a major automobile company. We use it as an example to demonstrate how a BN model and the causal inference can help improve system design.

The device is used for converting exhaust emissions, usually toxic, into less-toxic substances by creating chemical reactions of exhaust and additives inside the device. The device has to be redesigned, as it is driven by the tightened regulations from the Environmental Protection Agency (EPA). To achieve this, the design team starts from the device currently in production, compares its function to the new requirement, and twists the current design to satisfy the new requirement. In addition, the design team also needs to assess the reliability of the new design; that is, how likely the newly designed device may fail its intended functions.

Because data for the new model was scarce, the reliability team proposed a graphical model to depict the relationships between various required functions and subfunctions of this device and all possible consequences due to the failures of these functions. From this conceptual graph, a BN model is built, as shown in Fig. 3.

Fig. 3 represents the BN model of the new design. Each node in the graph is a device function and the directional arcs provide the causal relationships between these functions. It depicts the relationship between the different functional structures. The state of a node is binary: 1 for the dysfunctional state and 0 for the functional state. For convenience, these functions are represented by the following letters:

- Node G: Flow of exhaust gas.
- Node H: High temperature of element.
- Node F: Injection of fluid.
- Node R: Residence time of catalysis.
- Node T: Heat/mass transfer of elements.
- Node C: Amount of contaminants.
- Node B: Backpressure at outlet.
- Node S: Saturation of filters.

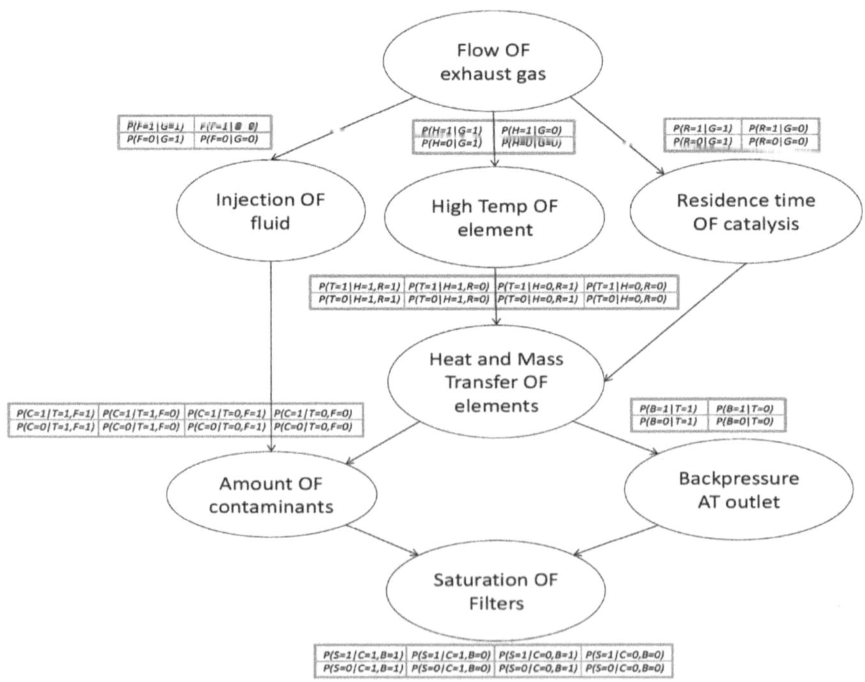

Fig. 3 The BN model for the hazard reduction device design.

Flow OF exhaust gas (G)	Nonfunctional	$P(G=1)=0.074$			
	Functional	$P(G=0)=0.926$			
Flow OF exhaust gas		Nonfunctional	Functional		
High temperature OF elements (H)	Nonfunctional	$P(H=1\|G=1)=0.714$	$P(H=1\|G=0)=0.044$		
	Functional	$P(H=0\|G=1)=0.286$	$P(H=1\|G=0)=0.956$		
Flow OF exhaust gas		Nonfunctional	Functional		
Injection OF fluids (F)	Nonfunctional	$P(F=1\|G=1)=0.979$	$P(F=1\|G=0)=0.012$		
	Functional	$P(F=0\|G=1)=0.021$	$P(F=0\|G=0)=0.988$		
Flow OF exhaust gas		Nonfunctional	Functional		
Residence time OF catalysis (R)	Nonfunctional	$P(R=0\|G=1)=0.963$	$P(R=1\|G=0)=0.027$		
	Functional	$P(R=0\|G=1)=0.037$	$P(R=0\|G=0)=0.973$		
High temperature OF elements		Nonfunctional		Functional	
Residence time OF catalysis		Nonfunctional	Functional	Nonfunctional	Functional
Heat and mass transfer OF elements	Nonfunctional	$P(T=1\|H=1,R=1)=0.993$	$P(T=1\|H=1,R=1)=0.834$	$P(T=1\|H=0,R=1)=0.982$	$P(T=1\|H=0,R=0)=0.005$
	Functional	$P(T=0\|H=1,R=1)=0.007$	$P(T=0\|H=1,R=0)=0.166$	$P(T=0\|H=0,R=1)=0.018$	$P(T=0\|H=0,R=0)=0.995$
Heat and mass transfer OF elements		Nonfunctional		Functional	
Injection OF fluids		Nonfunctional	Functional	Nonfunctional	Functional
Amount OF contaminants (C)	Nonfunctional	$P(C=1\|T=1,F=1)=0.998$	$P(C=1\|T=1,F=0)=0.862$	$P(C=1\|T=0,F=1)=0.334$	$P(C=1\|T=0,F=0)=0.059$
	Functional	$P(C=0\|T=1,F=1)=0.002$	$P(C=0\|T=1,F=0)=0.138$	$P(C=0\|T=0,F=1)=0.666$	$P(C=0\|T=0,F=0)=0.941$
Heat and mass transfer OF elements		Nonfunctional	Functional		
Backpressure AT outlet (B)	Nonfunctional	$P(B=1\|T=1)=0.918$	$P(B=1\|T=0)=0.069$		
	Functional	$P(B=0\|T=1)=0.082$	$P(B=0\|T=0)=0.931$		
Backpressure AT outlet		Nonfunctional		Functional	
Amount OF Contaminants		Nonfunctional	Functional	Nonfunctional	Functional
Saturation OF filters (S)	Nonfunctional	$P(S=1\|B=1,C=1)=0.984$	$P(S=1\|B=1,C=0)=0.743$	$P(S=1\|B=0,C=1)=0.946$	$P(S=1\|B=0,C=0)=0.022$
	Functional	$P(S=0\|B=1,C=1)=0.016$	$P(S=0\|B=1,C=0)=0.257$	$P(S=0\|B=0,C=1)=0.054$	$P(S=0\|B=0,C=0)=0.978$

Fig. 4 The marginal distribution parameters and the CPT parameters for each node of the BN model in Fig. 3.

To obtain their marginal distributions and the CPTs, both the expert opinions from product engineers and reliability engineers and the enhanced parenting process are applied during the redesign. The information is provided in Fig. 4.

These CPTs completed the quantitative part of the BN model. The fully defined BN model is able to provide insights to the reliability of the newly designed device. Different uses or insight angles toward reliability assessment give different queries made to the BN. For example, for the design team, the following scenario has been investigated.

The design team is interested in whether or not the current concept is capable of meeting the specific requirement of measured emission compliance with the standard by 90%. The node of interest is C, amount of contaminants. Its predecessor nodes are Node F, Injection of fluid, and Node T, heat/mass transfer of elements. It is found that, given the current design, Node C is functional only about 82% of the time, that is, $P(C=0)=0.82$. Therefore it did not meet the stated requirement.

A sensitivity analysis is conducted so that the team can evaluate a constraint such as $P(C=1)=0.1$. The sensitivity analysis was conducted by using the Shenoy-Shafer algorithm because it is one of the main methods for probability propagation in a joint tree [15]. Two recommendations on the CPTs for Node C and Node T are provided, as depicted in Figs. 5 and 6, respectively. The recommended changes are highlighted.

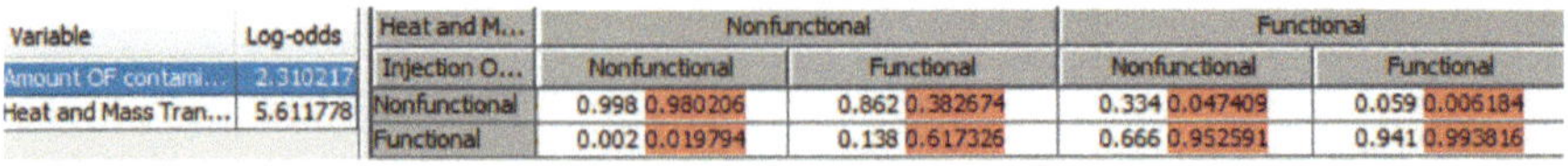

Variable	Log-odds
Amount OF contami...	2.310217
Heat and Mass Tran...	5.611778

Heat and M...	Nonfunctional		Functional	
Injection O...	Nonfunctional	Functional	Nonfunctional	Functional
Nonfunctional	0.998 0.980206	0.862 0.382674	0.334 0.047409	0.059 0.006184
Functional	0.002 0.019794	0.138 0.617326	0.666 0.952591	0.941 0.993816

Fig. 5 Recommended changes for Node C.

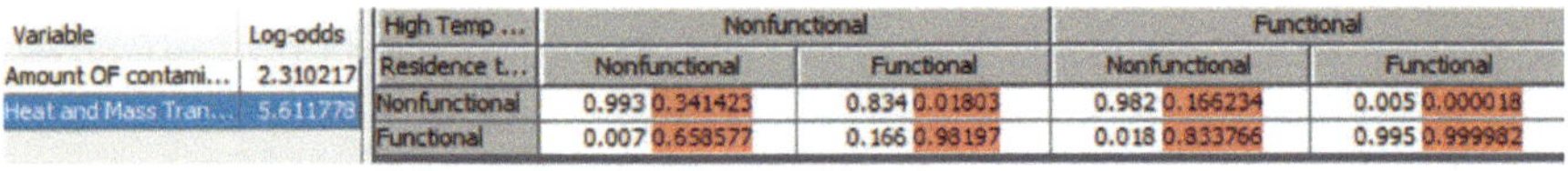

Variable	Log-odds
Amount OF contami...	2.310217
Heat and Mass Tran...	5.611778

High Temp ...	Nonfunctional		Functional	
Residence t...	Nonfunctional	Functional	Nonfunctional	Functional
Nonfunctional	0.993 0.341423	0.834 0.01803	0.982 0.166234	0.005 0.000018
Functional	0.007 0.658577	0.166 0.98197	0.018 0.833766	0.995 0.999982

Fig. 6 Recommended changes for Node T.

Using the log-odds statistics, which is the difference in the natural logarithm of the odds before and after applying the change of CPT parameter values, we can evaluate these two options [16]. The greater the value of log-odds, the greater the required change is. Eventually, it is decided to adopt the recommendations for Node C.

This decision can be further validated by examining the highlighted recommendations on the CPT of Node T. For example, on Fig. 6 it is recommended that $P(T=0|H=1, R=1)=0.007$ needs to be changed to $P(T=0|H=1, R=1)=0.658$, and $P(T=0|H=1, R=0)=0.166$ to be changed to $P(T=0|H=1, R=0)=0.981$. These suggested changes are too dramatic to be feasible.

Through this exercise, the team is able to obtain an insight into the system reliability related to the functionality of a particular node. Furthermore, sensitivity analysis provides a more objective decision-making process. The experts involved in this study are able to understand that a more robust design is needed to change the way Node C is affected by its predecessor nodes. After a further analysis, it is found that the attention of redesign should concentrate on $P(C=0|T=1, F=0)$. In other words, Node C needs to be functional even if Node T was nonfunctional. Thus it is necessary to increase the independence between Node T and Node C.

In summary, this example demonstrates how a BN analysis can help a designer to better choose robust components for key system functions, with the aim to meet the established system reliability requirement.

5 SOME NEW DEVELOPMENTS

In the past two decades, BN models for modeling system reliability structures have received a lot of attention from academia. Weber et al. [17] provided a comprehensive review of BN applications in complex system modeling. Appling BN models to analyze large system reliability, such as power systems, military vehicles, and semiconductor manufacturing systems, etc., can be found in the literature [18–22].

However, the application of Bayesian inference techniques in BN models has been relatively few. Sigurdsson et al. [23] addressed the issue of incorporating expert opinions into BN modeling. Li et al. [24] studied the BN in which the probability models for nodes were derived from the information from multiple sources. Yontay et al. [25] discussed the use of discrete BN for estimating system reliability, whereas CPT parameters were estimated by combining historical failure data (parent products) and expert opinions through Bayesian inference. This discussion was further extended in Yontay and Pan [26] for the scenarios where failure data were simultaneously generated from system level and component level.

Although the discrete BN models for modeling reliability structure have been successfully applied on many applications as previously mentioned, the use of discrete BN has become infeasible for big, complex systems because the parameterization of a large BN model can be too complicated to be manageable. A disorganized quantification process will result in poor probabilistic inference. To overcome these hurdles, a recent development in this field is to explore the utility of nonparametric BN (NPBN) for system reliability assessment [27].

In brief, NPBN can handle continuous random variables, yet it does not require these variables to be normally distributed. A NPBN model can depict the entire reliability configuration of the system and the strength of the relationship between the system and its components without using CPTs, thus saving the efforts for computing and managing CPT parameters. A drawback of NPBN is that it relies on Monte Carlo simulation for updating the joint and marginal distributions in the model, but it can be avoided by using the normal copula vine approach presented by Hanea [28]. In a recent publication [29], the utility of NPBN was studied for evaluating and analyzing system reliability at its early design stages. Given that no physical products and product tests are available at these early design stages, historical data from parent products and expert opinions on the system design are the main sources for obtaining design knowledge.

6 CONCLUSIONS

In this chapter, we introduce the BN modeling method as a generalization of RBD or FT for modeling system reliability with the uncertainty of system structure. The top-down and bottom-up inferences by BNs are discussed. More importantly, we demonstrate how to use information from different sources to quantify the model parameters in BN models, thus the system reliability is comprehensively assessed. Furthermore, some recent developments of the BN modeling for system reliability evaluation are summarized in the last section. Nevertheless, it has to be pointed out that this chapter is only a brief introduction to this research topic. For most technical details, readers should refer to the references cited herein.

REFERENCES

[1] I. Ben-Gal, Bayesian networks, in: Encyclopedia of Statistics in Quality and Reliability, John Wiley & Sons, Hoboken, New Jersey, 2007.
[2] H. Langseth, L. Portinale, Bayesian networks in reliability, Reliab. Eng. Syst. Saf. 92 (1) (2007) 92–108.
[3] G.F. Cooper, The computational complexity of probabilistic inference using Bayesian belief networks, Artif. Intell. 42 (2) (1990) 393–405.
[4] C.M. Bishop, Pattern Recognition and Machine Learning, vol. 1, Springer, New York, 2006.
[5] L.M. Sanchez, R. Pan, An enhanced parenting process: predicting reliability in product's design phase, Qual. Eng. 23 (4) (2011) 378–387.
[6] A. Gelman, J.B. Carlin, H.S. Stern, D.B. Rubin, Bayesian Data Analysis, Francis, Taylor &, 2014.
[7] V.E. Johnson, T.L. Graves, M.S. Hamada, C.S. Reese, in: A hierarchical model for estimating the reliability of complex systems, Bayesian Statistics 7: Proceedings of the Seventh Valencia International Meeting, Oxford University Press, USA, 2003, pp. 199–214.
[8] M. Hamada, H.F. Martz, C.S. Reese, T. Graves, V. Johnson, A.G. Wilson, A fully Bayesian approach for combining multilevel failure information in fault tree quantification and optimal follow-on resource allocation, Reliab. Eng. Syst. Saf. 86 (3) (2004) 297–305.
[9] C.S. Reese, M. Hamada, D. Robinson, Assessing system reliability by combining multilevel data from different test modalities, Qual. Technol. Quant. Manag. 2 (2005) 177–188.
[10] A.G. Wilson, T.L. Graves, M.S. Hamada, C.S. Reese, Advances in data combination, analysis and collection for system reliability assessment, Stat. Sci. 21 (2006) 514–531.
[11] T.L. Graves, M.S. Hamada, R. Klamann, A. Koehler, H.F. Martz, A fully Bayesian approach for combining multi-level information in multi-state fault tree quantification, Reliab. Eng. Syst. Saf. 92 (10) (2007) 1476–1483.
[12] A.G. Wilson, A.V. Huzurbazar, Bayesian networks for multilevel system reliability, Reliab. Eng. Syst. Saf. 92 (10) (2007) 1413–1420.
[13] T.L. Graves, M.S. Hamada, R. Klamann, A. Koehler, H.F. Martz, Using simultaneous higher-level and partial lower-level data in reliability assessments, Reliab. Eng. Syst. Saf. 93 (8) (2008) 1273–1279.
[14] J. Guo, Bayesian Methods for System Reliability and Community Detection (Ph.D. Thesis), Iowa State University, 2011.
[15] G.-J. Park, T.-H. Lee, K.H. Lee, K.-H. Hwang, Robust design: an overview, AIAA J. 44 (1) (2006) 181–191.

[16] H. Chan, A. Darwiche, When do numbers really matter? in: Proceedings of the Seventeenth Conference on Uncertainty in Artificial Intelligence, Morgan Kaufmann Publishers Inc., 2001, pp. 65–74.
[17] P. Weber, G. Medina-Oliva, C. Simon, B. Iung, Overview on Bayesian networks applications for dependability, risk analysis and maintenance areas, Eng. Appl. Artif. Intell. 25 (4) (2012) 671–682.
[18] S. Mahadevan, R. Zhang, N. Smith, Bayesian networks for system reliability reassessment, Struct. Saf. 23 (3) (2001) 231–251.
[19] D.C. Yu, T.C. Nguyen, P. Haddawy, Bayesian network model for reliability assessment of power systems, IEEE Trans. Power Syst. 14 (2) (1999) 426–432.
[20] T. Daemi, A. Ebrahimi, M. Fotuhi-Firuzabad, Constructing the Bayesian network for components reliability importance ranking in composite power systems, Int. J. Electr. Power Energy Syst. 43 (1) (2012) 474–480.
[21] M. Neil, N. Fenton, S. Forey, R. Harris, Using Bayesian belief networks to predict the reliability of military vehicles, Comput. Control. Eng. 12 (1) (2001) 11–20.
[22] M.F. Bouaziz, E. Zamai, F. Duvivier, Towards Bayesian network methodology for predicting the equipment health factor of complex semiconductor systems, Int. J. Prod. Res. 51 (15) (2013) 4597–4617.
[23] J. Sigurdsson, L. Walls, J. Quigley, Bayesian belief nets for managing expert judgement and modelling reliability, Qual. Reliab. Eng. Int. 17 (3) (2001) 181–190.
[24] M. Li, J. Liu, J. Li, B.U. Kim, Bayesian modeling of multi-state hierarchical systems with multi-level information aggregation, Reliab. Eng. Syst. Saf. 124 (2014) 158–164.
[25] P. Yontay, L.M. Sanchez, R. Pan, in: Bayesian network for reliability prediction in functional design stage, Reliability and Maintainability Symposium (RAMS), Annual, IEEE, 2015, pp. 1–6.
[26] P. Yontay, R. Pan, A computational Bayesian approach to dependency assessment in system reliability, Reliab. Eng. Syst. Saf. 152 (2016) 104–114.
[27] D. Kurowicka, Distribution-free continuous Bayesian belief, in: Modern Statistical and Mathematical Methods in Reliability, Series on Quality, Reliability and Engineering Statistics, 10, World Scientific, Singapore, 2005, p. 309.
[28] A.M. Hanea, Algorithms for Non-parametric Bayesian Belief Nets (Ph.D. Thesis), TU Delft, Delft University of Technology, 2008.
[29] D. Lee, R. Pan, A nonparametric Bayesian network approach to assessing system reliability at early design stages, Reliab. Eng. Syst. Saf. 171 (2018) 57–66.

CHAPTER 10

Multistate Multifailures System Analysis With Reworking Strategy and Imperfect Fault Coverage

Monika Manglik*, Mangey Ram†
*Department of Mathematics, University of Petroleum & Energy Studies, Dehradun, India
†Department of Mathematics, Computer Science & Engineering, Graphic Era University, Dehradun, India

Abstract

In designing a system, reliability prediction has an important role. When predicting the reliability of a system, the two key factors are considered: failure distribution of the component and configuration of the system. For the development of this rapidly developing field, this chapter provides an overview of a multicomponent repairable system and discusses the effect of the coverage factor on designed systems with various types of failures such as partial, catastrophic, and human failures. In this chapter, a Markov model has been developed for the system; also, the state transition probabilities have been determined for evaluating the various reliability measures. Supplementary variable techniques and Laplace transformations have been used for evaluating these reliability measures. Some specific cases along with a comparative study of different failure rates are also discussed. Finally, the practical utility of the model is discussed with the help of some graphical illustrations.

Keywords: Imperfect fault coverage, Multistate system, Reworking strategy, System performance

1 INTRODUCTION

Today, reliability has become the main concern due to the modern industrial process, and with an increasing level of sophistication, it comprises the highest number of manufacturing and industrial systems. Every system usually follows a failure-repair cycle. It means a system can exist in binary states: working (upstate) and failed state (downstate). These failures can be covered or uncovered. Because it directly impacts the system's reliability, it is therefore important to demonstrate the proportion of recovered fault for accurately determining the coverage factor of the equipment and to analyze

Advances in System Reliability Engineering
https://doi.org/10.1016/B978-0-12-815906-4.00010-5

the impact of the parameter on the reliability of the overall system, according to Ram and Manglik [1].

In a fault-tolerant system, the proportion of handled faults and total faults is called the coverage factor. As far as the coverage probability is concerned, it is the probability of successful detection and recovery from a failure. Klauwer reinterpreted the definition of the coverage factor as a mathematical ratio [2]:

Coverage factor = all faults of a system can be exposed so that are covered / all possible faults a system can be exposed to coverage

The coverage factor of the perfect coverage model is unity; any desired reliability of the system can be achieved by adding the replacements. However, it is not possible in the case of imperfect fault coverage. To improve the reliability of the system, not only is additional redundancy required but also improvement in the coverage factor. For an optimal level of redundancy, an accurate analysis is required; otherwise, due to imperfect fault coverage, the reliability of the system can be decreased with an increment in redundancy [3].

For the improvement of reliability of the systems, various authors have previously used the concept of coverage factor. Pham [4] examined a high voltage system with imperfect fault coverage by taking the failure rate of fault coverage as a constant. Akhtar [5] and Moustafa [6] analyzed the reliability of k-out-of-n:G system with imperfect fault coverage. The impact of fault detection and imperfect fault coverage on a repairable system has been considered by Trivedi [7]. Myers [8] studied the reliability of an k-out-of-n:G system with imperfect fault coverage. Ke et al. [9, 10] extended the model developed by Trivedi [7] with the help of asymptotic estimation and imperfect fault coverage; authors used Bayesian approach for predicting the performance measures of a repairable system with detection, imperfect coverage, and reboot. A comparative analysis for the availability of two systems with warm standby units has been given by Wang et al. [11]. In this study, authors assumed that the coverage factor of the active unit failure is different from that of the standby unit failure. Powell et al. [12] revealed that the coverage factor is used to compute the efficiency of their fault-tolerance mechanisms. For estimating the effectiveness of a reconfiguration scheme in a fault-tolerant network, Reibman and Zaretsky [13] discussed a modeling approach in which a higher-level model and lower-level model represented the occurrence of failures in the network and the network reconfiguration system, respectively. Furthermore, coverage factor has been used by a number of authors in various reliability purposes, such as Cai et al.

[14], Prabhudeva and Verma [15], and Kumar et al. [16]. Ram et al. [17] analyzed a parallel redundant system under two types of failures incorporating coverage factor. Cost analysis of three-state standby redundant electronic equipment has been done by Gupta and Sharma [18]. Manglik and Ram [19] analyzed a hydroelectric production power plant, and various reliability measures have also been discussed by the authors. Kumar and Ram [20] analyzed the system reliability measures under common cause failure. Ram and Manglik [21, 22] analyzed a multistate manufacturing system with common cause failure and waiting repair strategy; also, in addition to that, the authors analyzed the various reliability measures of an industrial system under standby modes and catastrophic failure. Ram and Manglik [23] proposed the mathematical modeling of a biometric system. Ram and Goyal [24] analyzed a flexible manufacturing system under the Copula-Coverage approach. Chopra and Ram [25] analyzed two nonidentical unit parallel systems with the concept of waiting repair. Goyal et al. [26] analyzed the sensitivity of three-unit series system under k-out-of-n redundancy, and Goyal and Ram [27] also studied a series-parallel system under warranty and preventive maintenance.

Human performance plays an important role in the study of human-machine interaction and system design [28]. Due to human error, the efficiency, security, and system concert is also reduced [29]. According to Dhillon and Yang [30] and Dhillon and Liu [31], human error is defined as the failure in execution of a particular task, due to which the planned operation or outcome is interrupted. Yang and Dhillon [32, 33] analyzed the availability of a repairable standby human-machine system and a general standby system with constant human error. To further the previously mentioned research, the authors designed a mathematical model for a standby complex system having two subsystems in the series by Markov process, and they obtained the state transition probability of each possible state of the system. The first subsystem has one main and one standby unit; after the failure of the main unit, the standby unit automatically takes the load. Further, the second subsystem has n units in series with partial failure property in its component. In this chapter, the effect of the coverage factor on reliability measures is discussed, and the chapter also analyzes how the fault coverage affects the expected profit of the system. The structure of the current chapter is as follows: Section 2 consists of the details related to the model including the nomenclature, system description, and assumptions. This section also describes the formulation and solution to the proposed model. In Section 3, the particular cases of reliability measures are discussed. Section 4 includes the results and discussion. Finally, in Section 5 the conclusion of the proposed analysis is discussed.

2 MATHEMATICAL MODEL DETAILS

2.1 Model Description and Assumptions

In this chapter, the authors developed the mathematical model of a standby complex system with three types of failures by using the concept of coverage factor. The considered system consists of two subsystems A and B connected in series. Subsystem A consists of two units in which one is the main and the second one is in standby mode. On the other hand, subsystem B consists of n identical units in series. Initially, the system works with full capacity. After the failure of the main unit of subsystem A, the standby unit takes over its functioning immediately through the switch-over device. Failure of both units of subsystem A results as a complete failure of the system. The partial failure of any jth unit (where $j=1, \ldots, n$) of subsystem B brings the system to a reduced state from where the system can be automatically repaired (perfect coverage). Here, a reduced state means that the system state works less efficiently. Failure of both units of subsystem A, due to human failure and catastrophic failure, brings the system to complete failure mode. Furthermore, the authors have assumed that the system cannot be automatically recovered (imperfect fault coverage) due to catastrophic failure (a complete, sudden, often unexpected breakdown in any system). The configuration of the system and state transition diagram of the designed model are shown in Fig. 1A and B. Failure rates are inherently constant in general, whereas the repairs follow a general distribution. Markov process, supplementary variable technique, and Laplace transforms are used to evaluate the reliability measures.

2.2 Nomenclature

The following nomenclature are used throughout the designing:

λ_A	The constant failure rate of the main and standby units of subsystem A.
λ_j	The constant partial failure rate of a jth unit of subsystem B.
λ_h, λ_c	Failure rates of human failure and catastrophic failures, respectively.
$P_{0,0}(t)/P_{0,j}(t)/P_{1,0}(t)/P_{1,j}(t)$	The probability at time t when the system is in full working condition/reduced state due to the partial failure of a jth unit of subsystem B/due to the failure of the main unit of subsystem A/due to the partial failure of a jth unit of subsystem B with the failure of the main unit of subsystem A, respectively.

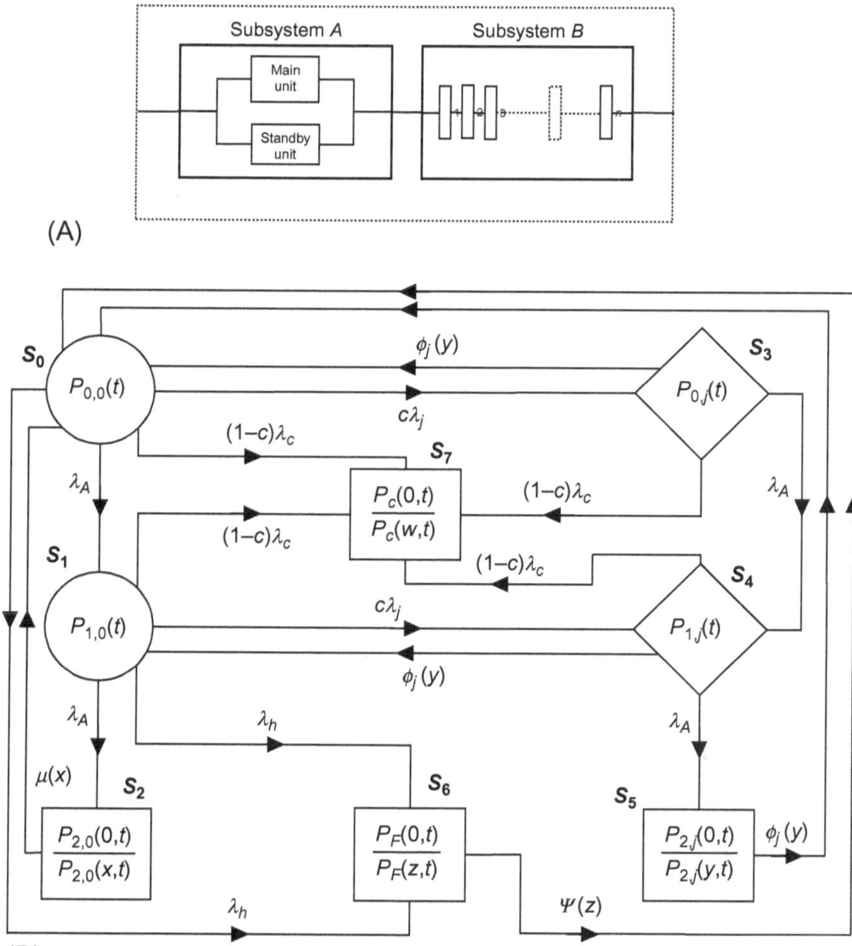

Fig. 1 (A) System configuration. (B) State transition diagram of the proposed model.

$\mu(x)/\psi(z)/\varphi_j(y)$ Repair rates of the system from failed state due to standby unit failure/human error/jth unit partial failure of subsystem B with the failure of the standby unit of subsystem A.

$P_{2,0}(x,\ t)$ The probability of state that both the units of subsystem A have failed and the system is under repair, elapsed repair time is x; t.

$P_F(z,\ t)$ The probability of state that the system has failed due to human error and under repair, elapsed repair time is z; t.

$P_{2,j}(y, t)$	The probability of state that the system has failed due to the failure of subsystem A with subsystem B is degraded and under repair, elapsed repair time is y; t.
$P_c(w, t)$	The probability of state that the system has failed due to catastrophic failure.
C	Coverage factor of the system.
$E_p(t)$	Expected profit during up time.
K_1, K_2	Revenue cost and service cost per unit time, respectively.
S	Laplace parameter.

2.3 State Description and Transition Diagram

The state description of the system is given as:

State	State description
S_0	The system is working with full capacity.
S_1	The system is in a good state after the failure of the main unit of subsystem A.
S_2	It is in failed state after the failure of the standby unit subsystem A.
S_3	It is in a reduced state due to the partial failure of any jth unit subsystem B.
S_4	It is in a reduced state due to the failure of the main unit of subsystem A and the partial failure of any jth unit of subsystem B.
S_5	It is in failed state due to the failure of both units of subsystem A and the partial failure of any jth unit of subsystem B.
S_6	The system has stopped working due to human failure.
S_7	Failed state due to catastrophic failure.

The assumptions associated with the model are mentioned here:

(i) Primarily, the system is working with full capacity.
(ii) The three states of the system are good, reduced, and failed states.
(iii) Catastrophic failure, human failure, and failure of a standby unit of subsystem A results the complete failure of the system.
(iv) The system can be automatically repaired in case of covered faults (partially failed units of subsystem B).
(v) Repairing is not possible after the sudden (catastrophic) failure (uncovered fault).
(vi) After repairing the system, it works with full capacity.

2.4 Formulation and Solution of the Mathematical Model

The following set of differential equations govern the present mathematical model:

$$\left[\frac{\partial}{\partial t}+\lambda_A+c\lambda_j+\lambda_h+(1-c)\lambda_c\right]P_{0,0}(t)=\int_0^\infty P_{2,0}(x,t)\mu(x)dx$$

$$+\int_0^\infty P_{2,j}(y,t)\phi_j(y)dy+\int_0^\infty P_F(z,t)\psi(z)dz+\phi_j(y)P_{0,j}(t) \tag{1}$$

$$\left[\frac{\partial}{\partial t}+\lambda_A+c\lambda_j+\lambda_h+(1-c)\lambda_c\right]P_{1,0}(t)=\varphi_j(y)P_{1,j}(t)+\lambda_A P_{0,0}(t) \tag{2}$$

$$\left[\frac{\partial}{\partial t}+\frac{\partial}{\partial x}+\mu(x)\right]P_{2,0}(x,t)=0 \tag{3}$$

$$\left[\frac{\partial}{\partial t}+\lambda_A+\varphi_j(y)+(1-c)\lambda_c\right]P_{0,j}(t)=c\lambda_j P_{0,0}(t) \tag{4}$$

$$\left[\frac{\partial}{\partial t}+\lambda_A+\varphi_j(y)+(1-c)\lambda_c\right]P_{1,j}(t)=c\lambda_j P_{0,0}(t)+\lambda_A P_{0,j}(t) \tag{5}$$

$$\left[\frac{\partial}{\partial t}+\frac{\partial}{\partial y}+\varphi_j(y)\right]P_{2,j}(y,t)=0 \tag{6}$$

$$\left[\frac{\partial}{\partial t}+\frac{\partial}{\partial z}+\psi(z)\right]P_F(z,t)=0 \tag{7}$$

$$\left[\frac{\partial}{\partial t}+\frac{\partial}{\partial w}\right]P_c(w,t)=0 \tag{8}$$

Boundary conditions:

$$P_{2,0}(0,t)=\lambda_A P_{1,0}(t) \tag{9}$$

$$P_{2,j}(0,t)=\lambda_A P_{1,j}(t) \tag{10}$$

$$P_F(0,t)=\lambda_h[P_{0,0}(t)+P_{1,0}(t)] \tag{11}$$

$$P_c(0,t)=(1-c)\lambda_c\left[P_{0,0}(t)+P_{1,0}(t)+P_{0,j}(t)+P_{1,j}(t)\right] \tag{12}$$

$$\begin{cases} P_0(t)=1, & t=0 \\ P_0(t)=0, & t>0 \end{cases} \tag{13}$$

Taking the Laplace transformation of Eqs. (1)–(12), we get:

$$\left[s+\lambda_A+c\lambda_j+\lambda_h+(1-c)\lambda_c\right]\overline{P}_{0,0}(s)=1+\int_0^\infty \overline{P}_{2,0}(x,s)\mu(x)dx$$
$$+\int_0^\infty \overline{P}_{2,j}(y,s)\phi_j(y)dy+\int_0^\infty \overline{P}_F(z,s)\psi(z)dz+\phi_j(y)\overline{P}_{0,j}(s) \tag{14}$$

$$\left[s+\lambda_A+c\lambda_j+\lambda_h+(1-c)\lambda_c\right]\overline{P}_{1,0}(s)=\varphi_j(y)\overline{P}_{1,j}(s)+\lambda_A\overline{P}_{0,0}(s) \tag{15}$$

$$\left[s+\frac{\partial}{\partial x}+\mu(x)\right]\overline{P}_{2,0}(x,s)=0 \tag{16}$$

$$\left[s+\lambda_A+\phi_j(y)+(1-c)\lambda_c\right]\overline{P}_{0,j}(s)=c\lambda_j\overline{P}_{0,0}(s) \tag{17}$$

$$\left[s+\lambda_A+\phi_j(y)+(1-c)\lambda_c\right]\overline{P}_{1,j}(s)=c\lambda_j\overline{P}_{0,0}(s)+\lambda_A\overline{P}_{0,j}(s) \tag{18}$$

$$\left[s+\frac{\partial}{\partial y}+\varphi_j(y)\right]\overline{P}_{2,j}(y,s)=0 \tag{19}$$

$$\left[s+\frac{\partial}{\partial z}+\psi(z)\right]\overline{P}_F(z,s)=0 \tag{20}$$

$$\left[s+\frac{\partial}{\partial w}\right]\overline{P}_c(w,s)=0 \tag{21}$$

$$\overline{P}_{2,0}(0,s)=\lambda_A\overline{P}_{1,0}(s) \tag{22}$$

$$\overline{P}_{2,j}(0,s)=\lambda_A\overline{P}_{1,j}(s) \tag{23}$$

$$\overline{P}_F(0,s)=\lambda_h\left[\overline{P}_{0,0}(s)+\overline{P}_{1,0}(s)\right] \tag{24}$$

$$\overline{P}_c(0,s)=(1-c)\lambda_c\left[\overline{P}_{0,0}(s)+\overline{P}_{1,0}(s)+\overline{P}_{0,j}(s)+\overline{P}_{1,j}(s)\right] \tag{25}$$

Solving Eqs. (14)–(21) with the help of Eqs. (13) and (22)–(25), one may get various state probabilities:

$$\overline{P}_{0,0}(s)=\frac{1}{d_2} \tag{26}$$

$$\overline{P}_{1,0}(s)=\frac{1}{(s+d)}\left[\varphi_j(y)\left\{\frac{c\lambda_j}{(s+d_1)}\left[1+\frac{\lambda_A}{(s+d_1)}\right]\right\}+\lambda_A\right]\overline{P}_{0,0}(s) \tag{27}$$

$$P_{0,j}(s)=\frac{1}{(s+d_1)}c\lambda_j\overline{P}_{0,0}(s) \tag{28}$$

$$\overline{P}_{1,j}(s)=\frac{c\lambda_j\overline{P}_{0,0}(s)}{(s+d_1)}\left[1+\frac{\lambda_A}{(s+d_1)}\right] \tag{29}$$

$$\overline{P}_{2,0}(s)=\left[\frac{\lambda_A\varphi_j(y)}{(s+d)}\left\{\frac{c\lambda_j}{(s+d_1)}\left[1+\frac{\lambda_A}{(s+d_1)}\right]\right\}\left[\frac{1-\overline{S}_\mu(s)}{s}\right]\right.$$
$$\left.+\frac{1}{(s+d)}\lambda_A{}^2\left[\frac{1-\overline{S}_\mu(s)}{s}\right]\right]\overline{P}_{0,0}(s) \tag{30}$$

$$\overline{P}_{2,j}(s)=\frac{\lambda_A c\lambda_j}{(s+d_1)}\left[1+\frac{\lambda_A}{(s+d_1)}\right]\left[\frac{1-\overline{S}_\varphi(s)}{s}\right]\overline{P}_{0,0}(s) \tag{31}$$

$$\overline{P}_F(s)=\lambda_h+\left[\frac{\lambda_h}{(s+d)}\varphi_j(y)\left\{\frac{c\lambda_j}{(s+d_1)}\left[1+\frac{\lambda_A}{(s+d_1)}\right]\right\}\left[\frac{1-\overline{S}_\psi(s)}{s}\right]\right.$$
$$\left.+\frac{\lambda_A\lambda_h}{(s+d)}\left[\frac{1-\overline{S}_\psi(s)}{s}\right]\right]\overline{P}_{0,0}(s) \tag{32}$$

$$\overline{P}_c(s)=\frac{1}{s}(1-c)\lambda_c\overline{P}_{0,0}(s)\left[\begin{array}{l}1+\frac{1}{(s+d)}\left[\varphi_j(y)\left\{\frac{c\lambda_j}{(s+d_1)}\left[1+\frac{\lambda_A}{(s+d_1)}\right]\right\}+\lambda_A\right]+\\ \frac{1}{(s+d_1)}c\lambda_j+\frac{c\lambda_j}{(s+d_1)}\left[1+\frac{\lambda_A}{(s+d_1)}\right]\end{array}\right] \tag{33}$$

Where $\overline{S}_\mu(s)=\int_0^\infty\mu(x)\{\exp\{-sx-\int_0^x\mu(x)dx\}\}dx$,
$\overline{S}_\psi(s)=\int_0^\infty\psi(z)\{\exp\{-sz-\int_0^x\psi(z)dz\}\}dx$,
$\overline{S}_\varphi(s)=\int_0^\infty\varphi_j(y)\left\{\exp\left\{-sy-\int_0^x\varphi_j(y)dy\right\}\right\}dy$, $d=\lambda_A+c\lambda_j+\lambda_h+(1-c)\lambda_c$,
$d_1=\lambda_A+(1-c)\lambda_c+\varphi_j(y)$, and

$$d_2=\left[\begin{array}{l}(s+d)-\frac{\lambda_A\phi_j(y)}{(s+d)}\left(\frac{c\lambda_j}{(s+d_1)}\right)\left(1+\frac{\lambda_A}{(s+d_1)}\right)\overline{S}_\mu(s)-\frac{\lambda_A{}^2}{(s+d)}\overline{S}_\mu(s)\\ -\lambda_A\frac{c\lambda_j}{(s+d_1)}\left[1+\frac{\lambda_A}{(s+d_1)}\right]\overline{S}_\phi(s)-\lambda_h\overline{S}_\psi(s)\\ -\frac{\lambda_h}{(s+d)}\overline{S}_\psi(s)\phi_j(y)\frac{c\lambda_j}{(s+d_1)}\left[1+\frac{\lambda_A}{(s+d_1)}\right]-\frac{1}{(s+d)}\lambda_A\lambda_h\overline{S}_\psi(s)-\phi_j(y)\frac{c\lambda_j}{(s+d_1)}\end{array}\right]$$

$$\overline{P}_{up}(s)=\overline{P}_{0,0}(s)+\overline{P}_{1,0}(s)+\overline{P}_{0,j}(s)+\overline{P}_{1,j}(s)$$
$$=\left[\begin{array}{l}1+\frac{1}{(s+d)}\left[\varphi_j(y)\left\{\frac{c\lambda_j}{(s+d_1)}\left[1+\frac{\lambda_A}{(s+d_1)}\right]\right\}+\lambda_A\right]\\ +\frac{1}{(s+d_1)}c\lambda_j+\frac{c\lambda_j}{(s+d_1)}\left[1+\frac{\lambda_A}{(s+d_1)}\right]\end{array}\right]\overline{P}_{0,0}(s) \tag{34}$$

$$\overline{P}_{down}(s)=\overline{P}_{2,0}(s)+\overline{P}_F(s)+\overline{P}_{2,j}(s)+\overline{P}_c(s)$$

$$=\left[\begin{array}{l}\left[\dfrac{\lambda_A\varphi_j(\gamma)}{(s+d)}\left\{\dfrac{c\lambda_j}{(s+d_1)}\left[1+\dfrac{\lambda_A}{(s+d_1)}\right]\right\}\left[\dfrac{1-\overline{S}_\mu(s)}{s}\right]+\dfrac{1}{(s+d)}\lambda_A{}^2\left[\dfrac{1-\overline{S}_\mu(s)}{s}\right]\right]\\ +\lambda_h+\left[\dfrac{\lambda_h}{(s+d)}\varphi_j(\gamma)\left\{\dfrac{c\lambda_j}{(s+d_1)}\left[1+\dfrac{\lambda_A}{(s+d_1)}\right]\right\}\left[\dfrac{1-\overline{S}_\psi(s)}{s}\right]+\dfrac{\lambda_A\lambda_h}{(s+d)}\left[\dfrac{1-\overline{S}_\psi(s)}{s}\right]\right]\\ +\dfrac{\lambda_A c\lambda_j}{(s+d_1)}\left[1+\dfrac{\lambda_A}{(s+d_1)}\right]\left[\dfrac{1-\overline{S}_\varphi(s)}{s}\right]\\ +\dfrac{1}{s}(1-c)\lambda_c\left[\begin{array}{l}1+\dfrac{1}{(s+d)}\left[\varphi_j(\gamma)\left\{\dfrac{c\lambda_j}{(s+d_1)}\left[1+\dfrac{\lambda_A}{(s+d_1)}\right]\right\}+\lambda_A\right]\\ +\dfrac{1}{(s+d_1)}c\lambda_j+\dfrac{c\lambda_j}{(s+d_1)}\left[1+\dfrac{\lambda_A}{(s+d_1)}\right]\end{array}\right]\end{array}\right]\overline{P}_{0,0}(s) \tag{35}$$

3 PARTICULAR CASES

3.1 Availability Analysis

Consider the values of various parameters as $\lambda_A=0.020$, $\lambda_c=0.025$, $\lambda_h=0.030$, $\lambda_j=0.035$, $\varphi_j(\gamma)=1$, $c=0.1$, 0.2, 0.3, 0.4, 0.5. Using all these values in Eq. (34), then after taking the inverse Laplace transform and varying time unit t from 0 to 15, we obtain Table 1 and correspondingly Fig. 2, which represents the behavior of availability of the system with respect to time.

Table 1 Availability versus time

	Availability $P_{up}(t)$				
Time (t)	**$c=0.1$**	**$c=0.2$**	**$c=0.3$**	**$c=0.4$**	**$c=0.5$**
0	1.00000	1.00000	1.00000	1.00000	1.00000
1	0.96267	0.96837	0.97409	0.97981	0.98555
2	0.93788	0.94878	0.95971	0.97069	0.98171
3	0.91771	0.93338	0.94915	0.96501	0.98096
4	0.89935	0.91946	0.93973	0.96017	0.98078
5	0.88179	0.90695	0.93056	0.95532	0.98034
6	0.86470	0.89284	0.92134	0.95019	0.97942
7	0.84795	0.87973	0.91199	0.94474	0.97798
8	0.83149	0.86669	0.90251	0.93896	0.97604
9	0.81531	0.85373	0.89291	0.93288	0.97366
10	0.79941	0.84085	0.88322	0.92655	0.97087
11	0.78378	0.82805	0.87344	0.91999	0.96772
12	0.76843	0.81536	0.86361	0.91323	0.96425
13	0.75334	0.80278	0.85375	0.90631	0.96050
14	0.73852	0.79031	0.84387	0.89924	0.95650
15	0.72397	0.77797	0.83398	0.89205	0.95228

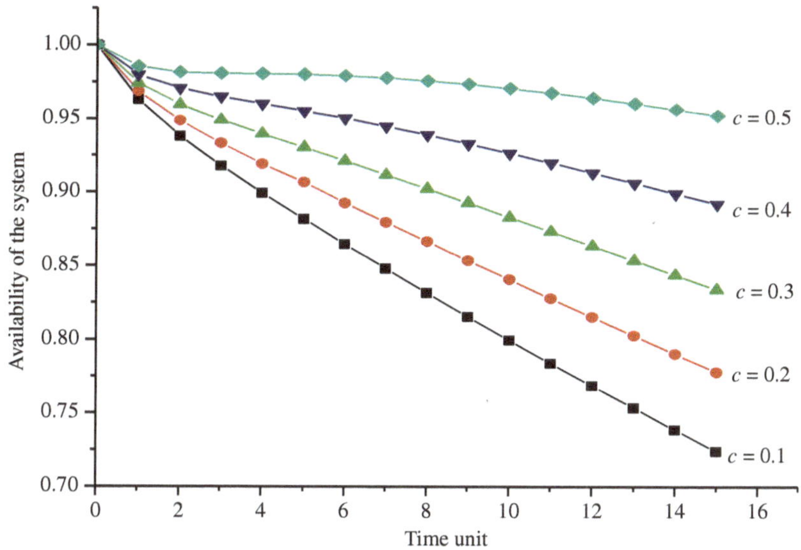

Fig. 2 Availability versus time.

3.2 Reliability Analysis

The time-dependent system reliability discussed by Yang and Dhillon [33] without repair can be obtained by inverting Eq. (34). Taking the same set of the parameter as mentioned in Section 3.1 and varying time unit t from 0 to 15, one can obtain the system's reliability without repair as shown in Table 2 and Fig. 3, respectively.

3.3 Mean Time to Failure (MTTF) Analysis

MTTF of a system is given as the average time between the failures of a system. By considering all the repairs equal to zero in Eq. (34), as s tends to zero, we can obtain the MTTF as:

$$\mathrm{MTTF} = \frac{1 + \dfrac{\lambda_A}{\lambda_A + c\lambda_j + \lambda_h + (1-c)\lambda_c} + \dfrac{c\lambda_j}{\lambda_A + (1-c)\lambda_c} + \dfrac{c\lambda_j\left(1 + \dfrac{\lambda_A}{\lambda_A + (1-c)\lambda_c}\right)}{\lambda_A + (1-c)\lambda_c}}{\lambda_A + c\lambda_j + \lambda_h + (1-c)\lambda_c} \tag{36}$$

Setting $\lambda_A = 0.020$, $\lambda_c = 0.025$, $\lambda_h = 0.030$, $\lambda_j = 0.035$, $c = 0.1, 0.2, 0.3, 0.4, 0.5$ and varying λ_A, λ_c, λ_h, λ_j one by one, respectively, as 0.1, 0.2, 0.3, 0.4, 0.05, 0.6, 0.7, 0.8, 0.9 in Eq. (36), one may obtain the variation of MTTF with respect to failure rates as mentioned in Table 3 and Fig. 4.

Table 2 Reliability versus time

Time (*t*)	Reliability *R*(*t*)				
	c = 0.1	*c* = 0.2	*c* = 0.3	*c* = 0.4	*c* = 0.5
0	1.00000	1.00000	1.00000	1.00000	1.00000
1	0.95198	0.95767	0.96338	0.96910	0.97483
2	0.90591	0.91672	0.92758	0.93849	0.94944
3	0.86174	0.87716	0.89266	0.90826	0.92395
4	0.81945	0.83897	0.85865	0.87850	0.89851
5	0.77897	0.80216	0.82559	0.84927	0.87320
6	0.74026	0.76672	0.79351	0.82064	0.84811
7	0.70328	0.73263	0.76241	0.79264	0.82333
8	0.66797	0.69986	0.73231	0.76532	0.79890
9	0.63427	0.66840	0.70320	0.73870	0.77490
10	0.60213	0.63821	0.67510	0.71280	0.75135
11	0.57150	0.60927	0.64797	0.68764	0.72830
12	0.54232	0.58154	0.62183	0.66323	0.70577
13	0.51454	0.55498	0.59664	0.63956	0.68378
14	0.48809	0.52957	0.57240	0.61665	0.66235
15	0.46293	0.50526	0.54909	0.59447	0.64149

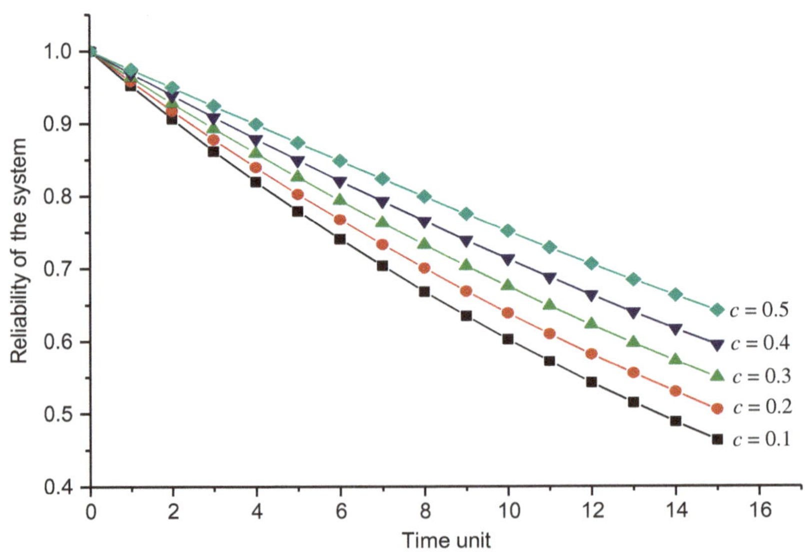

Fig. 3 Reliability versus time.

Table 3 MTTF as a function of failure rates

Failure rates of λ_A, λ_c, λ_h, λ_j	MTTF											
	$c = 0.1$				$c = 0.2$				$c = 0.3$			
	λ_A	λ_c	λ_h	λ_j	λ_A	λ_c	λ_h	λ_j	λ_A	λ_c	λ_h	λ_j
0.1	11.03520	8.42365	9.18114	22.10592	11.47911	9.48894	10.70445	27.46913	11.94738	10.82837	12.46262	33.14661
0.2	7.13613	4.80686	5.22260	25.71725	7.27926	5.41135	6.14765	33.47107	7.42688	6.18056	7.21765	40.86428
0.3	5.26544	3.35948	3.64527	28.67382	5.33499	3.78201	4.30875	37.72189	5.40608	4.32240	5.07702	45.75862
0.4	4.17018	2.58107	2.79888	31.13810	4.21111	2.90567	3.31597	40.88888	4.25274	3.32179	3.91512	49.13777
0.5	3.45156	2.09519	2.27122	33.22318	3.47851	2.35862	2.69481	43.33910	3.50580	2.69674	3.18582	51.61051
0.6	2.94405	1.76313	1.91086	35.01005	2.96308	1.98473	2.26957	45.29085	2.98234	2.26937	2.68549	53.49821
0.7	2.56654	1.52186	1.64915	36.55828	2.58071	1.71307	1.96020	46.88208	2.59502	1.95879	2.32095	54.98644
0.8	2.27480	1.33864	1.45047	37.91256	2.28575	1.50678	1.72504	48.20415	2.29680	1.72290	2.04353	56.18983
0.9	2.04258	1.19478	1.29450	39.10714	2.05130	1.34480	1.54025	49.32000	2.06009	1.53768	1.82534	57.18299

$c = 0.4$				$c = 0.5$			
λ_A	λ_c	λ_h	λ_j	λ_A	λ_c	λ_h	λ_j
12.44194	12.54064	1.79471	39.32620	12.96489	14.82247	16.94411	46.23507
7.57917	7.19011	8.46944	48.38263	7.73632	8.56940	9.95313	56.43331
5.47875	5.03675	5.97672	53.64559	5.55303	6.02274	7.04407	61.95352
4.29508	3.87376	4.61718	57.08440	4.33815	4.63958	5.45050	65.41281
3.53348	3.14609	3.76138	59.50697	3.56154	3.77141	4.44481	67.78349
3.00183	2.64809	3.17317	61.30566	3.02154	3.17609	3.75238	69.50947
2.60948	2.28596	2.74402	62.69391	2.62408	2.74264	3.24660	70.82221
2.30795	2.01081	2.41711	63.79779	2.31920	2.41304	2.86095	71.85425
2.06895	1.79471	2.15979	64.69656	2.07788	2.15400	2.55719	72.68691

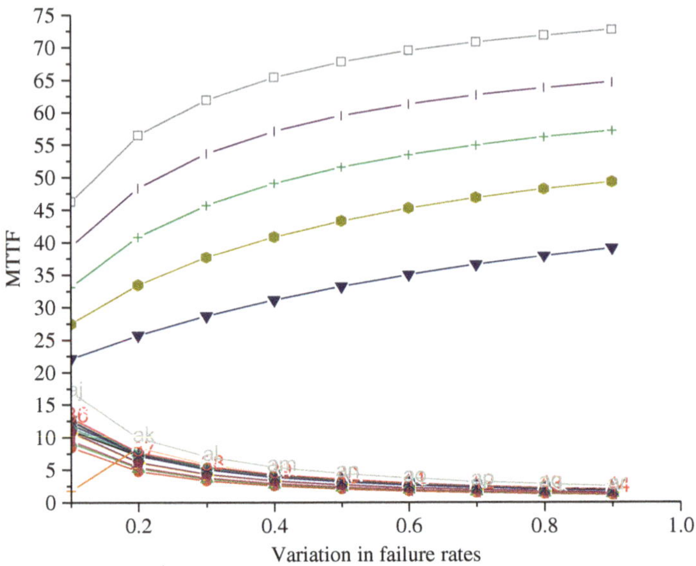

Fig. 4 MTTF as a function of failure rates.

3.4 Expected Profit

To maintain the product reliability, cost control is very important. Undoubtedly, only the reliability will not guarantee the viability of the product or system. Similarly, when the relating systems' reliabilities are too low, the arbitrary cost-cutting can be harmful to profit [19]. The expected profit during the interval [0, t) is given as:

$$E_p(t) = K_1 \int_0^t P_{up}(t)\,dt - tK_2 \tag{37}$$

Using the availability equation, expected profit for the same values of the parameters is given by setting $K_1 = 1$ and $K_2 = 0.1, 0.3, 0.6$, respectively, and one gets Table 4 and Fig. 5, respectively.

3.5 Busy Period Analysis or Mean Time to Repair (MTTR)

MTTR is the average time taken by a system to recover from any failure. By taking all values of all repair rates to zero in Eq. (35), as s tends to zero, one can obtain the MTTR as:

$$\text{MTTR} = \frac{\lambda_h}{\lambda_A + c\lambda_j + \lambda_h + (1-c)\lambda_c} \tag{38}$$

Table 4 Expected profit versus time

	Expected profit $E_p(t)$								
	$c=0.1$			$c=0.2$			$c=0.3$		
Time (t)	$K_2=0.1$	$K_2=0.3$	$K_2=0.6$	$K_2=0.1$	$K_2=0.3$	$K_2=0.6$	$K_2=0.1$	$K_2=0.3$	$K_2=0.6$
0	0	0	0	0	0	0	0	0	0
1	0.87969	0.67969	0.37969	0.88259	0.68259	0.38259	0.88550	0.68550	0.38550
2	1.72938	1.32938	0.72938	1.74062	1.34062	0.74062	1.75188	1.35188	0.75188
3	2.55696	1.95696	1.05696	2.58151	1.98151	1.08151	2.60615	2.00615	1.10615
4	3.36540	2.56540	1.36540	3.40786	2.60786	1.40786	3.45055	2.65055	1.45055
5	4.15593	3.15593	1.65593	4.22060	3.22060	1.72060	4.28570	3.28570	1.78570
6	4.92914	3.72914	1.92914	5.02004	3.82004	2.02004	5.11166	3.91166	2.11166
7	5.68545	4.28545	2.18545	5.80632	4.40632	2.30632	5.92834	4.52834	2.42834
8	6.42515	4.82515	2.42515	6.57953	4.97953	2.57953	6.73561	5.13561	2.73561
9	7.14853	5.34853	2.64853	7.33974	5.53974	2.83974	7.53333	5.73333	3.03333

$c=0.4$			$c=0.5$		
$K_2=0.1$	$K_2=0.3$	$K_2=0.6$	$K_2=0.1$	$K_2=0.3$	$K_2=0.6$
0	0	0	0	0	0
0.88840	0.68840	0.38840	0.89132	0.69132	0.39132
1.76318	1.36318	0.76318	1.77450	1.37450	0.77450
2.63089	2.03089	1.13089	2.65573	2.05573	1.15573
3.49346	2.69346	1.48346	3.53660	2.73660	1.53660
4.35123	3.35123	1.85123	4.41720	3.41720	1.91720
5.20402	4.00402	2.20402	5.29713	4.09713	2.29713
6.05152	4.65152	2.55152	6.17587	4.77587	2.67587
6.89340	5.29340	2.89340	7.05293	5.45293	3.05293
7.72934	5.92934	3.22934	7.92782	6.12782	3.42782

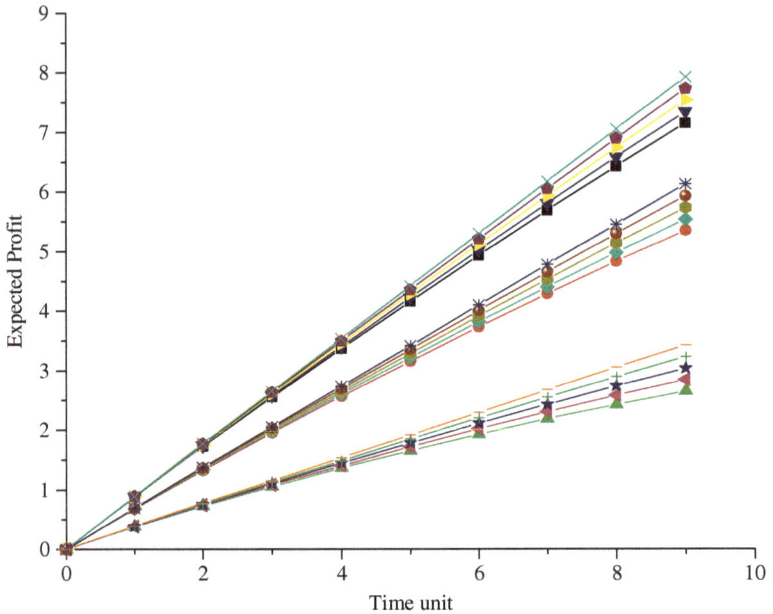

Fig. 5 Expected versus time.

Setting $\lambda_A = 0.020$, $\lambda_c = 0.025$, $\lambda_h = 0.030$, $\lambda_j = 0.035$, $c = 0.1$, 0.2, 0.3, 0.4, 0.5 and varying λ_A, λ_c, λ_h, λ_j one by one, respectively, as 0.1, 0.2, 0.3, 0.4, 0.05, 0.6, 0.7, 0.8, 0.9 in Eq. (38), the variation of mean time to repair with respect to failure rates may be easily obtained, as seen in Table 5 and Fig. 6.

4 RESULTS AND DISCUSSION

In this chapter, the authors have analyzed a multistate, repairable system under the concept of covered faults with three types of failure. Also, various reliability measures by using Markov model and supplementary variable techniques have been discussed by the authors.

The availability of any system lies between 0 and 1. Fig. 2 shows the impact of coverage factor on the availability of the system. It can be easily observed from the figure that, with an increase in time, the availability of the system decreases, whereas an increment in coverage factor results in an increase in availability of the system.

Fig. 3 shows the behavior of reliability of the system with rest to time. From the graph, it can be observed that the reliability of the system decreases

Table 5 MTTR as a function of failure rates

Failure rates of λ_A, λ_c, λ_h, λ_j	MTTR											
	$c=0.1$				$c=0.2$				$c=0.3$			
	λ_A	λ_c	λ_h	λ_j	λ_A	λ_c	λ_h	λ_j	λ_A	λ_c	λ_h	λ_j
0.1	0.19230	0.20905	0.68493	0.36363	0.19108	0.21897	0.68027	0.33333	0.18987	0.22988	0.67567	0.30769
0.2	0.11718	0.12847	0.81300	0.32432	0.11673	0.13824	0.80971	0.27272	0.11627	0.14962	0.80645	0.23529
0.3	0.08426	0.09273	0.86705	0.29268	0.08403	0.10101	0.86455	0.23076	0.08379	0.11090	0.86206	0.19047
0.4	0.06578	0.07255	0.89686	0.26666	0.06564	0.07957	0.89485	0.20000	0.06550	0.08810	0.89285	0.16000
0.5	0.05395	0.05958	0.91575	0.24489	0.05385	0.06564	0.91407	0.17647	0.05376	0.07308	0.91240	0.13793
0.6	0.04573	0.05054	0.92879	0.22641	0.04566	0.05586	0.92735	0.15789	0.04559	0.06243	0.92592	0.12121
0.7	0.03968	0.04389	0.93833	0.21052	0.03963	0.04862	0.93708	0.14285	0.03957	0.05449	0.93582	0.10810
0.8	0.03504	0.03878	0.94562	0.19672	0.03500	0.04304	0.94451	0.13043	0.03496	0.04834	0.94339	0.09756
0.9	0.03138	0.03474	0.95137	0.18461	0.03134	0.03861	0.95036	0.12000	0.03131	0.04344	0.94936	0.08888

$c=0.4$				$c=0.5$			
λ_A	λ_c	λ_h	λ_j	λ_A	λ_c	λ_h	λ_j
0.18867	0.24193	0.67114	0.28571	0.18750	0.25531	0.66666	0.26666
0.11583	0.16304	0.80321	0.20689	0.11538	0.17910	0.80000	0.18461
0.08356	0.12295	0.85959	0.16216	0.08333	0.13793	0.85714	0.14117
0.06535	0.09868	0.89086	0.13333	0.06521	0.11214	0.88888	0.11428
0.05366	0.08241	0.91074	0.11320	0.05357	0.09448	0.90909	0.09600
0.04552	0.07075	0.92449	0.09836	0.04545	0.08163	0.92307	0.08275
0.03952	0.06198	0.93457	0.08695	0.03947	0.07185	0.93333	0.07272
0.03492	0.05514	0.94228	0.07792	0.03488	0.06417	0.94117	0.06486
0.03128	0.04966	0.94836	0.07058	0.03125	0.05797	0.94736	0.05853

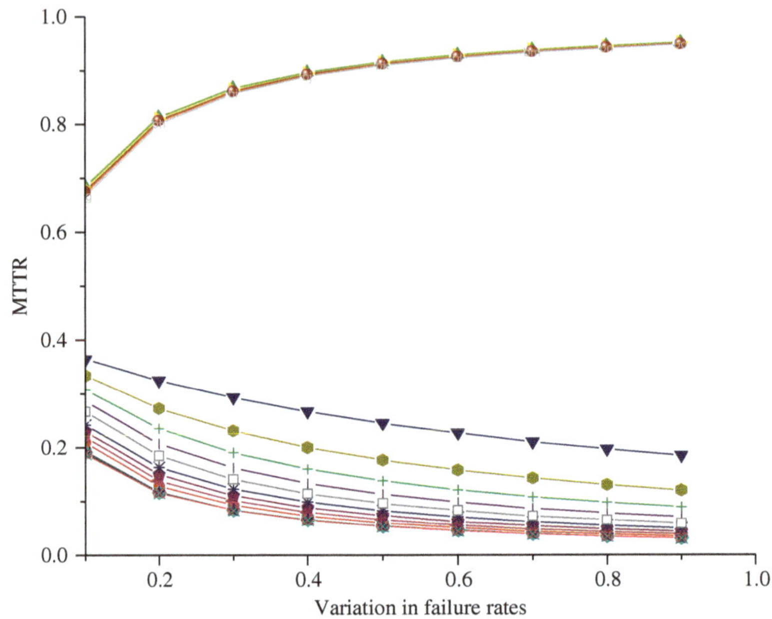

Fig. 6 MTTR as a function of failure rates.

constantly as the time increases, whereas the reliability of the system increases with the increment in coverage factor.

Fig. 4 is the study of MTTF of the system with respect to various failures. From the figure, it can be observed that mean time to failure of the system increases due to the partial failure of the *j*th unit of subsystem B, and decreases due to the failure of the main and standby units of subsystem *A*, human failure, and catastrophic failure. From the graph, it can be concluded that, at the time of system operation, both the failures (human and catastrophic) are controlled comparative to partial failure of any *j*th unit of subsystem *B*. It can also be seen that mean time to failure of the system increases with the increment in coverage factor.

Table 4 is obtained by varying service cost, taking the revenue cost per unit time at value 1 with coverage factor values at 0.1, 0.2, 0.3, 0.4, 0.5. From the figure, it can be observed that expected profit decreases with increase in service cost. This analysis shows that maximum profit can be attained by minimizing the service cost. Also, it can be seen that an increment in coverage factor results in an increase in expected profit.

Mean time to repair of the system with respect to failure rates at different values of coverage factor has been shown in Fig. 6. It can be easily observed from the figure that it increases with increment in human failure rate, whereas it decreases with an increment in unit failure rate and catastrophic failure rate. Hence, the study shows that the system takes more recovery time due to human failure in comparison to unit failure and catastrophic failure. Also, from Fig. 6, we determined that MTTR decreases in each case with an increment in coverage factor.

5 CONCLUSION

In this work, the authors have investigated a two-unit standby system with multiple failures, including the concept of fault coverage, and derived the explicit expressions as a function of time and coverage factor for the system performance measures. The numerical illustrations of these reliability measures show the effect of coverage factor, revealing that the performance of the system is improved significantly. The availability and reliability of the system and reduction in cost can be improved by introducing the concept of coverage factor. The results can be applied to the analysis, design, and operation of various systems subjected to fault coverage involving different reliability measures. Future work will be open to involving the performance analysis of various systems using the concept of coverage.

APPENDIX 1

At any time t, if the system is in state S_i, then the probability of the system to be in that state is defined as: the probability that the system is in state S_i at time t and remains there in the interval $(t, t+\Delta t)$ or/and if it is in some other state at time t then it should transit to the state S_i in the interval $(t, t+\Delta t)$ provided transition exist between the states and $\Delta t \rightarrow 0$)

Accordingly, Eqs. (39)–(46) are interpreted as:

The probability of the system to be in state S_0 in the interval $(t, t+\Delta t)$ is given by:

$$
\begin{aligned}
P_{0,0}(t+\Delta t) = &(1-\lambda_A \Delta t)\left(1-c\lambda_j \Delta t\right)(1-\lambda_h \Delta t)(1-(1-c)\lambda_c \Delta t)P_0(t) \\
&+\int_0^{\infty} P_{2,0}(x,t)\mu(x)\Delta t dx \\
&+\int_0^{\infty} P_{2,j}(y,t)\phi_j(y)\Delta t dy \\
&+\int_0^{\infty} P_F(z,t)\psi(z)\Delta t dz + \phi_j(y)\Delta t P_{0,j}(t)
\end{aligned}
$$

$$\Rightarrow \lim_{\Delta t \to 0} \frac{P_{0,0}(t+\Delta t) - P_{0,0}(t)}{\Delta t} + \left(\lambda_A + c\lambda_j + \lambda_h + (1-c)\lambda_c\right) P_{0,0}(t)$$
$$= \int_0^\infty P_{2,0}(x,t)\mu(x)dx + \int_0^\infty P_{2,j}(y,t)\phi_j(y)dy + \int_0^\infty P_F(z,t)\psi(z)dz$$
$$+ \phi_j(y)P_{0,j}(t) \Rightarrow \left[\frac{\partial}{\partial t} + \lambda_A + c\lambda_j + \lambda_h + (1-c)\lambda_c\right] P_{0,0}(t) \tag{39}$$
$$= \int_0^\infty P_{2,0}(x,t)\mu(x)dx + \int_0^\infty P_{2,j}(y,t)\phi_j(y)dy + \int_0^\infty P_F(z,t)\psi(z)dz$$
$$+ \phi_j(y)P_{0,j}(t)$$

For state S_1,

$$P_{1,0}(t+\Delta t) = (1-\lambda_A\Delta t)\left(1-c\lambda_j\Delta t\right)(1-\lambda_h\Delta t)(1-(1-c)\lambda_c\Delta t P_{1,0}(t)$$
$$+ \phi_j(y)\Delta t P_{1,j}(t) + \lambda_A \Delta t P_{0,0}(t) \Rightarrow \lim_{\Delta t \to 0} \frac{P_{1,0}(t+\Delta t) - P_{1,0}(t)}{\Delta t}$$
$$+ \left(\lambda_A + c\lambda_j + \lambda_h + (1-c)\lambda_c\right) P_{1,0}(t) = \varphi_j(y)P_{1,j}(t) + \lambda_A P_{0,0}(t)$$
$$\Rightarrow \left[\frac{\partial}{\partial t} + \lambda_A + c\lambda_j + \lambda_h + (1-c)\lambda_c\right] P_{1,0}(t) = \varphi_j(y)P_{1,j}(t) + \lambda_A P_{0,0}(t) \tag{40}$$

For state S_2,

$$P_{2,0}(x+\Delta x, t+\Delta t) = \{1-\mu(x)\Delta t\} P_{2,0}(x,t)$$
$$\Rightarrow \lim_{\substack{\Delta x \to 0 \\ \Delta t \to 0}} \frac{P_{2,0}(x+\Delta x, t+\Delta t) - P_{2,0}(x,t)}{\Delta t} + \mu(x)P_{2,0}(x,t)$$
$$= 0 \Rightarrow \left[\frac{\partial}{\partial t} + \frac{\partial}{\partial x} + \mu(x)\right] P_{2,0}(x,t) = 0 \tag{41}$$

For state S_3,

$$P_{0,j}(t+\Delta t) = (1-\lambda_A\Delta t)\left(1-\varphi_j(y)\Delta t\right)(1-(1-c)\lambda_c\Delta t)P_{0,j}(t)$$
$$+ c\lambda_j\Delta t P_{0,0}(t) \Rightarrow \lim_{\Delta t \to 0} \frac{P_{0,j}(t+\Delta t) - P_{0,j}(t)}{\Delta t} + \lambda_A P_{0,j}(t) + \varphi_j(y)P_{0,j}(t)$$
$$+ (1-c)\lambda_c P_{0,j}(t) = c\lambda_j P_{0,0}(t)$$
$$\Rightarrow \left[\frac{\partial}{\partial t} + \lambda_A + \varphi_j(y) + (1-c)\lambda_c\right] P_{0,j}(t) = c\lambda_j P_{0,0}(t) \tag{42}$$

For state S_4,

$$P_{1,j}(t+\Delta t)=(1-\lambda_A\Delta t)\Big(1-\varphi_j(y)\Delta t\Big)(1-(1-c)\lambda_c\Delta t)P_{1,j}(t)$$
$$+c\lambda_j\Delta tP_{0,0}(t)+\lambda_A\Delta tP_{0,j}(t)\Rightarrow\lim_{\Delta t\to 0}\frac{P_{0,j}(t+\Delta t)-P_{0,j}(t)}{\Delta t}$$
$$+\lambda_AP_{0,j}(t)+\varphi_j(y)P_{0,j}(t)+(1-c)\lambda_cP_{0,j}(t)=c\lambda_jP_{0,0}(t)$$
$$+\lambda_AP_{0,j}(t)\Rightarrow\left[\frac{\partial}{\partial t}+\lambda_A+\varphi_j(y)+(1-c)\lambda_c\right]$$
$$P_{0,j}(t)=c\lambda_jP_{0,0}(t)+\lambda_AP_{0,j}(t) \tag{43}$$

For state S_5,

$$P_{2,j}(y+\Delta y,t+\Delta t)=\Big\{1-\varphi_j(y)\Delta t\Big\}P_{2,j}(y,t)$$
$$\Rightarrow\lim_{\substack{\Delta y\to 0\\ \Delta t\to 0}}\frac{P_{2,j}(y+\Delta y,t+\Delta t)-P_{2,j}(y,t)}{\Delta t}+\varphi_j(y)P_{2,j}(y,t)$$
$$=0\Rightarrow\left[\frac{\partial}{\partial t}+\frac{\partial}{\partial y}+\varphi_j(y)\right]P_{2,j}(y,t)=0 \tag{44}$$

For state S_6,

$$P_F(z+\Delta z,t+\Delta t)=\{1-\psi(z)\Delta t\}P_F(z,t)$$
$$\Rightarrow\lim_{\substack{\Delta z\to 0\\ \Delta t\to 0}}\frac{P_F(z+\Delta y,t+\Delta t)-P_F(y,t)}{\Delta t}+\psi(z)P_F(z,t)=0$$
$$\Rightarrow\left[\frac{\partial}{\partial t}+\frac{\partial}{\partial z}+\psi(z)\right]P_F(z,t)=0 \tag{45}$$

For state S_7,

$$P_c(w+\Delta w,t+\Delta t)=0$$
$$\Rightarrow\lim_{\substack{\Delta w\to 0\\ \Delta t\to 0}}\frac{P_c(w+\Delta w,t+\Delta t)-P_c(w,t)}{\Delta t}=0$$
$$\Rightarrow\left[\frac{\partial}{\partial t}+\frac{\partial}{\partial w}\right]P_c(z,t)=0 \tag{46}$$

APPENDIX 2

Boundary conditions of the system are obtained corresponding to transitions between the states where the transition from a state with elapsed repair time exists [32]. From Fig. 1B, the boundary conditions are:

$$P_{2,0}(0,t) = \lambda_A P_{1,0}(t) \tag{47}$$

$$P_{2,j}(0,t) = \lambda_A P_{1,j}(t) \tag{48}$$

$$P_{2,j}(0,t) = \lambda_h [P_{0,0}(t) + P_{1,0}(t)] \tag{49}$$

$$P_c(0,t) = (1-c)\lambda_c \left[P_{0,0}(t) + P_{1,0}(t) + P_{0,j}(t) + P_{1,j}(t)\right] \tag{50}$$

When the system is perfectly good, that is, in initial state S_0, then

$$P_0(0) = 1 \tag{51}$$

and other state probabilities are zero at $t=0$.

REFERENCES

[1] M. Ram, M. Manglik, Performance evaluation of a multi-state system covering imperfect fault coverage, Commun. Stat. Simul. Comput. 45 (9) (2016) 3259–3280.
[2] L.A.T. Klauwer, Application of Formal Methods to Fault Injection and Coverage Factor Calculation (Master Thesis in Computer Science), Chalmers Tekniska Hogskolas INSA'de Lyon, 2006.
[3] T.F. Arnold, The concept of coverage and its effect on the reliability model of a repairable system, IEEE Trans. Reliab. C-22 (1973) 251–254.
[4] H. Pham, Reliability analysis of a high voltage system with dependent failures and imperfect coverage, Reliab. Eng. Syst. Saf. 37 (1992) 25–28.
[5] S. Akhtar, Reliability of *k*-out-of-*n*:G system with imperfect fault coverage, IEEE Trans. Reliab. 43 (1994) 101–106.
[6] M. Moustafa, Reliability analysis of *k*-out-of-*n*:G systems with dependent failures and imperfect coverage, Reliab. Eng. Syst. Saf. 58 (1997) 15–17.
[7] K.S. Trivedi, Probability and Statistics With reliability. Queuing and computer Science Applications, John Wiley & sons, Inc., New York, 2002.
[8] A. Myers, *k*-out-of-*n*:G system reliability with imperfect fault coverage, IEEE Trans. Reliab. 56 (2007) 464–473.
[9] J.C. Ke, S.L. Lee, Y.L. Hsu, On a repairable system with detection, imperfect coverage and reboot: Bayesian approach, Simul. Model. Pract. Theory 16 (2008) 353–367.
[10] J.C. Ke, Z.-L. Su, K.-H. Wang, Y.L. Hsu, Simulation inferences for an availability system with general repair distribution and imperfect fault coverage, Simul. Model. Pract. Theory 18 (3) (2010) 338–347.
[11] K.-H. Wang, T.-C. Yen, Y.-C. Fang, Comparison of availability between two systems with warm standby units and different imperfect coverage, Qual. Technol. Quant. Manag. 9 (3) (2012) 265–282.
[12] D. Powell, E. Martins, J. Arlat, Y. Crouzet, Estimators for fault tolerance coverage evaluation, IEEE Trans. Comput. 44 (2) (1995) 261–274.

[13] A. Reibman, H. Zaretsky, in: Modeling fault coverage and reliability in a fault-tolerant network, Global Telecommunications Conference, vol. 2, 1990, pp. 689–692.
[14] K.-Y. Cai, C.-Y. Wen, M.-L. Zhang, Fuzzy reliability modeling of gracefully degradable computing systems, Reliab. Eng. Syst. Saf. 33 (1991) 141–157.
[15] S. Prabhudeva, A.K. Verma, Coverage modeling and reliability analysis using multistate function, Int. J. Autom. Comput. 4 (4) (2007) 380–387.
[16] K. Kumar, J. Singh, P. Kumar, Fuzzy reliability and fuzzy availability of the serial process in butter- oil processing plant, J. Math. Stat. 5 (1) (2009) 65–71.
[17] M. Ram, R.G. Varshney, S.B. Singh, Performance improvement of a parallel redundant system with coverage factor, J. Eng. Sci. Technol. 8 (3) (2013) 350–358.
[18] P.P. Gupta, R.K. Sharma, Cost analysis of a three-state standby redundant electronic equipment, Microelectron. Reliab. 25 (6) (1985) 1029–1033.
[19] M. Manglik, M. Ram, Behavioural analysis of a hydroelectric production power plant under reworking scheme, Int. J. Prod. Res. 53 (2) (2015) 648–664.
[20] A. Kumar, M. Ram, System reliability measures in the presence of common cause failures, Int. J. Ind. Syst. Eng. 24 (1) (2016) 44–61.
[21] M. Ram, M. Manglik, An analysis to multi-state manufacturing system with common cause failure and waiting repair strategy, Cogn. Eng. 3 (1) (2016) 1266185.
[22] M. Ram, M. Manglik, Reliability measures analysis of an industrial system under standby modes and catastrophic failure, Int. J. Oper. Res. Inf. Syst. 7 (3) (2016) 36–56.
[23] M. Ram, M. Manglik, Stochastic biometric system modelling with rework strategy, Int. J. Math. Eng. Manag. Sci. 1 (1) (2016) 1–17.
[24] M. Ram, N. Goyal, Stochastic design exploration with rework of flexible manufacturing system under copula-coverage approach, Int. J. Reliab. Qual. Saf. Eng. 25 (2) (2017) 1850007.
[25] G. Chopra, M. Ram, Stochastic analysis of two non-identical unit parallel system incorporating waiting time, Int. J. Qual. Reliab. Manag. 34 (6) (2017) 817–832.
[26] N. Goyal, M. Ram, S. Amoli, A. Suyal, Sensitivity analysis of a three-unit series system under *k*-out-of-*n* redundancy, Int. J. Qual. Reliab. Manag. 34 (6) (2017) 770–784.
[27] N. Goyal, M. Ram, Series-parallel system study under warranty and preventive maintenance, in: M. Ram, J.P. Davim (Eds.), Mathematics Applied to Engineering, Elsevier, 2017, pp. 97–113, ISBN: 978-0-12-810998-4.
[28] B.S. Dhillon, Human errors: a review, Microelectron. Reliab. 29 (3) (1989) 299–304.
[29] M. Ram, S.B. Singh, V.V. Singh, Stochastic analysis of a standby system with waiting repair strategy, IEEE Trans. Syst. Man Cybern. Syst. Hum. 43 (3) (2013) 698–707.
[30] B.S. Dhillon, N. Yang, Probabilistic analysis of a maintainable system with human error, J. Qual. Maint. Eng. 1 (2) (1995) 50–59.
[31] B.S. Dhillon, Y. Liu, Human error in maintenance: a review, J. Qual. Maint. Eng. 12 (1) (2006) 21–36.
[32] N. Yang, B.S. Dhillon, Availability analysis of a repairable standby human-machine system, Microelectron. Reliab. 35 (11) (1995) 1401–1413.
[33] N. Yang, B.S. Dhillon, Stochastic analysis of a general standby system with constant human error and arbitrary system repair rates, Microelectron. Reliab. 35 (7) (1995) 1037–1045.

CHAPTER 11

Software Patch Scheduling Policy Incorporating Functional Safety Standards

Adarsh Anand*, Mohini Agrawal†, Navneet Bhatt*, Mangey Ram‡
*Department of Operational Research, University of Delhi, New Delhi, India
†School of Business, Galgotias University, Greater Noida, India
‡Department of Mathematics, Computer Science & Engineering, Graphic Era University, Dehradun, India

Abstract

Of late, the attacks or breaching of software systems has increased compared with accounts in the past. Vulnerabilities present in software have become a serious threat to security, and many times, intruders exploit known (as well as unknown) vulnerabilities. There is a current need to protect computer systems from exploits and develop systems that are potentially robust from such security breechings. Thus it is important to analyze the optimal time when vulnerabilities should be made public and also deduce an optimal time to fix certain vulnerabilities before they can be exploited. Based on the scientific technique of calculating software safety integrity level (SIL) in the functional safety standards, we have proposed the quantitative method for ensuring SIL within software systems to determine an optimal patch release policy for fixing the vulnerabilities after the software has been launched into the field. Validation of the proposed policy has been done using real-life vulnerability data of Microsoft Windows, and obtained results are well-versed in explaining the proposition. Further sensitivity analysis has been performed to form a better understanding.

Keywords: Dangerous failure ratio, Patch, Risk, Safety integrity level, Vulnerability

1 INTRODUCTION

Advancement of technology and expansion of Internet-based applications have led to a rapid shift in the way computers are linked with software applications, leading to a momentous increase in software security. Intensive dependence on computer systems and skyscraping network connectivity pave way to enhanced susceptibility of attack from exploiters. At this point in time, the utmost concern for any software-developing firm is software security. There is a current need to protect computer systems from exploits and develop systems that are potentially robust from such security

Advances in System Reliability Engineering
https://doi.org/10.1016/B978-0-12-815906-4.00011-7

breechings. Software plays a significant role in providing security but is also a major source of security problems [1]. It is the weakest link in the security chain with the possible exception to the "human factor," implying that security aspects should be an integral part of software design, right from the beginning of the development process. In the development of software, an essential objective is to offer some new functionalities or services; at the same time, overseeing related dangers is an inferred/auxiliary concern. Software is built using complicated languages that run on huge and complicated infrastructures using various tools. Thus it is imperative to acknowledge that there can be instances when things can go wrong or would not perform a desired task (flaws or weakness), and these flaws can actually be exploited by attackers (vulnerability).

Vulnerability is nothing but a weak breach of an asset that can be exploited by one or more threats [2]. A loophole in a system's design, implementation, or operation and management could be exploited to violate the system's security policy. It is an intersection of three core elements: the existence of flaw or loophole(s), access of attackers to these weak points, and the attacker's potential to exploit these flaws. Vulnerability exposure is assumed to follow a hypothetical pattern, which is termed as its life cycle [3]. There are certain events important in the discovery process of vulnerability, such as [4, 5]:

- *Injection time*: Time point at which the vulnerability was introduced into the software code prior to release in the field.
- *Release time*: The day the software system is released into the field for its operation, which contains certain loopholes lying latent.
- *Time of discovery*: Point of time at which flaws that can be exploited are initially identified.
- *Vulnerability disclosure time*: At this moment, the loophole detector plans to notify software firms.
- *Public release*: Firms plan to inform the public about the existence of a vulnerability.
- *Release of patch*: The date the corrective piece of code is released to fix the loopholes that were left in the software either at design or as a code-level defect.

The presence of one or more vulnerabilities in any parts of the software systems makes it prone to attacks. Generally, the defects that arise are of two kinds: imperfections at configuration-level issues prompting a security hazard, or bugs arising due to execution-level issues prompting security risk. An essential requirement to uncover these flaws and minimize the risk is to anticipate the need to have proficient human expertise. Failure in safety-critical systems may lead to several forms of physical harm, for example, the failure of the software designed for the London ambulance system that led to service disruption [6]. Another example of failure was found in the Therac-25 Radiotherapy Machine, which led to several deaths from overdose [6]. Such vulnerable systems need strong protection over the entire life of the software, making it important to identify the safety risk and locating measures for these risks. Large numbers of approaches exist and are used to develop safety-critical software with the common intention of developing software that is compliant with stated standards and is safe for its intended purpose [7]. A significant pattern that should be followed throughout the software development lifecycle is to have verification integrated within each stage of development, that is, a software development plan should be driven using a V-model development process [6]. As developers become aware of software security's importance, they are gradually implementing and developing various sets of practices to address these problems. One such practice is to incorporate the security aspect deeply into the various software stages during software development, which is shown in Fig. 1 [8].

The concept frequently used to manage software system risk is known as functional safety [8]. It is comprised of the overall safety of a system that depends upon the system's performance based on its response to different inputs, which are dependent upon operator errors and environmental changes that may exist. It is achieved when all the predefined safety functions are carried out, and the performance level of each safety function is met. Also, it is expected that the attributes of safety can be acquired by enhancing the dependability characterized by institutionalized software quality attributes ISO/IEC 9126 [9]. In the first part of functional safety development, a hazard and risk assessment of the target application is performed that determines the level of acceptable risk for an application, which can be termed as the safety integrity level (SIL). The lower the value of acceptable risk, the more the SIL target will be achieved (and vice versa). Thus software development has become more challenging [10].

Many times, to achieve these safety standards, a firm will plan to release a patch to fix those loopholes that might occur when the software is in field or to enhance its functionalities. Several factors determine the timing when a firm should release a patch. For example, under certain situations, a firm may not have the provision of developing a patch before disclosure of the vulnerabilities present in the software or the amount of time before the patch development begins is further dependent upon the resources available from the vendor's side. Furthermore, the patch release schedule adopted by a firm also influences the application of patch, for example, certain large

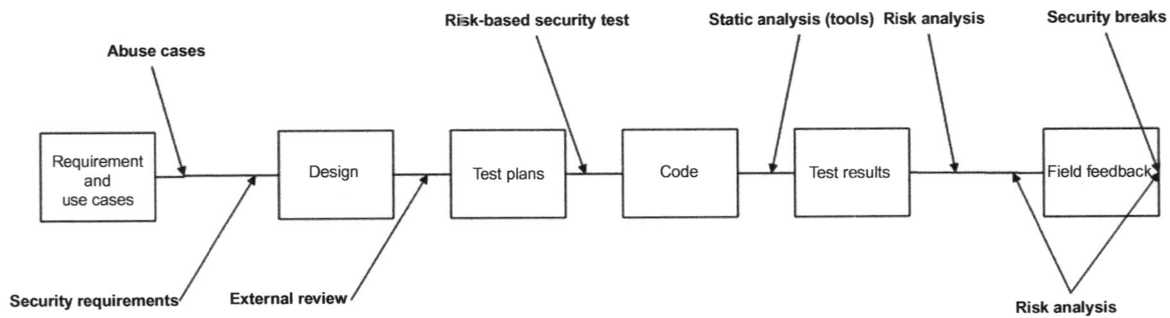

Fig. 1 Software security practice.

organizations, such as Oracle and Microsoft, have a fixed quarterly or monthly schedule of releasing patches [2].

Working on the basis of these concepts, in this chapter, an optimal patch release policy for fixing loopholes that may trigger after the software has been launched should be determined based on the concept of SIL in the functional safety standards. The rest of this chapter is organized as follows: Section 2 gives a brief literature review. Section 3 consists of notations, assumptions, and mathematical formulation of our proposed model. The numerical illustration for better understanding of the proposed model is provided in Section 4. Sensitivity is presented in Section 5. Further conclusion and references are given in Section 6 and in reference section, respectively.

2 BRIEF LITERATURE

With a logical presentation of the security designs and loopholes within the software, developers can get better insight of the potential breaching sites and mitigate the menace prior to exploitation by intruders. Anand and Bhatt [2] defined vulnerability as "a defect that enables an attacker to bypass security measures". The examination of vulnerabilities can assist designers to gain better understanding of their past mistakes and also showcase the shortcomings of the software. It can also indicate inadequate and wrongful practices due to which some vulnerability may appear in the software. Vulnerability discovery models (VDM) may turn out to be a valuable apparatus for assessing and anticipating weak attributes in software. Nevertheless, the present literature on vulnerability overlooks critical uncertainty and hypothetical concerns [11].

As of late, analysts have begun exploring how the weakness disclosure process can be portrayed by utilizing some VDMs. Several researchers have modeled different set of VDMs by utilizing different cases viz. Anderson [12], Rescorla [13], Alhazmi and Malaiya [14, 15], which uses an alternate approach and different set of parameters. However, these works have risen as of late, and a considerable number of their confinements have not yet been examined. Making use of Brady et al. [16] model for software reliability, Anderson [12] studied the trend in vulnerability discovery, whereas Rescorla [13] has made efforts to categorize the different trends in vulnerability data by utilizing the linear and exponential models. Alhazmi and Malaiya [14] laid emphasis on modeling the relationship between the cumulative vulnerabilities over time wherein they proposed a logistic and effort-based exponential model by isolating the efforts required to discover different vulnerabilities. Alhazmi and Malaiya [17] tried to develop a basis for systematic quantitative approaches for characterizing security by focusing upon vulnerability discovery. In 2015, Kapur et al. [18] demonstrated the application of a logistic rate for discovering vulnerabilities. Anand and Bhatt [2] developed a novel approach to model the vulnerability discovery process that follows a hump-shaped pattern. Recently, Anand et al. [4, 5] deduced a multiversion VDM to capture the discovery influenced by the code-sharing phenomenon. Bhatt et al. [19] suggested a VDM that accounts for the discovery of vulnerabilities discovered additionally due to the influence of previously identified vulnerabilities.

Besides having an understanding behind different VDMs, to have better knowledge about the manner in which these vulnerabilities can be fixed, it is important to have brief awareness on patching. Several researchers have worked on the concept of patching in the field of software reliability growth modeling. Some well-known works include: Jiang and Sarkar [20] claimed that, after the software has been delivered to users, the job of testing is not complete and the software is being tested; once flaws are identified, they are fixed by issuing patches. This discovery of faults can be from either the user's end or the testers' end. Going with the widely acceptable fact that "It is not possible to remove all the bugs from the software in first instance," Arora et al. [21] made a claim that patches are an integral part of postsales services offered by software developers, which is most important for the overall security of the software. Several researchers claim that the timely release of patches is crucial for preventing losses. Anand and Bhatt [2] proposed a scheduling policy for a software product and showed the importance of patching in lowering the system's failure rate and making it more economical.

Moreover, the literature on VDMs has tried to model the pattern of vulnerability discovery, but little or no attention has been laid upon when to fix these vulnerabilities by releasing patches. With this aim, the emphasis here has been given to understand the vulnerability discovery process based on which cost analysis has been performed to determine the optimal time when a firm should release its first patch. The concept of supportability has been introduced, which is the ability of the system to provide relevant information helpful for identifying and resolving issues when the software fails to work correctly. Furthermore, the scheduling policy has been designed in such a fashion that it caters to both the minimum value of SIL and dangerous failure ratio (DFR) for better decision-making.

3 THE PROPOSED MODELING FRAMEWORK

The significance of the thought is to secure vulnerable things (in this situation, software) from assault, yet the issue is that vulnerabilities in the product let vindictive programmers skirt standard security innovations with exemption. In the event that this was not the situation, at that point, the security defenselessness issue would not be extended. Securing a system under the advancing programming era has become troublesome, even if it is patched regularly. If the software system was in, some sense, self-ensuring (by being defensive, and all the more appropriately tested from a security point of view) or possibly less loaded with vulnerabilities, It would be easier to run and, at the same time, be cost-effective.

On the road toward achieving a high security level for software, it is important to focus on software security at the ground level. It is quite crucial for both the developer as well as the practitioner to become aware of security importance; they should adopt and evolve best practices to address security issues. Security testing must include two procedures: testing security usefulness with standard functional testing methods, and hazard construct security testing with respect to threat prototypes and danger models. Moving with the concept of maintaining a certain threshold of SIL, the presented proposal highlight the time to issue the first patch in such a manner that the overall cost incurred is minimized. The following are notations and assumptions.

3.1 Notations

$\Omega(t)$: Cumulative number of vulnerabilities.
N: The total number of vulnerabilities.
λ: Constant rate of vulnerability detection.
c_1: Cost of fixing the vulnerability before patch release.
c_2: Cost of fixing the vulnerability after patch release.
c_3: Cost of fixing vulnerability per unit time.
T_{LC}: Time point representing the lifecycle of the software.

3.2 Assumptions

i. Vulnerabilities are discovered in independent fashion.
ii. The rate of vulnerability discovery is directly dependent on remaining number of undiscovered vulnerability.
iii. The best that VDM can model is the discovery process given the vulnerability detection environment remains constant.
iv. The cost of fixing vulnerability after patch release will be higher compared to the cost of fixing before patch release.

3.3 Rescorla Model: Revisited

For the formulation of release time problem, we have considered the model given by Rescorla [13], which is based on the Goel-Okumoto model (1979) [22] that describes an exponential vulnerability detection curve, which makes it useful in many situations. The exponential model to quantify the cumulative number of vulnerabilities can be given as follows:

$$\Omega(t) = N \cdot \left(1 - e^{-\lambda t}\right) \tag{1}$$

where N is the total number of vulnerabilities present in the software, and "λ" is the constant rate for vulnerability discovery.

3.4 Bhatt et al. [19]: Revisited

Another model that has been used for modeling the count of vulnerabilities discovered and additional vulnerabilities discovered due to the influence of previously discovered vulnerabilities is given by Bhatt et al. [19]. The model captures the proportion of vulnerabilities discovered due to the influence of previously discovered vulnerabilities. The vulnerability discovery rate follows an s-shaped growth curve.

The mean value function describing the number of vulnerabilities discovered until time point "t" can be given as follows:

$$\Omega(t) = N \cdot \left(\frac{1 - e^{-(r+s)t}}{1 + \frac{s}{r} e^{-(r+s)t}} \right) \tag{2}$$

where "r" represents the rate of vulnerability detection, and "s" represents the rate that constitutes the discovery of some additional vulnerabilities influenced by discovered vulnerabilities.

3.5 Designing of Cost Structure Under SIL

Software security is the notion of designing software with the goal that it keeps on working accurately under pernicious assault. Most technologists recognize this current endeavor's significance; however, they require help in how to handle it. This new approach gives that assistance by investigating software security best practices by fixing vulnerabilities that minimize the associated risk with that exploitation of vulnerabilities. New terminology called supportability is introduced that can be defined as the probability that no vulnerability has been identified and resolved in $(t, t+x)$ provided that the last vulnerability has been identified and resolved before "t.". To obtain the software reliability discovery process focusing on dangerous vulnerabilities under the user's operation environment, we can characterize supportability $V(x|t)$ based on the severity coefficient (SC) and DFR as portrayed by Yamada and Nishikawa [10] for defining reliability:

$$V(x|\,t) = \mathrm{Exp}\{-\mathrm{DFR}\{\Omega(t + (\mathrm{SC} \times x)) - \Omega(t)\}\} \tag{3}$$

Eq. (3) has been derived in an analogous sense as reliability in the field of software reliability. Moreover, it is quite important to understand the associated risk factor. It is eventually the probability that surely some vulnerability will be discovered, which can be modeled as:

$$\mathrm{Risk} = 1 - V(x|\,t) = 1 - \mathrm{Exp}\{-\mathrm{DFR}\{\Omega(t + (\mathrm{SC} \times x)) - \Omega(t)\}\} \tag{4}$$

Making use of these defined models for vulnerability, a scheduling policy for patch release has been discussed. To understand the scheduling policy, it is important to first create the structure for total cost in detecting vulnerabilities present in the software. The total cost can be studied componentwise as follows:

i. Cost of fixing the vulnerability prior to patch release in the software can be given as follows:

$$C_1(t) = c_1\Omega(t) \tag{5}$$

ii. Cost of fixing the vulnerability after the release of a patch can be given as:

$$C_2(t) = c_2\{\Omega(T_{LC}) - \Omega(t)\} \tag{6}$$

iii. Cost of fixing vulnerability per unit time:

$$C_3(t) = c_3 \cdot t \tag{7}$$

Therefore the total expected cost of testing the software using Eqs. (5)–(7) can be defined as follows:

$$\begin{aligned} C(t) &= C_1(t) + C_2(t) + C_3(t) \\ &= c_1\Omega(t) + c_2\{\Omega(T_{LC}) - \Omega(t)\} + c_3 \cdot t \end{aligned} \tag{8}$$

The optimization model with the objective to minimize the total expected cost of fixing the vulnerability and the condition of minimum threshold limit on supportability as an important constraint, we define the optimal problem as:

$$\begin{array}{ll} \text{Min} & C(t) \\ \text{s.t.} & V(x|\,t) \geq V_0 \end{array} \tag{9}$$

where V_0 is the minimum threshold value for supportability.

Optimization model as given in Eq. (9) can be rewritten by using Eqs. (3), (8) as:

$$\begin{array}{ll} \text{Min} & C(t) = c_1\Omega(t) + c_2\{\Omega(T_{LC}) - \Omega(t)\} + c_3 \cdot t \\ \text{s.t.} & \text{Exp}\{-\text{DFR}\{\Omega(t + (\text{SC} \times x)) - \Omega(t)\}\} \geq V_0 \end{array} \tag{10}$$

The problem given in Eq. (10) is solved to obtain the optimal time point to release the first patch to fix the loopholes that can be exploited.

4 NUMERICAL ILLUSTRATION

For the applicability of the proposed methodology of determining the optimal patch release time, we considered the vulnerability data set of Microsoft Windows, primarily XP Windows, for a period of 15 years from 2000 to 2014 [23]. Two different scheduling policies based on different models as given in Eqs. (1), (2) have been solved and presented as two different cases.

Case 1: Using the previously mentioned data set, the estimated parameters of the Rescorla model [2] obtained by using SAS software are $N=186541.4$, $\lambda=0.000233$ [24]. To supplement the release time of the first patch, we have assumed the cost parameters to be $c_1=200$, $c_2=250$, and $c_3=50$; and other assumed parameters include$x=1$, $\text{DFR}=0.01$, and $\text{SC}=1$. It is important to note that all of these values are taken hypothetically, keeping in mind the nature of associated components. Making use of the optimization solver LINGO, and by solving the optimization problem as given in Eq. (10) under the threshold limit of 10% and incorporating Eq. (1), using the values of assumed parameters, optimal time $T^*=0.425\times10^{-1}=0.0425$ year, which is equal to 15.51 days (i.e., 16 days), which is the time point at which a firm should release its first patch. Furthermore, the objective value is 0.4663526E+08, which is the total expected cost of fixing the discovered vulnerabilities.

Case 2: The estimated parameters of the Bhatt et al. [19] model as obtained by using SAS software [24] are $N=942.148$, $r=0.0107$, and $s=0.3059$.

For the calculation of the patch release time, we have assumed the cost parameters to be $c_1 = 190$, $c_2 = 240$, and $c_3 = 50$; and other assumed parameters include $x = 1$, DFR = 0.1, and SC = 1. Moving onto similar lines as for Case 1, the optimal time $T^* = 0.9634 \times 10^{-1} = 0.09634$ year, which is equal to 35.19 days (i.e., 35 days), which is the time point at which a firm should release its first patch. Furthermore, the objective value is 226,070.8, which is the total expected cost.

5 SENSITIVITY ANALYSIS

Sensitivity analysis, also known as "simulation analysis" or the "what-if analysis," determines the robustness of a model's outcome. It studies the effect of independent parameters on dependent parameters. The independent variables are varied over a range, and its effect on the outcome is observed. If the output varies noticeably when changing the input variable from minimum to maximum over a range, then the output is said to be sensitive. If the output doesn't change much, it is said to be insensitive or robust. Outcomes that remain robust while changing the input values of the parameters help strengthen the credibility of the model. Sensitivity analysis also helps in quantifying the risk and identifying the optimal parameter settings of the model. Repeated application of sensitivity analysis aids in better understanding the pros and cons of the model. This information is very crucial when recommending an approach or model. To have in-depth interpretation of a supportability index, we carried out the sensitivity analysis by varying the threshold value V_0, and the obtained results are summarized in Table 1.

From Table 1, it can be clearly visualized that, when the threshold limit on level of supportability increases, then a patch has to be issued as soon as possible.

Table 1 Sensitivity analysis on the variable V_0

	Case 1		Case 2	
V_0	Time (in days)	Cost	Time (in days)	Cost
0.08	17.38707	46,635,250	42.51695	226,061.9
0.1	15.53168	46,635,260	35.61209	226,070.8
0.15	12.16039	46,635,280	22.96654	226,086.9
0.2	9.76841	46,635,290	13.91566	226,098.4
0.25	7.91305	46,635,300	6.84952	226,107.3
0.3	6.39712	46,635,310	1.04605	226,114.5

6 CONCLUSION

Here, we have modeled a release policy considering the effects of vulnerability, as they become serious hazards to software security. Several cases have been reported in which intruders exploit vulnerabilities. Of great importance is the need to shield software systems from these endeavors and grow such frameworks that are conceivably strong from such security breechings. In this manner, it is critical to investigate the ideal time when vulnerabilities should be disclosed and furthermore to decide upon an ideal time to settle certain vulnerabilities before they can be misused. In view of the logical strategy of computing programming security integrity level in the practical well-being benchmarks, we have proposed a quantitative technique for guaranteeing SIL inside the programming framework to decide an ideal fix discharge arrangement for settling the vulnerabilities after the product has been propelled into the field. The results obtained after the verification of the proposed strategy are knowledgeable in clarifying the recommendation.

REFERENCES

[1] J. Ren, R.N. Taylor, A secure software architecture description language, In: Workshop on Software Security Assurance Tools, Techniques, and Metrics, CreateSpace Independent Publishing Platform, Long Beach, California, USA, 2005, pp. 82–89.
[2] A. Anand, N. Bhatt, Vulnerability discovery modeling and weighted criteria based ranking, J. Indian Soc. Probab. Stat. 17 (1) (2016) 1–10.
[3] A. Ozment, in: Improving vulnerability discovery models, Proceedings of the 2007 ACM Workshop on Quality of Protection, ACM, October 2007, pp. 6–11.
[4] A. Anand, S. Das, D. Aggrawal, Y. Klochkov, Vulnerability discovery modelling for software with multi-versions. in: Advances in Reliability and System Engineering, Springer International Publishing, Cham, 2017, pp. 255–265. https://doi.org/10.1007/978-3-319-48875-2_11. Print ISBN: 978-3-319-48874-5, Online ISBN: 978-3-319-48875-2.
[5] A. Anand, M. Agarwal, Y. Tamura, S. Yamada, Economic impact of software patching and optimal release scheduling, Qual. Reliab. Eng. Int. 33 (1) (2017) 149–157.
[6] C. Ebert, Implementing functional safety, IEEE Softw. 32 (5) (2015) 84–89.
[7] G. McGraw, Software Security: Building Security, vol. 1, Addison-Wesley Professional, Boston, USA, 2006.
[8] G. McGraw, Software security, IEEE Secur. Priv. 2 (2) (2004) 80–83.
[9] ISO/IEC 9126, Software Engineering—Product Quality, 1991.
[10] S. Yamada, T. Nishikawa, Quantitative assessment for software safety integrity level with functional safety standards and risk costs, Int. J. Reliab. Qual. Saf. Eng. 21 (06) (2014). 1450030.
[11] J. Kim, Y.K. Malaiya, I. Ray, in: Vulnerability discovery in multi-version software systems, 10th IEEE High Assurance Systems Engineering Symposium (HASE'07), IEEE, November 2007, pp. 141–148.

[12] R.J. Anderson, Security in opens versus closed systems-the dance of Boltzmann, Coase and Moore, in: Open Source Software: Economics, Law and Policy, Toulouse, France, 2002.

[13] E. Rescorla, Is finding security holes a good idea? IEEE Secur. Priv. 3 (1) (2005) 14–19.

[14] O.H. Alhazmi, Y.K. Malaiya, Modeling the vulnerability discovery process, Proceedings of 16th IEEE International Symposium on Software Reliability Engineering (ISSRE'05), 2005, pp. 129–138.

[15] O.H. Alhazmi, Y.K. Malaiya, in: Quantitative vulnerability assessment of systems software, Proc. Annual Reliability and Maintainability Symposium, 2005, pp. 615–620.

[16] R.M. Brady, R.J. Anderson, R.C. Ball, Murphy's Law, the Fitness of Evolving Species, and the Limits of Software Reliability (No. UCAM-CL-TR-471), University of Cambridge, Computer Laboratory, Cambridge, UK, 1999.

[17] O.H. Alhazmi, Y.K. Malaiya, Application of vulnerability discovery models to major operating systems, IEEE Trans. Reliab. 57 (1) (2008) 14–22.

[18] P.K. Kapur, N. Sachdeva, S.K. Khatri, in: Vulnerability discovery modeling, International Conference on Quality, Reliability, Infocom Technology and Industrial Technology Management, 2015, pp. 34–54.

[19] N. Bhatt, A. Anand, V.S.S. Yadavalli, V. Kumar, Modeling and characterizing software vulnerabilities, Int. J. Math. Eng. Manag. Sci. 2 (4) (2017) 288–299.

[20] S. Jiang, S. Sarkar, in: Optimal software release time with patching considered, Proc. 13th Annual Workshop Information Technologies and Systems, Seattle, 2003, pp. 61–66.

[21] A. Arora, R. Telang, H. Xu, Optimal policy for software vulnerability disclosure, Manag. Sci. 54 (4) (2008) 642–656.

[22] A.L. Goel, K. Okumoto, Time-dependent error-detection rate model for software reliability and other performance measures, IEEE Trans. Reliab. 28 (3) (1979) 206–211.

[23] X.P. Windows, Vulnerability Statistics, http://www.cvedetails.com/product/739/Microsoft-Windows.html?vendor_id=26. Accessed 20 February 2016.

[24] SAS Institute Inc, SAS/ETS User's Guide Version 9.1, SAS Institute Inc., Cary, NC, 2004.

CHAPTER 12

Tools and Techniques in Software Reliability Modeling

Preeti Malik*, Lata Nautiyal*, Mangey Ram†

*Department of Computer Science and Engineering, Graphic Era Deemed to be University, Dehradun, India

†Department of Mathematics, Computer Science & Engineering, Graphic Era University, Dehradun, India

Abstract

The primary aim of software engineering is to deliver the best quality product for low cost and within a specified time. Every day, more complex software is produced, along with highly scientific tools and technologies to fulfill the latest need of the modern world. As our dependency for all our activities has shifted from manual to automated systems, the complexity of software has also increased; therefore reliability has become one of the fundamental attributes during the software development process. This chapter presents reliability measurement tools and techniques used in the modern software industry. An evaluation parameter for evaluating reliability models is also included. A hierarchy of software reliability models is also included in this chapter. How can we perform reliability modeling during different phases of software development? This question is also answered in the chapter.

Keywords: Software, Reliability, Reliability model, Probability

1 INTRODUCTION

Information technology plays an important role in every aspect of our lives. All of the daily appliances we use, for instance, air conditioners, televisions, refrigerators, washing machines, and other apparatus, are controlled by CPUs and software. A primary fact about software is that it never wears out or distorts; also, the natural environment cannot harm it, and it can operate for years and years as long as its hardware components allow it to operate. The driving force behind software is software engineering. The primary aim of software engineering is to deliver the best quality product for low cost and within a specified time [1, 2]. Every day, more complex software is produced, along with highly scientific tools and technologies to fulfill the latest need of the modern world. As our dependency for all our activities has shifted from manual to automated systems, the complexity of software has also increased; therefore reliability has become one of the fundamental attributes during the

Advances in System Reliability Engineering
https://doi.org/10.1016/B978-0-12-815906-4.00012-9

software development process. A well-defined and most popular description for *software reliability* [3] is the likelihood of execution without any failure for a given time interval, called the *mission time*. Software reliability [4] are the integral characteristics of software quality, along with other important characteristics, for instance, functionality, usability, presentation, serviceability, competence, installation, maintainability, and well-defined documentation. In layman's terms, reliability is regarded as the property of mentioning "how well the software meets its prerequisites" and "the probability of failure-free procedure for the given period of time."

If we define this in a more specialized manner, then:

> *The term Software reliability is actually the probability of the given software under the specified circumstances.*

By applying the given formula, the probability of failure can be considered by testing an example of all existing input conditions:

$$\text{Mean Time Between Failure} = \text{Mean Time To Failure} + \text{Mean Time To Repair}$$

$$\text{Probability} = \text{Number of failing cases} / \text{Total number of cases under consideration}$$

The given group of all likely input states is known as the input space. To find out exactly the reliability of any software, it is also important to find the output space from a given input space and software. When further described, reliability testing is an important parameter of software reliability. Before actually going through the process of reliability testing, data is collected by numerous stages of development, for instance, design and operating stages. The tests are limited due to restrictions such as cost and time restrictions. Studying statistical samples of software products tests reliability of software. Time constraints are handled by applying deadlines. After this design phase, the implementation phase starts. Each datum is gathered precisely according to restrictions on cost and time. To achieve satisfactory outcomes of reliability testing, reliability characteristics must be considered, For instance, mean time to failure (MTTF) is evaluated based on three important factors: operating time, number of on/off cycles, and calendar time.

2 RELIABILITY MODELING BUZZWORDS

The word cloud shown in Fig. 1 shows the various elements related to software reliability. The following section discusses some of these buzzwords.

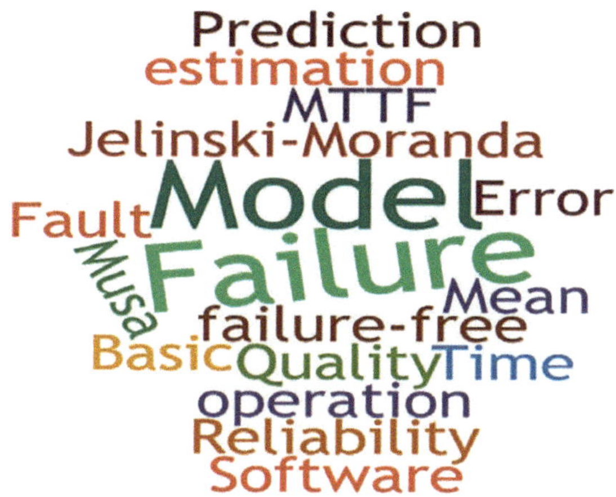

Fig. 1 Word cloud for software reliability buzzwords.

2.1 Errors, Faults, Failures, and Reliability

When the work in the direction of software reliability begins, the first thing that needs to be considered is what we need to achieve exactly, and what various failures may hamper the success of the project. Underneath, IEEE terminology [5] will be followed to discover the various buzzwords.

The term *fault* is also referred to as a *bug* or both; the concepts can be used complementary to each other. It is assumed that, whenever there is a fault, it may always cause a *failure,* also defined as "when a component or the entire software system does not function as per the desired output by the user is also categorized as a failure." The ultimate relationship between correctness and reliability can be shown by Fig. 2.

Among the terms faults and failures, the fault is quiescent mode. A large number of different significances are related to failure, for instance, some are harmless and others are perhaps fatal. There are various categories of failures, for example, catastrophic, severe, significant, and minor. How one defines severity varies as the needs of the system. Table 1 presents an example of classification taken from Donnelly et al. [6].

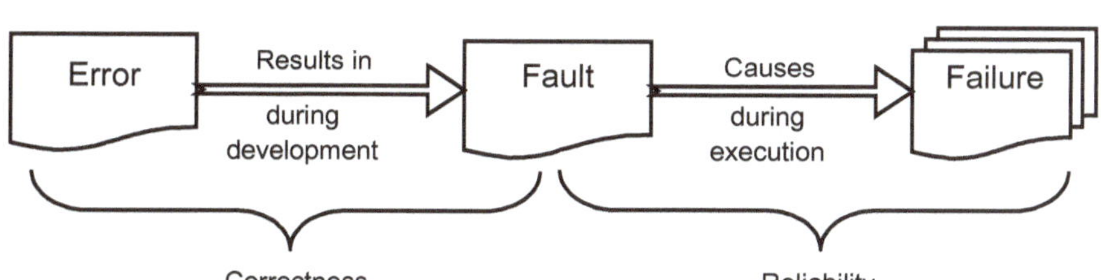

Fig. 2 Relationship between correctness and reliability.

Table 1 Classification of severity based on service degradation

Classification	Definition	
Catastrophic	Complete software failure; nothing left.	The robot does not work at all.
Severe	Component of the software having highest priority is not working.	The robot must be operated manually.
Significant	Customers change how to use the system.	The robot's configuration can't be changed distantly, if they require alteration, then they must be configured through the control panel.
Minor	Problem is not noticeable by the customer.	Some maintenance functions are not currently performed.

Error can be defined as inconsistency between computed and measured value and the specified or true value or condition. Errors are useful in fault-tolerant computing to portray in-between stages of faults and failures [7]. A human actor or component may also cause failures by inserting wrong input. Hardware failing or some other equipment failing may also cause failure of system. *Reliability* is:

> *The ability of a system or component to perform its required functions under stated conditions for a specified period of time.*

Sommerville [8] defined reliability in somewhat different terms:

> *Reliability is the probability that the system or components performs its required functions under stated conditions for the specified period of time.*

Some measures of reliability depend on particular situation and include [9]:

- *MTTF*: This measure indicates that the software is reliable enough that it is ready to release.
- *Rate of occurrence of failures*: It defines the expected number of failures in a fixed time. This kind of measure is suitable for systems that control potentially dangerous processes.
- *Probability of failure on demand*: This measure is appropriate for safety systems. These systems only come into existence when another system is stuck into some dangerous circumstances.
- *Availability*: This measure is used in situations when failures in a system depend on the length of time that system was unobtainable. For instance: airline reservation systems, telephone switches.

3 RELIABILITY CHARACTERISTICS OF SOFTWARE

It is well known that the complexity of any software is quite high and, hence, achieving software reliability tends to be a trying task. Unless and until a replica of the software is efficiently and accurately developed, we cannot call that particular software "reliable software." The following are a few ideas that contribute to the possibility of software failure (see also Kene [10] and Herrmnn [11]):

- Software from various applications and domains has distinctive properties. Any change, big or small, in the source code of the software can result in entirely different software.
- Imperfections that are not on view in the code are prone to more faults, rather than those that are visible. They are constant and invariable unless and until they are fixed.
- Rather than time, software reliability relies on the quality of amendment the software receives.
- It is rarely heard that the cause behind the failure of software is due to the fault. This is possible as the bugs can only appear during the testing phase of the software. An article by Pham and Zhang [12] shows that, in any commercial software product, 350,000 lines of code will have more than 2000 programming errors, that is, an average of six software faults for every 1000 lines of code written. Although, it is depicted that a commercial software product may quietly encompass numerous errors because of its complexity.
- Thanks to the various testing phases of software, bugs are effortlessly revealed and hence the faults rarely ever cause the software to fail or crash.
- External conditions such as the physical environment do not affect the software. Internal conditions like memory available, speed of clock, etc. certainly do affect software reliability.
- The greater the age of the software, the fewer the number of bugs in it. Because they are rectified as soon as they are found, they are removed at the earliest time. Only newer versions of the software could invite potential bugs, which, again, are detected as soon as possible and hence corrected.
- There are a few conditions in which software can make its own faults quite obvious and apparent.
- It is possible that the occurrence of even a single error could cause the entire software to crash. This continues until the fault is detected and hence corrected.

- Until the fault is corrected, the software could face a straggle or delay in its working. This delay is analyzed statistically but depends on the essence of the fault.
- It is always observed that the MTTF of a software product in considerably sized systems is inversely proportional to size of program; it means that the count of faults in each line is not uniformly constant [13].
- At the time of implementation, the software can be installed on various platform devices. However, the modus operandi of each platform may vary from each another. So the collected data of faults will be addressed from various locations.

3.1 Reliability and Software Life-Cycle

Various phases of SDLC shows the variation in the methodologies, requirements for evaluation, and prediction of software reliability [13, 14]:

- Early prediction models are used in the requirement analysis and designing phase before the deployment of the software product. Reliability should be properly evaluated based on the architecture and specified prerequisites.
- While discussing the implementation and testing phases, software reliability evaluation is necessary to make decisions, such as when to stop testing and debugging. For example, when the MTTF is long enough, the software is ready to release.
- At the time of launching of software, generally it is assumed that all the bugs have been properly debugged and amended. Therefore after the launching process, a *reliability model* is used to forecast the MTTF that can be projected.
- If the software product is failure-free, or in case the software product is not maintained, a reliability model is most relevant.
- It is also assumed that if there are no failures and the software is up-to-date, it is sure that the reliability growth model is in order.

4 RELIABILITY MODELS

Reliability models are used in the debugging phase of the software development lifecycle [15]. They analyze the quality of the software by testing its various modules in certain phases. Once a fault is found, the code is modified, and the corrected version of the software is found; the code is modified

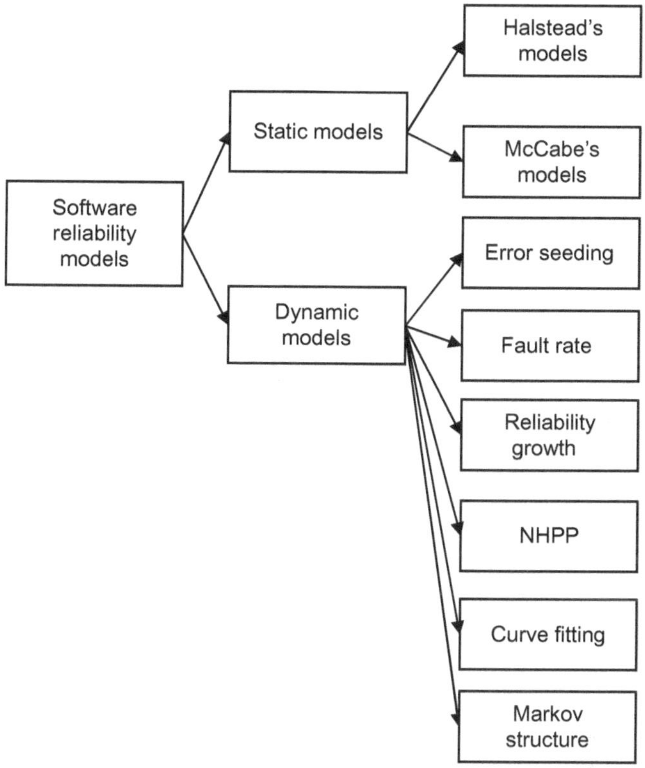

Fig. 3 Software reliability model's hierarchy.

and the corrected version of the software is tested all over again. Software reliability models are either deterministic or probabilistic. Halstead's software metric model and McCabe's cyclomatic complexity metric model are the two most popular deterministic models (refer to Fig. 3). The probabilistic models include the following:

- Failure rate model
- Failure count model
- Fault seeding model
- Software reliability growth model

4.1 Failure Rate Model

The process involved is to gauge the duration between each detected fault. This time is then analyzed and parameters are established, hence making an estimate judgment of the reliability of the software.

4.1.1 Jelinski-Moranda Model

The model proposed by Jelinski-Moranda is the most frequently used model [16]. Let N be the number of software faults that could potentially cause the entire system to crash. The software failure rate is calculated during time t_1 between the $(i-1)$th and ith failure. The formula to calculate failure rate is:

$$F(t_i) = \alpha[N - (i-1)]$$

where α is proportionality constant. This function is called a hazard function, and it remains static between failures. Its size decreases when a fault is eradicated.

4.1.2 Schick and Wolverton Model

The presumptions of the Jelinski-Moranda model are a basis for the software model with the exception of the proportionality of the hazard function with respect to the fault content of the software program along with the time before a most recently found failure. Its formula is:

$$F(t_i) = \alpha[N - (i-1)]t_i$$

Hazard rate is one-dimensional (linear) at each interval of failure. A moderated formula could be as follows [17]:

$$F(t_i) = \alpha[N - (i-1)]\left(-at_i^2 + bt_i + c\right)$$

where a, b, c are constants. When failure time $t_i = 0$, the hazard function is proportional to the hazard function of the Jelinski-Moranda model.

4.1.3 Goel-Okumoto Imperfect Debugging Model

An extension of the Jelinski-Moranda model, namely the "imperfect debugging model", was suggested by Goel and Okumoto [18]. This model considers that, at any time, the number of faults in the system follows the Markov process. Its hazard function is:

$$F(t_i) = \alpha[N - p(i-1)]\lambda$$

where

p = imperfect debugging probability

λ = failure rate per fault

4.1.4 Littlewood-Verrall Bayesian Model

In this proposal, time instances between each failure are to be exponentially distributed. The criterion of this model is random variables with the formula of gamma distribution, which is as follows:

$$F(t_i|\lambda_i) = \lambda e^{-\lambda_i t_i}$$

4.2 Failure Count Model

4.2.1 Musa Execution Time Model

Based on execution time of the software, this model caters to the following formula:

$$F(T) = \alpha f(N - n_c)$$

where

T = execution time
f = linear execution frequency
n_c = corrected number of faults

4.2.2 Goel-Okumoto NHPP Model

Let us assume $N(t)$ to be the cumulative number of failures detected at a time instant t. $N(t)$ can then be modeled as:

$$P\{N(t) = y\} = m((t))^y e^{-m(t)}, \quad y = 0, 1, 2, 3, \ldots, y!$$

$$m(t) = a\left(1 - e^{-bt}\right)$$

$$\lambda(t) = m'(t) = abe^{-bt}$$

where

$m(t)$ = number of expected failures at time t
$\lambda(t)$ = failure rate
a = expected number of failures observed
b = fault detection rate per fault

We need to understand that, at any arbitrary interval, the software can be subjected to failure because of the number of faults in its modules.

4.2.3 Goel Generalized NHPP Model

At the time of testing, the rate of failure curve first goes up and then down. To handle this variable failure rate process, Goel advanced a generalized NHPP model with this particular formula:

$$P\{N(t) = y\} = m((t))^{y} e^{-m(t)}, \quad y = 0, 1, 2, 3, \ldots, y!$$
$$m(t) = a\left(1 - e^{-bt}\right)$$

The failure rate is found using the formula:

$$\lambda(t) = m'(t) = abe^{-bt} t^{c-1}$$

where

a = expected number of faults
b, c = constants

4.3 Fault Seeding Models

These models are so-called because of a known quantity of faults "seeded" into the software program. These faults are tallied at the time of testing of the software. Various fault estimation and computing algorithms and models are used to calculate the total number of faults and to gauge the reliability of the software.

4.3.1 Mills Seeding Model

Most simple, yet most popular, this model proposes to seed a certain quantity of faults in a program code and then run and test it for a particular interval of time. Original and genuine faults can be established through the group of aboriginal and seeded faults, which had been ignored at the time of testing. Another name for this model is a "tagging model" because a given fault is tagged as seeded.

4.3.2 Lipow Model

A model that advances the estimation of probability of faults in any testing phase of the software is the Lipow model [19]. These probabilities are calculated to tally the total number of original and seeded faults.

4.4 Input Domain-Based Models

In this particular model, various test cases are produced from an input domain/field. It is extremely complex to divide these input domains into uniform classes. The reliability is measured using the number of failures or execution of the generalized test cases.

4.4.1 Nelson Model

Nelson [20] brought forward the Nelson model. Here, the software is sampled with n inputs. These inputs are unsystematically picked from a certain input domain. This random sampling will establish the degree of reliability of the software.

4.4.2 Ramamurthy and Bastani Model

This model [21] predominantly prioritizes reliability of real-time, vital, and critical process control programs. This model generates an approximate probability that the program is accurate and acceptable for all random set of inputs.

5 EVALUATION OF MODELS

Iannino et al. [22] established numerous methods in which the integrity of a model could be assessed.

(i) *Predictive validity*: The ability of a reliability model to foresee subsequent behavior of failure either during the phases or during testing. It also needs to predict the accuracy of the model, its trend, or the structured change of bias from small to large chunks of failure, time instances, and noise, which can be gauged by the corresponding change in the predicted rate of failure. Lyu and Nikora [23] further separated it in:
  - Correctness is calculated by prudential likelihood
  - Bias, as measured by the U-plot
  - Trend, as measured by the Y-plot
  - Noise, as measured by the relative change in the predicted failure rate

(ii) *Capability*: Estimating the aggregate needed by the software managers, developers, and users, to plan, manage, and control operations of the software systems.

(iii) *Assumption quality*: Checking the integrity, clarity, and lucidity of any particular assumption made by a model.

(iv) *Applicability*: The functionality and practicality of the model across various software platforms, environments, and the lifestyle phases.

(v) *Simplicity*: The clarity and accessibility of the collected data to know the nature of the model and establish its implementation into the problem.

(vi) *Measuring parameters*: This involves making correct and accurate predictions regardless of the completeness or quality of data [23].

6 SELECTION OF RELIABILITY MODELS

A number of methods exist for selecting the perfect model to estimate reliability of software. One could do an elaborate comparative analysis or find a set of models that are more accurate and steady than other models [24]. Another approach was given by Lyu and Nikora [23], in which a basic set of models is chosen for acceptance, and later the models whose prediction biases don't cancel out are dropped. These models are then collaborated by calculating a total of probability distribution function. Another method proposed by Khoshgoftaar and Woodcock [25] was to find out that the model matches up with real operational data in a good way, then the best match is chosen based on some criterion. The last approach we are including was given by Stringfellow and Andrews [26]. This approach finds a group of accurate and stable models and conservatively sets the current failure probability as the maximum of the individual failure probabilities given by the models.

7 SOFTWARE RELIABILITY ASSESSMENT TOOLS

Software reliability tools are the software that can be used to estimate reliability of software. SMERFS, CASRE, SREPT, CASYS, GERT, and PISRAT are the name of such tools. CASYS, GERT, and PISRAT are common tools, and other tools are specialized assessment tools.

The Statistical Modeling and Estimation of Reliability Functions for Software (SMERFS) [27] uses a black-box technique, and a number of reliability models are supported by this tool. This tool assumes that the software is already tested. The latest version of this tool is SMERFA Cubed, developed for Windows operating systems. Computer-Aided Software Reliability Estimation (CASRE) is another black-box tool. The reliability model it supports is the same as in SMERFS. A negative point in this tool is that it doesn't include test coverage and also doesn't have an automated method for data collection. Test coverage criteria are not included in the reliability estimation process. This tool takes a cluster of several models into a single estimation process, and then a linear combination of weights from different models is taken in predictive analysis of reliability. This tool is also not suitable for large software products.

The Automated Test Analysis for C (ATAC) is a white-box tool. Software efficiency is checked using coverage criteria. The compiler of this tool is atacCC, and this compiler creates instrumented binary files. The size of test suit is reduced by this tool. It also assesses test completeness. Software Reliability and Estimation Prediction Tool (SREPT) is given by cycle [28]. This tool can be used across several stages of a software's lifecycle. This tool is able to predict reliability of a system as soon as its architecture is in place. It can also plan release time of software based on some defined criteria.

A tool designed by Li and Malaiya [29] named Reliability of Basic and Ultra-reliable Software systems (ROBUST) supports five growth models. One model among these includes test coverage. Two of the remaining uses are static metrics for reliability estimation. Accuracy of growth model is improved in this tool. Failure time, test coverage, and time intervals are used to operate on data. All of these tools are dependent on external data sources, and this needs significant effort done by the user before applying to software. GERT is an attempt to revolutionize this extra effort.

GERT was developed by Davidsson et al. [30], which utilizes multivariate regression model for estimating reliability within a number of metrics used for empirical estimation of reliability. While using GERT, one has to assume that a test suit is already available. The last tool discussed is Proportional Intensity-based Software Reliability Assessment Tool (PISRAT), developed by Shibata et al. [31]. Various testing metrics data and failure data are used by this tool. It is written in Java and has 54 classes with 8000 lines of code. JFreeChart library is used for generating charts. It is the first freeware tool for dynamic modeling of software reliability. This tool can support:

(i) Maximum likelihood method used for parameter estimation
(ii) The goodness-of-fit test under various optimization criteria
(iii) Quantitative software reliability is assessed

8 CONCLUSIONS

Reliability measurement is a way of measuring quality of software. There are a number of models presented by various authors for reliability estimation. Software reliability models are either deterministic or probabilistic. Halstead's software metric model and McCabe's cyclomatic complexity metric model are the two most popular deterministic models. The probabilistic models are failure rate model, failure count model, fault seeding model, and software reliability growth model. This chapter consists of these models. Readymade software products are the most suitable tools that are commonly used to assess the software reliability; CASYS, GERT, and PISRAT are the most common tools given in software reliability assessment tools, and other tools include specialized software reliability assessment tools.

REFERENCES

[1] C.-Y. Huang, et al., in: Software reliability modelling and cost estimation incorporating testing-effort and efficiency, Proceedings 10th International Symposium on Software Reliability Engineering (Cat. No. PR00443), 1999, pp. 62–72.

[2] P. Liggesmeyer, M. Trapp, Trends in embedded software engineering, IEEE Softw. 26 (3) (2009) 19–25.

[3] J. Pan, Software Reliability: Dependable Embedded Systems, Carnegie Mellon University, Springer, United States, 1999.

[4] H. Son, S. Koo, Software reliability improvement techniques, in: P. Seong (Eds.), *Reliability and Risk Issues in Large Scale Safety-critical Digital Control Systems*, Springer Series in Reliability Engineering, Springer, London, 2009, pp. 105–120.

[5] IEEE Standard, IEEE Standard Glossary of Software Engineering Terminology, The Institute of Electrical and Electronics Engineers, USA, 1990.

[6] M. Donnelly, et al., Best current practice of SRE, in: Handbook of Software Reliability Engineering, IEEE Computer Society Press and McGraw-Hill, Hightstown, NJ, USA, 1996, pp. 219–254.

[7] M.R. Lyu, Introduction, in: Handbook of Software Reliability Engineering, IEEE Computer Society Press and McGraw-Hill, Hightstown, NJ, USA, 1996.

[8] I. Sommerville, Software Engineering, sixth ed., Addison-Wesley, Boston, MA, USA, 2001.

[9] B. Littlewood, L. Strigini, Validation of ultra high dependability for software-based systems, Commun. ACM 36 (11) (1993) 69–80.

[10] S.J. Keene, Comparing hardware and software reliability, Reliab. Rev. 14 (4) (1994) 5–21.

[11] D.S. Herrmnn, Software Safety and Reliability, Wiley-IEEE Computer Society Press, US, February 2000.

[12] H. Pham, X. Zhang, A software cost model with warranty and risk costs, IEEE Trans. Comput. 48 (1) (1999) 71–75.

[13] D. Hamlet, Are we testing for true reliability? IEEE Softw. 9 (4) (1992) 21–27.

[14] C.A. Asad, M.I. Ullah, M.J.-U. Rehman, An approach for software reliability model selection, Proceedings of the 28th Annual International Computer Software and Applications Conference (COMPSAC'04), vol. 1, 2004, pp. 534–539.

[15] A.L. Goel, Sofware reliability models: assumptions, limitations and applicability, IEEE Trans. Softw. Eng. 11 (12) (1985) 1411–1423.

[16] Z. Jelinski, P. Moranda, Sofware reliability research, in: Statistical Computer Performance Evaluation, Academic Press, New York, 1972, pp. 465–484.

[17] G.J. Schick, R.W. Wolverton, An analysis of component sofware reliability models, IEEE Trans. Softw. Eng. 4 (2) (1978) 104–120.

[18] A.L. Goel, K. Okumoto, in: An analysis of recurrent software failure in a real-time control system, Proc. ACM Annual Tech Conf, ACM, Washington, DC, 1978, pp. 496–500.

[19] M. Lipow, Estimation of Sofware Package Residual Errors, TWR software series report, SS-72-09, 1972, pp. 359–371.

[20] E. Nelson, Estimating software reliability from test data, Microelectron. Reliab. 17 (1978) 67–74.

[21] C.V. Ramamurthy, F.B. Bastani, Sofware reliability: status and perspective, IEEE Trans. Softw. Eng. 8 (4) (1982) 359–371.

[22] A. Iannino, J.D. Musa, K. Okumoto, Criteria for software reliability model comparisons, ACM SIGSOFT Softw. Eng. Notes 8 (3) (1983) 12–16.

[23] M.R. Lyu, A. Nikora, Applying reliability models more effectively, IEEE Softw. 9 (4) (1992) 43–52.

[24] R.T. Clemen, Making Hard Decisions—An Introduction to Decision Analysis, second ed., Duxbury, Pacific Grove, 1996.
[25] T.M. Khoshgoftaar, T.G. Woodcock, in: Software reliability model selection: a case study software reliability model selection: a case study, Proceedings of the 1991 International Symposium on Software Reliability Engineering, 1991, pp. 183–191.
[26] C. Stringfellow, A. Andrews, An empirical method for selecting software reliability growth models, Empir. Softw. Eng. 7 (2002) 319–343.
[27] G.E. Stark, in: A survey of software reliability measurement tools, IEEE International Symposium on Software Reliability, 1991, pp. 90–97.
[28] S. Ramani, S. Gokhale, K.S. Trivedi, SREPT: software reliability estimation and prediction tool, in: 10th Intl. Conference on Modeling Techniques and Tools (Tools '98), Lecture Notes in Computer Science 1469, 1998, pp. 27–36.
[29] N. Li, Y. Malaiya, in: ROBUST: a next generation software reliability engineering tool, IEEE International Symposium on Software Reliability Engineering, 1995, pp. 375–380.
[30] M. Davidsson, et al., in: GERT: an empirical reliability estimation and testing feedback tool, Proceedings of the 15th International Symposium on Software Reliability Engineering (ISSRE'04), 2004.
[31] K. Shibata, K. Rinsaka, T. Dohi, in: PISRAT: proportional intensity-based software reliability assessment tool, 13th IEEE International Symposium on Pacific Rim Dependable Computing, 2007, pp. 43–52.

INDEX

Note: Page numbers followed by *f* indicate figures, and *t* indicate tables.

G

H

I

FSC
www.fsc.org
MIX
Papier | Fördert
gute Waldnutzung
FSC® C083411

Kolibri 360 GmbH
Ferdinand-Jühlke-Straße 7
99095 Erfurt, Deutschland
produktsicherheit@kolibri360.de